AF598026

METHODS IN MOLECULAR BIOLOGY™

Series Editor
John M. Walker
School of Life Sciences
University of Hertfordshire
Hatfield, Hertfordshire, AL10 9AB, UK

For further volumes:
http://www.springer.com/series/7651

Nanotechnology in Regenerative Medicine

Methods and Protocols

Edited by

Melba Navarro

Institute for Bioengineering of Catalonia (IBEC), Barcelona, Spain;
Networking Research Centre on Bioengineering, Biomaterials and Nanomedicine, CIBER-BBN, Barcelona, Spain

Josep A. Planell

Institute for Bioengineering of Catalonia (IBEC), Barcelona, Spain;
Networking Research Centre on Bioengineering, Biomaterials and Nanomedicine, CIBER-BBN, Barcelona, Spain

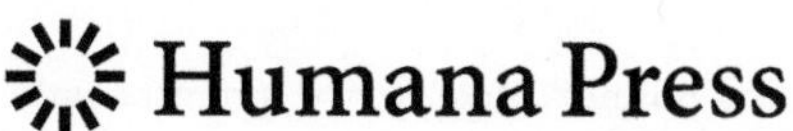

Editors
Melba Navarro
Institute for Bioengineering of Catalonia (IBEC)
Barcelona, Spain
and
Networking Research Centre on Bioengineering
Biomaterials and Nanomedicine, CIBER-BBN
Barcelona, Spain
mnavarro@ibecbarcelona.eu

Josep A. Planell
Institute for Bioengineering of Catalonia (IBEC)
Barcelona, Spain
and
Networking Research Centre on Bioengineering
Biomaterials and Nanomedicine, CIBER-BBN
Barcelona, Spain
japlanell@ibecbarcelona.eu

ISSN 1064-3745 e-ISSN 1940-6029
ISBN 978-1-61779-387-5 e-ISBN 978-1-61779-388-2
DOI 10.1007/978-1-61779-388-2
Springer New York Dordrecht Heidelberg London

Library of Congress Control Number: 2011940219

Printed on acid-free paper

Humana Press is part of Springer Science+Business Media (www.springer.com)

Preface

Nanotechnology encompasses the understanding of the fundamental physics, chemistry, biology, and technology of nanometer scale objects (*Institute of Physics – Nanotechnology Journal*, UK). Nanotechnology could also be defined as the study of manipulating matter on the atomic and molecular scale. Nanotechnology has become in recent years a popular and well-accepted term and a central pillar in many nationally funded research programs. The potential strength of nanotechnology is due to the fact that engineering materials and devices at the nanoscale implies not only the manipulation of the individual atoms and molecules to arrange them, forming bulk macroscopic bodies (bottom-up approach), but it also implies the possibility to analyze and evaluate matter at the nanolevel. Moreover, considerable funding has been allocated and invested in the development of this discipline in many countries such as the USA, Korea, Japan, Australia, and the European Union, including also several individual member states.

Life is organized at the level of cells, but it is well known that natural cellular events, interactions, and processes take place at the subcellular scale and at the molecular level. This is the reason why nanotechnology is meant to play a key leading role in developing tools able to identify, measure, and study such events at the nanometric level, as well as in contributing to the disclosure of unknown biological interactions and mechanisms. Moreover, it should also be key in engineering issues, such as producing material structures able to mimic the biological ones, efficient delivery systems, as well as devices aiming among other issues to identify and track proteins and cells.

The great existing interest in elucidating such unknown biological interactions has led to the convergence of disciplines, such as engineering, physics, chemistry, and molecular and cellular biology into a novel field known as nanobiotechnology. This new technology should allow detecting, evaluating, analyzing, and engineering biological nanostructures. This means to open wide the whole field of nanomedicine, including nanodevices for diagnostic and therapy, drug delivery systems, and regenerative medicine. The implication of nanotechnology to regenerative medicine is the objective of this book. In fact, regenerative medicine is meant to develop innovative in situ and cost-effective therapies by repairing and regenerating tissues for diseases and problems without solution at present, as well as for overcoming many present bionic solutions.

Taking the above point of view and given the importance and potential impact of nanotechnology in medicine, this book aims to provide an overview of a very wide range of the different currently used technologies and methods that involve nanotechnology principles and that may be used in tissue regeneration. Being that the application of nanotechnology to regenerative medicine is a very broad field, this book focuses its interests on particular areas, such as its use as a means to produce efficient platforms and structures for tissue engineering, delivery systems and biosensors, as well as the use of some techniques to study materials surfaces and the interactions between cells, biomolecules, and surfaces at the nanoscale.

Rather than a compilation of chapters, where the state-of-the-art of these technologies is reviewed, this book is a collection of experimental protocols, where an in-depth and step-by-step description of various nanotechnology involving methods is carried out.

The book is divided into 19 chapters. The first chapter is an introduction to the importance and the potential capacity of nanotechnology to develop new tools and means to have a better understanding of the biological interactions and processes with the ultimate aim to bring up new therapies for regenerative medicine. There are five overview chapters presenting a comprehensive review of very important topics in the field such as the development of novel strategies to engineer tissue in vitro, the design of diagnosis devices, modeling of bio/non-bio interactions, and also the ethical, legal, and social issues related to regenerative medicine.

The other chapters are dedicated to the full description of methodologies followed for the synthesis of new biomolecules and biomaterials, the fabrication of 3D scaffolds at the nanoscale, surface chemical modification through functionalization with biomolecules and protein patterns, and the detection and analysis of biological entities and events.

This volume provides established scientists, junior researchers, and students involved in the bioengineering, biotechnology, and biomedical fields with a sound foundation of a wide variety of nanotechnology approaches in regenerative medicine. Finally, we would like to acknowledge all the authors and colleagues that participated in the preparation of this book, not only for their outstanding contributions, but also for their effort and willingness in putting together a book with such a diversity of expertise and such an interdisciplinary approach.

Barcelona, Spain ***Melba Navarro***
Barcelona, Spain ***Josep A. Planell***

Contents

Contributors

JOANA C. ANTUNES • *Faculdade de Engenharia, Divisão de Biomateriais, INEB – Instituto de Engenharia Biomédica, Universidade do Porto, Porto, Portugal*
CONRADO APARICIO • *Department of Restorative Sciences, Minnesota Dental Research Center for Biomaterials and Biomechanics (MDRCBB), School of Dentistry, University of Minnesota, Minneapolis, MN, USA*
FRANCISCO JAVIER ARIAS • *G.I.R. Bioforge, University of Valladolid, CIBER-BBN, Valladolid, Spain*
GONÇAL BADENES • *ICFO Institute of Photonic Science, Castelldefels, Spain*
MARIA FERNANDA BAHIA • *Faculty of Pharmacy, Department of Pharmaceutical Technology, University of Porto, Porto, Portugal*
MÁRIO A. BARBOSA • *Faculdade de Engenharia, Divisão de Biomateriais, Departamento de Engenharia Metalúrgica e Materiais, INEB – Instituto de Engenharia Biomédica, Universidade do Porto, Porto, Portugal*
JEANANN S. BOYCE • *Computer Science and Business, Montgomery College, Takoma Park, MD, USA*
OSCAR CASTAÑO • *Institute for Bioengineering of Catalonia (IBEC), Barcelona, Spain; Networking Research Centre on Bioengineering, Biomaterials and Nanomedicine, CIBER-BBN, Barcelona, Spain*
GIANNI CIOFANI • *Smart Materials Lab, Center of MicroBioRobotics, Italian Institute of Technology, Scuola Superiore Sant'Anna, Pontedera, Pisa, Italy*
SERENA DANTI • *Otology – Cochlear Implants Unit, Azienda Ospedaliera Universitaria Pisana & University of Pisa, Pisa, Italy*
DENIS DOROKHIN • *Faculty of Science and Technology, Materials Science and Technology of Polymers, University of Twente, Enschede, The Netherlands; MESA+Institute for Nanotechnology, Enschede, The Netherlands*
MOHAMED ELTOHAMY • *Institute of Tissue Regeneration Engineering (ITREN), Dankook University, Cheonan, South Korea*
RAMON ERITJA • *Institute for Research in Biomedicine, IQAC-CSIC, CIBER de Bioingeniería, Biomateriales y Nanomedicina (CIBER-BBN), Barcelona, Spain*
MARIA F. GARCIA-PARAJO • *BioNanoPhotonics Group, IBEC – Institute for Bioengineering of Catalonia and CIBER-BBN, Barcelona, Spain; ICREA – Institució Catalana de Recerca i Estudis Avançats, Barcelona, Spain*
ALESSANDRA GIROTTI • *G.I.R. Bioforge, University of Valladolid, CIBER-BBN, Valladolid, Spain*
LINDA MACDONALD GLENN • *Alden March Bioethics Center, Albany Medical Center, Albany, NY, USA*
ANGELA GUACCIO • *Interdisciplinary Research Centre on Biomaterials (CRIB), University of Naples Federico II, Naples, Italy*

MING-YONG HAN • *Institute of Materials Research and Engineering, A*STAR (Agency for Science, Technology and Research), Singapore, Singapore; Division of Bioengineering, National University of Singapore, Singapore, Singapore*
GIORGIA IMPARATO • *Center for Advanced Biomaterials for Health Care@CRIB, Istituto Italiano di Tecnologia (IIT), Naples, Italy*
DOMINIK JAŃCZEWSKI • *Institute of Materials Research and Engineering, A*STAR (Agency for Science, Technology and Research), Singapore, Singapore*
HAE-WON KIM • *Institute of Tissue Regeneration Engineering (ITREN), Dankook University, Cheonan, South Korea*
STEFAN A. KOBEL • *Laboratory of Stem Cell Bioengineering (LSCB) and Institute of Bioengineering (IBI), Ecole Polytechnique Fédérale de Lausanne (EPFL), Lausanne, Switzerland*
MARK P. KREUZER • *ICFO Institute of Photonic Science, Castelldefels, Spain*
MATTHIAS P. LUTOLF • *Laboratory of Stem Cell Bioengineering (LSCB) and Institute of Bioengineering (IBI), Ecole Polytechnique Fédérale de Lausanne (EPFL), Lausanne, Switzerland*
YASSINE MAAZOUZ • *Biomaterials, Biomechanics, and Tissue Engineering Research Grroup, Technical University of Catalonia, ETSEIB, Barcelona, Spain*
BRENDAN MANNING • *Institute for Research in Biomedicine, IQAC-CSIC, CIBER de Bioingeniería, Biomateriales y Nanomedicina (CIBER-BBN), Barcelona, Spain*
M.-PILAR MARCO • *CIBER de Bioingeniería, Biomateriales y Nanomedicina (CIBER-BBN), Barcelona, Spain; Applied Molecular Receptors Group, IQAC-CSIC, Barcelona, Spain*
ELENA MARTINEZ • *Nanobioengineering Laboratory, Institute for Bioengineering of Catalonia (IBEC), Barcelona, Spain; Centro de Investigación Biomédica en Red en Bioingeniería, Biomateriales y Nanomedicina (CIBER-BBN), Barcelona, Spain*
M. CRISTINA L. MARTINS • *Divisão de Biomateriais, INEB – Instituto de Engenharia Biomédica, Universidade do Porto, Porto, Portugal*
ALVARO MATA • *The Nanotechnology Platform, Parc Científic Barcelona, Barcelona, Spain*
BENEDETTO MELE • *Department of Aeronautical Engineering, University of Naples Federico II, Naples, Italy*
MELBA NAVARRO • *Institute for Bioengineering of Catalonia (IBEC), Barcelona, Spain; Networking Research Centre on Bioengineering, Biomaterials and Nanomedicine, CIBER-BBN, Barcelona, Spain*
ANA PATRÍCIA NETO • *Faculty of Pharmacy, Department of Pharmaceutical Technology, University of Porto, Porto, Portugal*
PAOLO A. NETTI • *Italian Institute of Technology (IIT), Genoa, Italy; Interdisciplinary Research Centre on Biomaterials (CRIB), University of Naples Federico II, Naples, Italy*
JOSÉ DAS NEVES • *Faculty of Pharmacy, Department of Pharmaceutical Technology, University of Porto, Porto, Portugal*
LIAM PALMER • *Department of Chemistry, Northwestern University, Evanston, IL, USA*
MATEU PLA-ROCA • *Laboratory of Surface science and Technology, Swiss Federal Institute of Technology (ETH Zurich), Zurich, Switzerland*

JOSEP A. PLANELL • *Institute for Bioengineering of Catalonia (IBEC), Barcelona, Spain; Networking Research Centre on Bioengineering, Biomaterials and Nanomedicine, CIBER-BBN, Barcelona, Spain*

ROMAIN QUIDANT • *ICFO Institute of Photonic Science, Castelldefels, Spain*

ARTUR RIBEIRO • *G.I.R. Bioforge, University of Valladolid, CIBER-BBN, Valladolid, Spain*

CATARINA RIBEIRO • *Faculty of Pharmacy, Department of Pharmaceutical Technology, University of Porto, Porto, Portugal*

JOSÉ CARLOS RODRÍGUEZ-CABELLO • *G.I.R. Bioforge, University of Valladolid, CIBER-BBN, Valladolid, Spain*

J.-PABLO SALVADOR • *CIBER de Bioingeniería, Biomateriales y Nanomedicina (CIBER-BBN), Barcelona, Spain; Applied Molecular Receptors Group, IQAC-CSIC, Barcelona, Spain*

JOSEP SAMITIER • *Nanobioengineering Laboratory, Institute for Bioengineering of Catalonia (IBEC), Barcelona, Spain; Department of Electronics, University of Barcelona, Barcelona, Spain; Centro de Investigación Biomédica en Red en Bioingeniería, Biomateriales y Nanomedicina (CIBER-BBN), Barcelona, Spain*

BRUNO SARMENTO • *Faculty of Pharmacy, Department of Pharmaceutical Technology, University of Porto, Porto, Portugal*

HOLGER SCHÖNHERR • *Department of Physical Chemistry I, University of Siegen, Siegen, Germany*

SUSANA R. SOUSA • *Divisão de Biomateriais, INEB – Instituto de Engenharia Biomédica, Universidade do Porto, Porto, Portugal; REQUIMTE/Instituto Superior de Engenharia do Porto, Departamento de Engenharia Química, Porto, Portugal*

SAMUEL I. STUPP • *Department of Chemistry, Northwestern University, Evanston, IL, USA; Department of Materials Science and Engineering, Northwestern University, Evanston, IL, USA; Institute for BioNanotechnology in Medicine, Northwestern University, Evanston, IL, USA*

ESTHER TEJEDA-MONTES • *The Nanotechnology Platform, Parc Científic Barcelona, Barcelona, Spain*

ANTONIO TILOCCA • *Department of Chemistry and Thomas Young Centre, University College London, London, UK*

NIKODEM TOMCZAK • *Institute of Materials Research and Engineering, A*STAR (Agency for Science, Technology and Research), Singapore, Singapore*

FRANCESCO URCIUOLO • *Institute of Composite and Biomedical Materials (IMCB), National Research Council (CNR), Naples, Italy*

G. JULIUS VANCSO • *Faculty of Science and Technology, Materials Science and Technology of Polymers, University of Twente, Enschede, The Netherlands; MESA+Institute for Nanotechnology, Enschede, The Netherlands; Division of Bioengineering, National University of Singapore, Singapore, Singapore*

DEHUA YANG • *Ebatco, Eden Prairie, MN, USA*

SZCZEPAN ZAPOTOCZNY • *Faculty of Chemistry, Jagiellonian University, Krakow, Poland*

Chapter 1

Is Nanotechnology the Key to Unravel and Engineer Biological Processes?

Melba Navarro and Josep A. Planell

Abstract

Regenerative medicine is an emerging field aiming to the development of new reparative strategies to treat degenerative diseases, injury, and trauma through developmental pathways in order to rebuild the architecture of the original injured organ and take over its functionality. Most of the processes and interactions involved in the regenerative process take place at subcellular scale. Nanotechnology provides the tools and technology not only to detect, to measure, or to image the interactions between the different biomolecules and biological entities, but also to control and guide the regenerative process. The relevance of nanotechnology for the development of regenerative medicine as well as an overview of the different tools that contribute to unravel and engineer biological systems are presented in this chapter. In addition, general data about the social impact and global investment in nanotechnology are provided.

Key words: Regenerative medicine, Nanotechnology, Tissue engineering

1. The Paradigm of Regenerative Medicine

Regenerative medicine is an emerging field that combines tools from research in non-mammalian and human development, stem cell biology, genetics, materials science, bioengineering, and tissue engineering in order to come up with the required knowledge to use and trigger developmental pathways in order to rebuild the architecture of the original injured organ and take over its functionality.

The main aim of tissue regeneration consists of establishing reparative pathways in order to treat degenerative diseases, injury, and trauma.

Research conducted during recent years in the regenerative processes of different types of organisms has been usually focused

Melba Navarro and Josep A. Planell (eds.), *Nanotechnology in Regenerative Medicine: Methods and Protocols*, Methods in Molecular Biology, vol. 811, DOI 10.1007/978-1-61779-388-2_1, © Springer Science+Business Media, LLC 2012

on understanding the mechanisms of natural regeneration in lower species. More recently, stem cell biology has contributed to get a deeper insight on human developmental processes, and the identification and understanding of biological signals such as growth factors and their crucial role in tissue formation has been of paramount importance (1). Combination of stem cells biology together with novel biomaterials development, and basic knowledge of the developmental processes and growth factors potential have raised hopes in regenerative medicine, and this brought great expectations into providing clinicians with more efficient and functional response to tissue injury or disease.

There are different interpretations and points of view about the meaning of regenerative medicine and the processes that it involves. A really broad definition of regenerative medicine includes the repair, replacement, or regeneration of damaged tissues or organs with a combination of several technological approaches (2). According to Haseltine (2003), the term "regenerative medicine" encompasses four different evolutive phases (3). The first phase aims to emulate the body's own repair mechanisms by mimicking the actions of growth factors. The second phase involves implanting tissues or organs that have been generated outside the body with the help of the identified growth factors. The third phase involves technologies that can rejuvenate or renew old tissues, by resetting a cell's biological clock. The fourth phase is addressed to the exploitation of two emerging fields, namely, nanotechnology and materials science. Nanotechnology will provide the tools for engineering and studying biological entities and interactions to "subatomic physical tolerances." It will also provide the capability of engineering and characterizing artificial materials at the nanoscale.

A different definition of regenerative medicine describes it as a multidisciplinary field where various phases or disciplines are merged and some of them are currently in development. Some of these phases involve fields that already have a solid background such as the development of artificial organs, synthetic implants, and organ transplantation, while some others are more recent as in the cases of tissue engineering and gene therapy, and finally, some others have just started such as for stem cell therapy. In general, there is a common point in all these definitions, and this is the very close association of engineering with biology approaches. This combination is considered as an essential and integral characteristic of regenerative medicine (4).

In addition to the previous definitions where there is an important and key synergy between life sciences and engineering/technology in order to generate new tools and means for nanoscale study of bio/non-bio events, interactions, and structures (Fig. 1), some authors envision the "regenerative medicine revolution" as a more focused and narrow subject, based on a series of new exciting breakthrough discoveries in the field of stem cell biology.

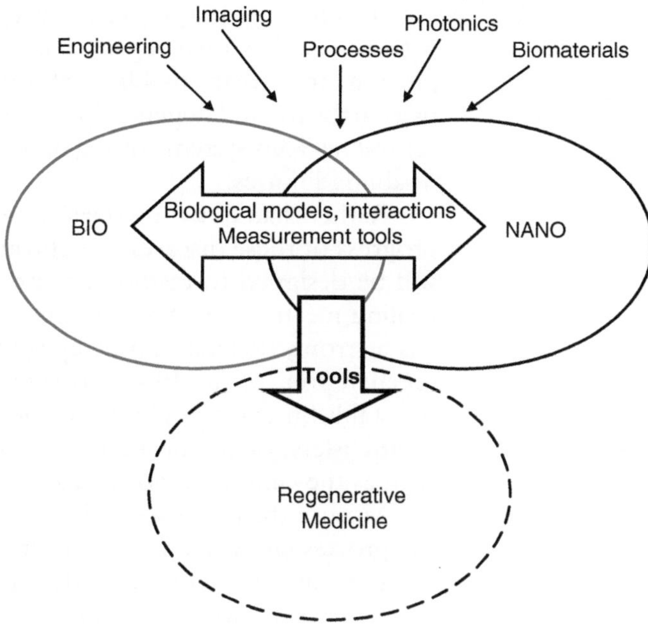

Fig. 1. Interactions between nanotechnology, biology, and regenerative medicine and the influence of other related fields.

In this case, regenerative medicine is described as an outstanding manifestation of the emergence of applied stem cell and developmental biology.

Whichever the approach is, regenerative medicine explores the repair and regeneration of organs and tissues using the natural signalling pathways and components such as stem cells, growth factors, and peptide sequences among other biomolecules, in combination with synthetic scaffolds (5). Furthermore, immediately after a triad of cells/scaffold/signals is implanted, a very complex cascade of biological events and processes is activated. Some of these processes such as angiogenesis and nutrients delivery are crucial to stimulate tissue growth. A key issue when trying to reproduce the regeneration events is to understand and translate the flow of information (signals) between stem cells that will differentiate, biosynthesize, and regenerate extracellular matrix (ECM), and their microenvironment. Biochemical cues dispersed in the extracellular space as well as other biophysical signals are interpreted by cells as a series of intracellular instructions that lead to changes in their behaviour and that activate certain cellular events in response to such signals. Cells may also reply to extracellular signals by sending their own messages into the surrounding medium either by biosynthesizing new ECM, or/and remodelling the existing one. This bidirectional communication of the cell with the extracellular medium is in fact the language that should be deeply understood, learnt, and used in order to control, guide, and engineer the regenerative process.

In summary, the objective of regenerative medicine is to work with the body's own repair capacity and healing mechanisms to prevent and treat disabling chronic diseases such as diabetes, osteoarthritis, and degenerative disorders of the cardiovascular and central nervous system among others, and to assist individuals with disabling injuries.

Rather than targeting the symptoms or attempting to delay the progress of these diseases and chronic conditions, future therapies will be designed to restore patients' health using the body's own healing mechanisms. Some examples where regenerative medicine has a strong societal impact by solving prevalent diseases are the regeneration of healthy cartilage in osteoarthritic joints, the re-establishment of a physiological release profile in diabetic pancreatic islets, or the promotion of self-repair mechanisms in areas such as the central nervous system and the heart.

Most of the processes and interactions involved in the regenerative process take place at subcellular scale. Nanotechnology provides the tools and technology to detect, to measure, or to image the interactions between the different biomolecules and biological entities. Moreover, nanotechnology allows the design and development of nanomaterials and nanodevices to interact directly with cells and tissues at a molecular level. This new technology in turns makes possible the investigation and transformation of biological systems, while biology offers inspiration models to nanotechnology.

2. Nanotechnology

There are different definitions in literature describing the term nanotechnology. In general it can be defined as the science and engineering involved in the design, synthesis, characterization, and application of materials and devices whose smallest functional organization in at least one dimension is on the nanometer scale (6). A different definition describes nanotechnology as "the ability to work at the atomic, molecular, and supramolecular levels (on a scale of approximately between 1–100 nm) in order to understand, create, and use material structures, devices, and systems with fundamentally new properties and functions resulting from their small structure" (7).

It is noteworthy to mention that nanotechnology is not in itself a single emerging scientific discipline but rather an interdisciplinary combination of traditional scientific disciplines such as chemistry, physics, materials science, and biology orchestrated to bring together the required collective expertise needed to develop these novel technologies.

The potential impact of nanotechnology stems directly from the fact that engineering materials and devices at the nanometer scale implies controlled manipulation of the individual atoms and molecules and their arrangement to form bulk macroscopic bodies (6).

For applications to medicine and physiology, these materials and devices can be designed to interact with cells and tissues at a molecular level with a high degree of integration between technology and biological systems not previously attainable.

At present, different methods for the synthesis of nanoengineered materials and devices are used. In general, however, most synthetic methods can be classified into two main approaches: "top-down" and "bottom-up" approaches, and their combination. The top-down approach consists in going from large scales to small scales, in this case, towards the nanoscale. A piece of the bulk material is gradually sculpted until the desired shape is achieved. That is, you start at the top of the bulk piece and work your way down removing material from where it is not required. Nanotechnology techniques for top-down fabrication vary but can be divided into mechanical and chemical fabrication. The most illustrative top-down fabrication technique is nanolithography. In this process, the required material is protected by a mask and the exposed material is etched away. Depending upon the level of resolution required for features in the final product, etching of the base material can be done chemically using acids or mechanically using ultraviolet light, X-rays, or electron beams. The bottom-up approach consists in going from very small scales towards larger ones. Bottom-up fabrication techniques place atoms or molecules one at a time to build the desired nanostructure. In this case, self-assembly techniques are used where the atoms arrange themselves as required.

In addition to synthesis and fabrication methods to obtain nanoengineered materials and nanostructures, there are other aspects where the use of nanotechnology is extremely valuable. The visualization, characterization, and manipulation of materials and devices require sophisticated imaging and quantitative techniques with spatial and temporal resolutions on the order of 10^{-6} and below the molecular level. These techniques are critical for understanding the relationship and interface between nanoscopic and mesoscopic/macroscopic scales, a particularly important objective for biological applications. As such, further nanotechnological advances will require parallel progress of these physical characterization techniques. These techniques are necessary not just to probe the structure of the materials themselves but also to study the interface between these materials and the cells and tissues they are designed to interact with, because they are the only available means to look at what is happening at these extremely small scales.

3. The Role of Nanotechnology in Regenerative Medicine

Nanobiotechnology is a recent term describing the convergence of the two existing but distant worlds of engineering and cellular and molecular biology. Engineers have been working for the past three decades on the development and fabrication of structures with dimensions within the nanoscale; mainly in the electronics industry, in order to enable faster and higher density electronic chips. Concomitantly, molecular biologists have a wide experience working with molecular and cellular dimensions which usually vary between several nanometers and several micrometres (8). It is believed that the combination of both disciplines will allow the development of a new type of devices and systems for biological and chemical analysis characterized by better sensitivity and specificity and higher rates of recognition compared with existing ones (9).

One of the most illustrative examples of the convergence of nanotechnology and molecular biology is the protein synthesis process. It works as a very efficient nanofabrication laboratory where atoms and molecules are combined and assembled following an extremely precise, reproducible, and organized bottom-up process. This is a highly controlled process where any single mistake may lead to important diseases. Nanobiotechnology will contribute to shed lights on such subcellular level processes and their translation into other bio and non-bio systems.

Nanobiotechnology is defined as a field that applies the nanoscale principles and techniques to understand and transform biosystems (living or non-living) and which uses biological principles and materials to create new devices and systems integrated from the nanoscale (10, 11). Furthermore, the convergence of nanoscale science with modern biology and medicine is a trend that should be reflected in science policy decisions (12, 13).

Thanks to nanobiotechnology, a cellular and molecular basis has been established for the development of innovative disease-modifying therapies for in situ tissue regeneration and repair, requiring only minimally invasive surgery.

Moreover, nanotechnology can play a critical role in the development of cost-effective therapies for in situ tissue regeneration. Thus, it is necessary to have a comprehensive insight of both basic biology of tissue regeneration and the effective ways to initiate and control the regenerative process. This nanobiomimetic strategy depends on three basic elements: smart biomaterials, bioactive signalling molecules, and cells which are the basic triad of Tissue Engineering. Nanotechnology can provide tools in the development of biomimetic, smart biomaterials by tailoring resorbable polymers at the molecular level for specific cellular responses. These biomaterials are specifically designed to mimic the ECM environment, thus, providing the right chemical cues such as attachment sequences and mechanical stimuli as to stimulate specific regenerative events at the molecular level, directing cell proliferation, cell differentiation, and ECM production and organization. Nanotechnology is a valuable tool for the development of systems for controlled supply and delivery of proteins, peptides, and genes to activate cell events and mimic nature's signalling cascades.

The third element of the Tissue Engineering triad are cells, thus, as previously mentioned, cells are in the spotlight of regenerative medicine and future efforts will be done towards the effective exploitation of the enormous self-repair potential that has been observed in adult stem cells. In fact, outstanding breakthroughs have been already achieved as in the case of the larynx transplant that took place in the Hospital Clinic of Barcelona (2008) where an allograft was seeded with stem cells from the own patient and successfully implanted avoiding any possible rejection of the immune system (14). Other example is the repair of heart muscle after heart attack carried out by researchers at the University of Minnesota (2010) where acute myocardial infarctions were treated by tagging stem cells with antibodies that recognize the cell surface proteins of damaged myocytes (15).

Nanotechnologies may assist in identifying signalling systems, in order to control the self-healing potential of endogenous adult stem cells, and in developing efficient targeting systems for stem cell therapies. The support of nanotechnology in the achievement of the proper triad will allow the attainment of cell-free, intelligent bioactive materials that would effectively provide signalling to stimulate the self-healing potential of the patient's own stem cells (16).

4. Nanotechnology Provides the Tools to Measure and Understand Biosystems and Biological Processes

Research of fundamental biological processes at the nanoscale has yielded relevant new knowledge in systems biology, cellular processes, and molecular self-assembly. In addition, significant advances have been made in the development of technology to measure biological entities and interactions at the subcellular level and in understanding the cell as a highly organized, self-repairing, self-replicating, information-rich molecular machine (17, 18). Single-molecule measurements are bringing knowledge and opening new pathways on the processes and mechanistic properties of molecular biological entities, both in vivo and in vitro, allowing the direct observation and investigation of molecular events, motors, enzyme reactions, protein dynamics, DNA transcription, and cell signalling.

Nanoscale instrumentation has allowed measurement of spatial and temporal interactions among cells, including intracellular forces, as well as the intermolecular binding strength between pairs of molecules in physiological solutions, providing the quantitative evidence of their cohesive function (19, 20).

Some examples of the success of the development of nanoscale instrumentation are the measurement of the intermolecular forces between single molecules such as proteins, polymer molecules, or other biomolecules performed by atomic force microscopy (AFM) (21) on the development of a nanoscale system that displays protein unfolding events as visible colour changes (22). This system will greatly improve the probability to visualize structural changes of proteins in complex synthetic and living systems; the quantitative measurement of interfacial and adhesion forces between living bacteria and mineral surfaces in situ by means of an atomic force microscope (23). Another contribution of nanotechnology has been the measurement of nanoRNAs and their significant effect on gene expression (24).

Furthermore, nanotechnology has facilitated the development of methods for detection of single cells or groups of few molecules. Nanoproteomics, the application of nanobiotechnology to proteomics, can enable detection of a single molecule or protein (25). Biobarcode assays enable detection of tiny quantities of proteins in body fluids that cannot be detected by conventional methods (26).

In general, nanoscience has allowed to understand self-organization, supramolecular chemistry, and assembly dynamics, and has moved forward our knowledge of the self-assembly of nanoscopic, mesoscopic, and even macroscopic components in living systems (27, 28). However, current understanding of the biosystem building blocks and their interactions at the nanoscale is rather limited and significant work has to be done to reach a more complete picture.

5. Nanotools in Biology and Medicine

In addition to instrumentation, nanotools such as nanoparticles, nanotubes, nanobiosensors, nanochannels, nanopores, nanowires, and nanopatterned surfaces are becoming of great importance in the study of biological events at the nanoscale.

Nanoparticles are among the most currently developed nanotools. Given the variety of nanoparticle technologies that are available, it is possible to tailor nanoparticle cores and surfaces with a high degree of specificity in order to use them in different applications such as cell-trackers, biomarkers, and diagnostics (29).

Nanoparticle technology offers significant advantages over other cell-labelling technologies under development. An example of this technology is the perfluorocarbon nanoparticles detected by MRI (magnetic resonance imaging). In this case, a MRI scanner is tuned to the specific frequency of the fluorine compound in the nanoparticles. Therefore, only the cells that contain the perfluorocarbon nanoparticles are visible and detected in the scan. Cells labelled with this type of nanoparticles have been studied and in vitro cell cultures have shown that the cells retained their usual surface markers and that their functionality was still retained after the labelling process. This cell-tracking technology is a useful and promising option in the diagnosis and monitoring of tumours as well as other pathologies.

In the case of biomarkers discovery, the physicochemical characteristics and high surface area of nanoparticles make them ideal candidates for developing biomarker-harvesting platforms. Nanoparticle surfaces can be modified to selectively bind a subset of biomarkers and sequester them for later study using high-sensitivity proteomic tests.

Although different types of nanoparticles have been used in the diagnostics field, the most currently used are gold nanoparticles, magnetic nanoparticles, and QDs (quantum dots). Gold nanoparticles can assemble onto a sensor surface in the presence of a complementary target. These nanoparticles are particularly good labels for sensors because a variety of analytical techniques can be used to detect them. Magnetic nanoparticles are used as labelling molecules for bioscreening. These particles usually consist in a magnetic core coated by a polymeric layer functionalized with antibodies for capturing cells or specific biomolecules. QDs are inorganic fluorophores that offer significant advantages over conventionally used fluorescent markers as their photobleach rate is much slower (30). QDs possess a combination of unique properties such as high sensitivity, broad excitation spectra, stable fluorescence with simple excitation, and no need for lasers. Quantum dots have an extensive variety of applications mainly for diagnostics; in particular, in viral diagnosis and in some diseases such as cancer they have proved to be a very promising tool. Their integration with therapeutics is also under study.

Carbon nanotubes (*CNTs*) have proved to serve as safer and more effective alternatives to previous drug delivery systems (31). Their dimensions within the nanoscale allow them to pass through the cell membrane, carrying therapeutic drugs, vaccines, peptides, and nucleic acids inside the cell and heading to very specific targets. CNTs can also be used as narrow conduits for flow-based assays, as reinforcement in composite materials, or as tips in AFM for high-resolution imaging. Besides CNTs, other compounds such as boron nitride, gallium nitride, boron carbide, and some organic polymers have been used in the fabrication of nanotubes for multiple applications (32). Though CNTs seem to be a very promising tool there are some doubts and controversy with respect to their in vivo behaviour and elimination through metabolic pathways.

Nanobiosensors are sensors used for the detection of chemical or biological signals at the nanoscale. These sensors can be electronically tuned to react to the binding of a single molecule. Some prototype sensors have demonstrated their capacity to detect nucleic acids, proteins, and ions. These sensors can operate in the liquid or gas phase, opening up an enormous variety of downstream applications. They are inexpensive to manufacture and are portable. It may even be possible to develop implantable detection and monitoring devices on the basis of these detectors.

Nanofluidic arrays and protein nanobiochips are also some examples of devices that incorporate nanotechnology-based biochips and microarrays. Nanotechnology on a chip is a new paradigm for total chemical analysis systems (33). One of the more promising uses of nanofluidic devices is isolation and analysis of individual biomolecules, such as DNA.

Nanopatterns are now being used in biomaterials science as a tool for controlling tissue regeneration, since it has been shown that nanotopography seems to have an essential role in guiding cell behaviour in vitro.

A wide variety of techniques have been used to produce specific nanotopography on biomaterial surfaces. Both, methods producing an ordered topography with a regular controlled pattern and methods producing unordered topography with random orientation and organization have been developed. Polymer demixing, chemical etching, and colloidal lithography are some of the most relevant techniques to obtain randomly organized patterns. Conversely, soft-lithography techniques and other methods using different sources of radiation (electrons, ions, or photons) to etch the substrate are among the most commonly used techniques to produce regular geometries.

Besides topography, surface chemistry is a critical parameter conditioning cell behaviour and consequently, the success of a biomaterial-based therapy for tissue regeneration.

All the techniques of nanostructuration cited here can be combined with different surface functionalization techniques where both chemical bonding and physical adsorption take place. This includes a variety of techniques ranging from rather simple ones such as dip-coating to more sophisticated "bottom-up" and "top-down" techniques such as self-assembly. Finally, dip-pen lithography is a simple method based on the AFM tip-assisted deposition of molecules (ink) to create chemical patterns at the nanoscale.

6. Social Impact and Global Investment

Nowadays, nanotechnology is considered as a worldwide priority. In fact, it is expected that the use of nanotechnological expertise will considerably contribute to accomplish several issues concerning future global demands such as how to meet energy needs, natural resources preservation, and how to provide comprehensive and preventive medical care, where regenerative medicine plays a relevant role.

Disease areas which can be expected to benefit the most from nanotechnology within the next 10 years are cancer, diabetes, diseases of a wide variety of systems such as cardiovascular, pulmonary,

Table 1
Global market for some of the most promising systems developed to tackle different defects and pathologies (42)

Market size (M€)	2006	2015	2020	2025
Spine devices	3,800	5,000	7,000	9,000
Wound Care (active dressing)		5,000	12,000	17,000
Bone fillers	115	240	300	380
Orthopaedic Biomaterials	180	260	320	430

Table 2
Global market size for cell therapies (42)

Market size (M€)	2015	2020
Non stem cell-based therapies (driven by wound healing, orthopaedics)	1,000	2,500
Tissue Engineering (driven by orthopaedics, wound healing, cardiac, neurological)	10,000	20,000
Stem cell therapies (driven by cardiovascular, diabetes)	1,000	7,000
Supporting technologies	3,500	8,000

blood, neurological, urinogenital, orthopaedic and traumatological problems, and inflammatory/infectious diseases (16). Tables 1 and 2 summarize the expected market size for some systems and devices dealing with different pathologies and for each disease ranked by research complexity. Since most of these diseases are age-related they are expected to grow with the increase of senior population.

Tissue repair and regeneration are other areas where nanotechnology could have great impact. This is the case of biodegradable nanoparticles which release appropriate biomolecules, namely morphogens, growth factors, peptide sequences, among others with a high specificity. An example of these nanoparticles as delivery systems is the release of angiogenic factors that could improve the regeneration of tissues.

Given the relevance of nanotechnology and its social and economical impact, more than 35 countries have developed programs in nanotechnology during the last decade. Moreover, there has been a significant increase in funding devoted to nanotechnology and nanobiosystems during recent years. The USA initiated a multidisciplinary strategy for the development of science and engineering by the creation of the National Nanotechnology Initiative

(NNI). The Federal Government in USA increased the investment in nanotechnology R&D from $116 million in fiscal year 1997 to about $697 million in fiscal year 2002 (34).

The proposed NNI budget for fiscal year 2011 is nearly $1.8 billion reflecting continued steady growth in the NNI investment from its beginnings in order to advance the understanding of nanoscale phenomena and the ability to engineer nanoscale devices and systems to cover national priorities and global challenges in areas such as energy conversion and storage, and medicine. The fact that US Federal government has devoted an important and increasing budget in nanotechnology research over the past 11 years of the NNI, is a clear demonstration of the strong support that US government gives to this program. This major investment is based on the potential that nanotechnology has to improve our fundamental understanding and control of matter at the nanoscale. This nanoscale control in turn leads to breakthroughs and outstanding changes in technology and industry for the benefit of society. Though the main focus of NNI is still directed towards the support and knowledge widening of fundamental research, infrastructure development, and technology transfer, investments for 2011 are not only allocated to pursue basic research but also to generate more applicable innovative tools. In this sense, significant emphasis is placed on accelerating the transition from basic R&D advances and capabilities into innovations that support national priorities such as sustainable energy technologies, healthcare, and environmental protection. There are also important investments in education and in research on ethical, legal, and other societal aspects of nanotechnology for a total over $260 million (35).

Research on nanotechnology has also received a significant support in the UK, most European countries, Japan, China, Israel, Finland, Singapore, Canada, South Korea, and Australia among others. Differences in funding allocation among countries are mainly observed in the research areas each country is aiming for, in the level of program integration into various industrial sectors, and in the time scale of their R&D targets.

In Australia, the 2007 report *Government Funding, Companies and Applications in Nanotechnology Worldwide* produced by the Technology Transfer Centre placed Australia eighth out of 16 nanotechnology leading countries, with a projected Australian Government funding in nanotechnology of $833 million for 2006–2010. Indeed, Australian Government support for nanotechnology in 2008–2009 was worth $107.7 million (36).

At present Europe has a strong position in the emerging field of Nanomedicine that has a high potential for technological and conceptual breakthroughs, innovation, and creation of employment. Close cooperation between industry, research centres, academia, hospitals, regulatory bodies, funding agencies, patient organizations, investors, and other stakeholders could dramatically boost this promising field.

In particular, the following three research areas have been identified as a basis of a Strategic Research Agenda (SRA) in this field: nanotechnology-based diagnostics including imaging, targeted drug delivery and release, and regenerative medicine.

In 2007, the first year of FP7 implementation, the European Commission allocated around EUR 600 million to projects for research and development in nanosciences and nanotechnologies. Around EUR 75 million was devoted to nanotechnology within the Marie Curie action. Around EUR 20 million was allocated to research on health, safety, and environmental aspects of nanomaterials, in particular on the possible impact that nanoparticles might have on health and the environment. Furthermore, some funding (EUR 8 million) was allocated to other issues also related to nanotechnology such as projects to promote debate and measures for the good supervision of nanotechnology, including the creation of an observatory on the development and use of nanotechnology (37).

In 2009 the European Technology Platform in Nanomedicine recognized once more the significant importance of a fluid communication between academia, clinics, and industry in order to translate and launch the generated knowledge into outstanding competitive products for regenerative medicine to boost up EU economy. As a result, ETP Nanomedicine proposed the creation of new programs and roadmaps involving all the key players namely, academia, industry, clinics, regulatory entities, and end users to drive the process.

The South Korean government has also recognized that nanotechnology is basic for the development in the areas of information, biology, materials, energy, environment, and military. In addition, South Korean government has considered that this newly emerging technology is definitely a driving force that will create new industries and high-tech products through technological breakthroughs. As a consequence, South Korean government has proposed a national nanotechnology initiative as the means to place the country at the top in certain competitive areas and to develop niche market for industry growth. Since 2000, funding devoted to nanosciences has considerably increased. In fact, the increase in government spending in nanotechnology for 2002 compared to 2000 was about 400%.

In 2002, the "Plan for Implementing Nanotechnology Development" was launched together with two new Frontier Research Programs, namely, Nanoscale Mechatronics & Manufacturing Technologies. Each of the programs was funded with \$100 million for the next 10 years. South Korea has committed \$2 billion over 10-year period (2001–2010) to nanoscience and technology R&D (38).

Singapore has also clearly recognized the importance of nanoscience R&D and the impact of nanotechnology on Singapore's key manufacturing industries, and ultimately the significance for the country's economy. Nanotechnology has been identified by the Economic Review Committee (ERC) as a key area for Singapore's pursuit of competitive advantages. The most important funding agency in Singapore for Nanoscience and Technology is the Agency for Science, Technology & Research (A*STAR). The A*STAR Nanotechnology Initiative started in September 2001. Its goal is to disseminate research focused on nanotechnology and generate basic background for the development of this new field. This is part of Singapore's continuing effort to invest in the development and in the promotion of innovation in areas that boost Singapore's industry (39).

Japanese government has also considered the development and exploitation of nanotechnology as one of their priorities in terms of worldwide competitiveness. In fact, it initially formulated a policy to promote nanotechnology in their second Science and Technology Basic Plan for 2001–2005. In the third plan for 2006–2010, nanotechnology was established as one of four priority research fields, with major investments targeted at reducing the cost of clean energy, electronics, and the development of green technologies.

In 2002, the total budget for nanotechnology research in Japan was about 82.5 billion yen and increased to about 90.4 billion in 2003. The main entities involved in nanotechnology research in Japan are the MEXT (Ministry of Education, Culture, Sports, Science and Technology) and METI (Ministry of Economy, Trade and Industry). MEXT has just less than two thirds of the total budget, most of which is associated with individual funds awarded competitively rather than non-competitive national ones. Most of the remaining total budget is used by METI, mainly in national projects (40).

The nanotech market in Japan has grown significantly in recent years. Indeed, in 2006, the Nomura Research Institute estimated that the domestic market for nanotechnology would grow to 6 trillion yen by 2010 and to 23 trillion yen by 2015. A different survey by the Japanese Ministry of Economy, Trade and Industries (METI) in 2008, estimated that the domestic market would grow to 13 trillion yen by 2020. Nearly 600 nanotech and related companies operated in Japan in 2004, according to a survey by the New Energy and Industrial Technology Development Organization (NEDO).

At present, basic research is being promoted by the MEXT, the Japan Society of the Promotion of Science (JSPS) and universities (41).

References

1. Gurtner, G.C., Calleghan, M.J., Longaker, M.T. (2007) Progress and Potential for Regenerative Medicine, *Annu. Rev. Med.* **58**, 299–312
2. Nanomedicine, Nanotechnology for health, (2006) ETP Strategic Research Agenda for Nanomedicine, Available online at: http://cordis.europa.eu/nanotechnology/nanomedicine.htm
3. Haseltine, W. (2003) Regenerative medicine 2003: An overview, *J Regener Med* **4**, 15–18
4. Mironov, V., Visconti, R.P., Markwald, R.R. (2004) What is regenerative medicine? Emergence of applied stem cell and developmental biology, *Expert Opin Biol Ther* **4**, 773–781
5. Hardouin, P., Anselme, K., Flautre, B., Bianchi, F., Bascoulenguet, G. & Bouxin, B. (2000) Tissue engineering and skeletal diseases. *Joint Bone Spine* **67**, 419–424
6. Silva, G.A., Introduction to nanotechnology an its applications to medicine, *Surg Neurol* **61**, 216–220
7. Craighead, H., Leong, K. (2000). Biological, medical and health applications. In: Research Directions, Roco, M.C., `Williams, R.S., Alivisatos, P (eds) Kluwer Academic Publishers
8. Whitesides, G.M. (2003) The "right" size in nanobiotechnology, *Nat Biotechnol* **18**, 760–763
9. Nienmeyer, C.M., Mirkin, C.A. eds (2004) Nanobiotechnology: Concepts, Applications and Perspectives, Wiley
10. Roco, M. (2005) Converging Technologies In: Biomedical Nanotechnology, Malsch, N.H. (ed) CRC Press
11. Roco, M.C., Bainbridge, W.S. (eds) Converging Technologies for Improving Human Performance. NSF-DOC Report, 2002; Kluwer: Boston. URL: http//wtec.org/ConvergingTechnologies
12. National Research Council: Small wonders-endless frontiers, a review of the National Nanotechnology Initiative. The National Academies Press, Washington DC; 2002. Available online at: URL: http//www.nsf.gov/home/cressprgm/nano/smallwonders_pdffiles.htm
13. National Research Council: Implications of emerging micro and nanotechnologies. The National Academies Press, Washington DC, 2002
14. Macchiarini, P., Jungebluth, P., Go, T., Asnaghi, M.A., Rees, L.E., Cogan, T.A., Dodson, A., Martorell, J., Bellini, S., Parnigotto, P.P., Dickinson, S.C., Hollander, A.P., Mantero, A., Conconi, M.T., Birchall, M.A. (2008) Clinical transplantation of a tissue-engineered airway, *Lancet* **372**, 2023–2030
15. Zhang, G., Hu, Q., Braunlin, E.A., Suggs, L.J., Zhang, J. (2010) Enhancing Efficacy of Stem Cell Transplantation to the Heart with a PEGylated Fibrin Biomatrix, *Advances in Tissue Engineering: Angiogenesis* **20**, 195–205
16. European Technology Platform on Nanomedicine, Nanotechnology for health (2005) European Commission, Luxembourg. Available online at: http://cordis.europa.eu/nanotechnology/nanomedicine.htm
17. Ishima, A., Yanagida, T. (2001) Single molecule nanobioscience, *Trends Biochem Sci* **26**, 438–444
18. Müller, D.J., Janovjak, H., Lehto, T., Kuerschner, L., Anderson, K. (2002) Observing structure, function and assembly of single proteins by AFM, *Prog Biophys Mol Biol* **79**, 1–43
19. Misevic, G.N. (2001) Atomic force microscopy measurements: binding strength between a single pair of molecules in physiological solutions, *Mol Biotechnol* **18**, 149–154
20. Bao, G. (2002) Mechanics of Biomolecules, *J Mechanics Phys Solids* **50**, 2237–2274
21. Ikai, A., Idiris, A., Sekiguchi, H., Arakawa, H., Nishida, S. (2002) Intra and intermolecular mechanics of proteins and polypeptides studies by AFM. *Appl Surf Sci* **188**, 506–512
22. Baneyx, G., Baugh, L., Vogel, V. (2001) Coexisting conformations of fibronectin in cell culture imaged using fluorescence resonance energy transfer, *Proc Natl Acad Sci USA* **98**; 14464–14468
23. Lower, S.K., Hochella, M.F.H., Beveridge, T.J. (2001) Bacterial recognition of mineral surfaces: nanoscale interactions between Shewanella and alpha-FEDOH, *Science* **292**, 1360–1363
24. Couzin, J. (2002) Breakthrough of the year: small RNAs make big splash, *Science* **298**, 2296–2297
25. Jain, K.K. (2007) Nanobiotechnology: Applications, Markets and Companies. Basel: Jain (Ed) PharmaBiotech Publications
26. Bao, Y.P., Wei, T.F., Lefebvre, P.A., An, H., He, L., Kunkel, G.T. (2006) Detection of

protein analytes via nanoparticle-based bio bar code technology. *Anal Chem* **78**, 2055–2059

27. Davis, A.V., Yeh, R.M., Raymond, K.N. (2002) Supramolecular assembly dynamics, *Proc Natl Acad Sci USA* **99**, 4793–4796
28. Whitesides, G., Boncheva, M. (2002) Beyond molecules: self assembling of mesoscopic and macroscopic component, *Proc Natl Acad Sci USA* **99**, 4769–4774
29. Geho, D.H., Jones, C.D., Petricoin, E.F., Liotta, L.A. (2006) Nanoparticles: potential biomarker harvesters. *Curr Opin Chem Biol* **10**, 56–61
30. Chan, W.C., Maxwell, D.J., Gao, X., Bailey, R.E., Han, M., Nie, S. (2002) Luminiscent quantum dots for multiplexed biological detection and imaging, *Curr Opin Biotechnol* **13**, 40–46
31. Lin, Y., Taylor, S., Huaping, L., Shiral, K. A., Qu, L., Wang, W., Gu, L., Zhou B., Sun Y-P., (2004) Advances toward bioapplications of carbon nanotubes, *J Mater Chem* **14**, 527–541 K
32. Ciofani, G., Raffa, V., Menciassi, A., and Cuschieri, A. (2009) Boron nitride nanotubes: an innovative tool for nanomedicine. *Nano Today* **4**, 8–10
33. Jain, K.K. (2005) Nanotechnology-based lab-on-a-chip devices. In: Fuchs J, Podda M. (eds). Encyclopedia of Diagnostic Genomics and Proteomics, New York: Marcel Dekkar Inc
34. Roco, M.C. (2001) International strategy for nanotechnology research and development, *J Nanoparticle Res* **3**, 353–360
35. The National Nanotechnology Initiative, Supplement to the president's 2011 budget, National Science and Technology Council, USA 2010. Available online at: http://www.nano.gov/NNI_2011_budget_supplement.pdf
36. National nanotechnology strategy annual report 2008–2009 (2009) Australian office of nanotechnology
37. Nanotechnology in Australia, Trends, applications and collaborative opportunities (2009) Australian Academy of Science http://www.science.org.au/reports/documents/nanotechnology09.pdf
38. European Commission, EU Policy for nanosciences and nanotechnologies. Available online at: ftp://ftp.cordis.europa.eu/pub/nanotechnology/docs/eu_nano_policy_2004-07.pdf
39. Nanotechnology Strategy in Korea, (2003) Asia Pacific Nanotechnology Weekly 1, Nanotechnology Research Institute, AIST. Available online at: http://www.nanoworld.jp/apnw/articles/library/pdf/27.pdf
40. Kishi, T., Bando, Y. (2004) Status and trends of nanotechnology R&D in Japan. *Nat Materials* **3**, 129–131
41. Okada, S. (2009) Big on nanotech: Japan is one of the world leaders in nanotechnology, but its position could be threatened by a lack of govern, *Entrepreneur*. Available online at: http://www.entrepreneur.com/tradejournals/article/print/201599863.html 2010
42. Roadmap in nanomedicine towards 2020. Expert report 2009, ETP Nanomedicine, European Commission. Available online at: http://www.etp-nanomedicine.eu/public/press-documents/publications/etpn-publications/091022_ETPN_Report_2009.pdf

Chapter 2

Synthesis of Genetically Engineered Protein Polymers (Recombinamers) as an Example of Advanced Self-Assembled Smart Materials

José Carlos Rodríguez-Cabello, Alessandra Girotti, Artur Ribeiro, and Francisco Javier Arias

Abstract

In this chapter, we describe two methods for bio-producing recombinant repetitive polypeptide polymers for use in biomedical devices. These polymers, known as elastin-like recombinamers (ELRs), are derived from the repetition of selected amino acid domains of extracellular matrix proteins with the aim of recreating their mechanical and physiological features. The proteinaceous nature of ELRs allows us to make use of the natural biosynthetic machinery of heterologous hosts to express advanced and large polymers or "recombinamers." Despite the essentially unlimited possibilities for designing recombinamers, the production of synthetic genes to encode them should allow us to overcome the difficulties surrounding bioproduction of these non-natural monotonous DNA and protein sequences. The aim of this work is to supply the biotechnologist with fine-tuning methods to biosynthesize advanced self-assembled smart materials.

Key words: Elastin-like recombinamers, Genetically engineered protein polymers, Tissue engineering biomaterials, Self-assembly

1. Introduction

The biosynthesis of peptide polymers using natural pathways allows monodisperse and extremely sophisticated polymers to be synthesized (1). Biomaterial properties such as composition, molecular weight, and monodispersity affect pharmacokinetics and cellular-transport phenomena, therefore complete sequence and architecture control allows such materials to be designed with an extremely high degree of both functional and structural complexity (2). This control is now possible by constructing synthetic genes, which can include any sequence that encodes functional peptide domains that confer

Melba Navarro and Josep A. Planell (eds.), *Nanotechnology in Regenerative Medicine: Methods and Protocols*, Methods in Molecular Biology, vol. 811, DOI 10.1007/978-1-61779-388-2_2, © Springer Science+Business Media, LLC 2012

mechanical, structural, or bioactive properties and also distribute and collocate them with the required ratio and consecutiveness (3). The resulting biopolymers, or recombinamers, possess enormous potential as advanced materials for a variety of biomedical applications, such as cell harvesting, drug delivery, and tissue engineering in the formation of different kinds of 2D or 3D scaffolds (4).

The two elastin-like recombinamers (ELRs) described herein have been designed to accomplish two specific biomedical purposes: to act as advanced scaffolds for tissue engineering and as self-assembled drug-delivery biomaterials (5). The first ELR contains different building blocks, namely the elastomeric pentapeptide VPGIG, which confers the desired mechanical behavior and biocompatibility, and a variation of VPGIG that contains a L-lysine instead of an L-isoleucine for cross-linking purposes. The polymer also contains two bioactive domains for cell adhesion and specific biodegradation. Thus, the CS5 fibronectin domain includes the REDV tetrapeptide, which is specifically recognized by the integrin α4β1 (6) present in endothelial cells selectively bound to REDV-coated surfaces (7). The other functional block (ValGlyValAlaProGly $(VGVAPG)_3$) is a recurring hexapeptide derived from the human elastin exon 24-encoded product (8). This sequence has been introduced to drive enzymatic hydrolysis of the synthetic scaffold by the same physiological pathways as natural elastin during extracellular matrix (ECM) remodeling (8).

The use of block copolymers in nanobiotechnology has aroused a great deal of interest over the past few years as their self-assembly properties enable the creation of different structures on a nanometer scale in an easy and inexpensive manner using a *bottom-up* approach (9). Amphiphilic block copolymers contain well-defined blocks of compositionally dissimilar monomers that have significantly different polarities and interaction affinities for aqueous solutions (10). Thus, amphiphilic ELR-based block copolymers comprise an apolar, hydrophobic block (in the second example above we used the L-alanine containing pentapeptide (A-block)) and a polar, hydrophilic block, in this case the pH-sensitive L-glutamic acid containing block (E-block). This type of molecule is ideally suited for applications involving the energetic and structural control of materials and biological interfaces, including emulsifiers, delivery agents, dispersants, gelation agents, compatibilizers, and foamants (11). Self-assembled micellar nanoparticles have recently been successfully employed in tissue engineering for targeted drug-delivery applications (12). The intrinsic self-assembly behavior of these amphiphilic molecules allows the synthesis of scaffolds suitable for cell culture in a layer-by-layer manner.

The ELR genes described herein are obtained by different approaches: concatemerization (13) and iterative–recursive methods (14). These procedures allow monomer genes to be "polymerized" in a seamless and unidirectional manner. This type of seamless cloning involves the use of type IIS restriction endonucleases that recognize asymmetric base sequences and cleave DNA outside their recognition site, thereby cleaving any DNA sequence that is at a defined distance from their recognition sequence. These endonucleases are ideal for a protein polymer biosynthesis approach as they guarantee unidirectional ligation and avoid the insertion of extraneous DNA at the ligation joints, thereby eliminating extraneous amino acid residues within the polymer sequence.

The purpose of this chapter is to describe different methods available for obtaining ELRs polymeric genes synthesis, or for any other proteinaceous modular polymers, their heterologous expression in a bacterial system and further purification (Fig. 1), and finally two examples of biomedical devices construction are based on these biomaterials.

2. Materials

2.1. Monomeric Gene Production

1. Monomeric ELRs and oligonucleotide primers (IBA Nucleic Acid Synthesis). Restriction sites are indicated in bold type. Overlapping sequences are underlined:

   REDV peptidic monomer: $(VPGIG)_2$VPGKG $(VPGIG)_2$ EEIQIGHIPREDVDYHLYP (VPGIG)$_2$VPGKG $(VPGIG)_2$ $(VGVAPG)_3$

   (a) *REDV For*:
   5′ACCA**CTCTTC**AGTACCGGGCATCGGTGTTCCGGGCATTGGTGTGCCGGGCAAAGGTGTTCCGGGCATTGGTGTGGCCGGCATTGGTGAAGAAATCCAGATCGGGCATATCCCACGCGAGGATGTGGACTACCACCTGTATCCGGTGCC 3′

   (b) *REDV Rev*:
   5′ACCA**CTCTTC**TTACACCCGGTGCAACGCCCACACCCGGGGCAACGCCCACACCCGGCGCAACGCCCAGGGGCAACGCCCACACCCGGCGCAACGCCCACTTTGCCCGGAACACCAATGCCCGGCACACCAATGCCTGGCACCGGATACAGGTGGTAGTCC 3′

   A-block peptidic monomer: $(VPAVG)_{20}$

   (c) *VPAVGfor*:
   5′AAC**GCTCTTC**AGTACCGGCCGTGGGCGTTCCAGCAGTGGGTGTTCCGGCTGTGGGCGTTCCGGCGGTGGGTGTTCCGGCGGTGGGCGTTCCGGCTGTGGGTGTTCCGGCGGTGGGCGTTCCTGCGGTGGGTGTTC CAGCAGTGGGTGTTCCTGCCGT CGGGGTCCCGGCCGTC 3′

   (d) *VPAVG rev*:
   5′AACA**CTCTTC**TTACGCCCACTGCCGGAACACCCACGGCTGGAACGCCCACAGCCGGAA

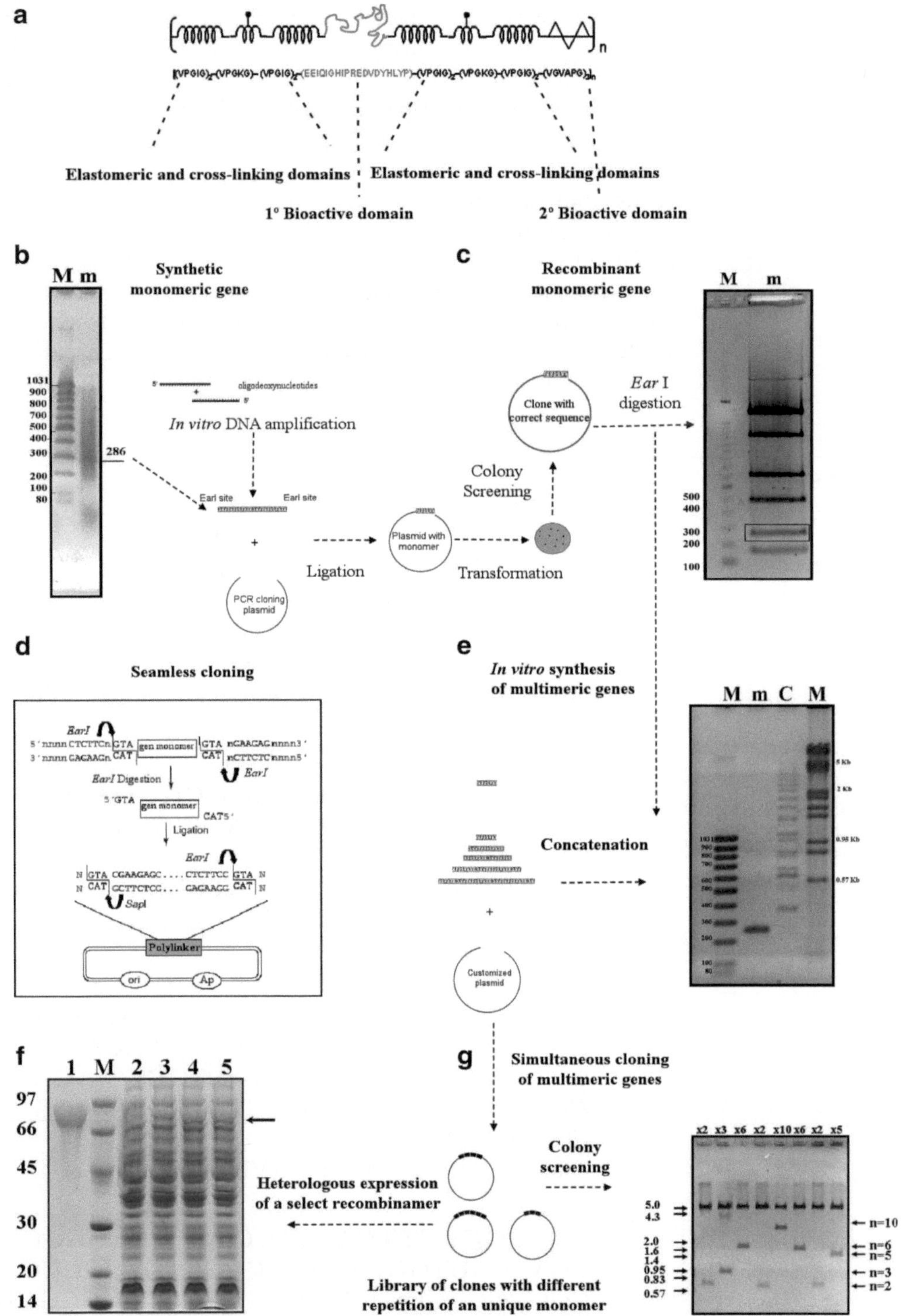

Fig. 1. Schematic representation of the concatemerization method for obtaining bioactive ELRs.

CACCCACCGCCGGAACGCCCACCGCCGGAACA CCCACCGCCGGAACGCCCACCGCCGGAACA CCCACCGCAGGAACGCCCACTGCTGGAACACC GACGGCCGGGACCCCGACG 3′

E-block peptidic monomer: VPGVG VPGVG VPGEG VPGVG VPGVG

(e) *VPGEG for*:
5′AGTTAGGATCC**CTCTTC**AGTACCAGG TGTTGGTGTTCCGGGTGTTGGCGTGCCGG GCGAAGGCGTGCCG 3′

(f) *VPGEG rev*:
5′AGTTAGAATTC**CTCTTC**TTACCCCTA CACCCGGAACACCAACACCCGGCACGCCT TCGCCCGGCACG 3

2. dNTP mix (10 mM) (Eppendorf) (see Note 1).
3. Proof reading *Pfu* DNA polymerase and buffer (Stratagene).
4. Sterile nuclease-free water.
5. Blunt-Ended PCR Cloning Kit (GE Healthcare). Store at –20°C.
6. 2–3% Agarose gel: 2–3% (w/v) agarose MetaPhor (Cambrex) in 1× TAE buffer (see Note 2).
7. SYBR Green I Nucleic Acid Gel Stain (Invitrogene).

2.2. Plasmid Customization: Site-Directed Mutagenesis

1. pET-25b(+) Expression plasmid (Novagen).
2. Oligonucleotides for mutagenesis (Thermo Electron Corporation):

   pET-25b(+) *Sap*I deletion

   (g) *Site 3153 for*: 5′ CCGCATCAGGCG*G*TCTTCCGCTTCC 3′

   (h) *Site 3153 rev*: 5′ GGAAGCGGAAGA*C*CGCCTGATGCGG 3′

   pET-25b(+) Polylinker mutation. Both primers were purchased 5′ phosphorylated.

   (i) *SitelinkerpETPhosF*: 5′GATATC**GCTCTTC**AGTATGAT CCGAATTCGAGCTCCGTCGAC3′.

   (j) *SitelinkerpETPhosR*: 5′GACCA**GCTCTTC**CTACCGG CAGCAGGG3′.
3. Quick Change Site-directed Mutagenesis Kit (Stratagene).
4. pDrive Cloning Vector (Qiagen).
5. XL1-blue *E. coli* competent cells (Stratagene) (see Note 3).
6. 1% Agarose gel: 1% (w/v) agarose in 1× TAE buffer (40 mM Tris–acetate, 1 mM pH 8 EDTA (ethylenediaminetetraacetic acid)).
7. Molecular weight markers: PCR DNA ladder (New England Biolabs), 100 bp DNA ladder (New England Biolabs); 1,000 bp DNA ladder (Novagen).

2.3. Cloning and Purification of ELR Genes

1. Luria–Bertani medium (LB-amp): LB + 100 μg/mL ampicillin (Apollo) (see Note 4).
2. Luria–Bertani medium agar plates: LB-amp + 1.5% agar. Stable at 4°C up to 2 weeks.
3. Quantum Prep Plasmid Mini/Midiprep Kits (Biorad).
4. Restriction endonucleases and buffers, T4 DNA ligase and buffer, shrimp alkaline phosphatase (SAP) and buffer (Fermentas).
5. QIAquick Gel Extraction Kit (Qiagen).
6. Glycogen (Roche).

2.4. ELR Expression and Purification

1. BLR(DE3) *E. coli* expression strain (Novagen) (see Note 5).
2. IPTG (isopropyl-β-D-thiogalactopyranoside) (Apollo Scientific Ltd). Dissolve in water at 100 mM, filter-sterilize, and store at 4°C.
3. 1× TBS (Tris-Buffered Saline): 20 mM Tris-base pH 8, 140 mM NaCl (Sigma-Aldrich). Store at 4°C.
4. 1× SB (Sonication Buffer): 20 mM Tris-base, 1 mM pH 8 EDTA, 1 mM PMSF (phenylmethylsulfonyl fluoride) (Apollo Scientific Ltd). Store at 4°C.
5. Millex-VV 0.1 μm filters (Millipore).

2.5. Biomedical and Nanotechnological Applications of ELRs

1. REDV-ELR.
2. Sterile type I water.
3. Phosphate-buffered saline (abbreviated PBS).
4. *N,N*-Dimethylformamide (DMF), dimethyl sulfoxide (DMSO), hexamethylene diisocyanate (HDI), and acetone.
5. Circular cover glasses (Thermo Scientific).
6. Teflon molds.
7. $E_{50}A_{60}$ ELR copolymer.
8. Varian Cary 50 UV–vis spectrophotometer with a thermostatized sample chamber.
9. JEM2200FS Electronic Microscope (JEOL, Europe) GATAN magnification system (Gatan Inc, France), CCD camera US400, Digital Micrograph software.

3. Methods

3.1. Monomeric Gene Production

The artificial nature of ELRs means that it is necessary to design and synthesize their genes de novo in order to obtain them. Due to current methodological and technical limitations it is impossible to obtain a complete synthetic gene; therefore, the only possible approach involves genetic engineering. The repetitive structure of recombinamers implies that a monomer DNA segment or "monomeric gene" encoding a specific polymer sequence can be created and subsequently seamlessly joined independently of the presence of restriction sites in the target DNA to generate multimerized genes that express the whole recombinamer.

3.1.1. Oligomer Design Guidelines

The ELR monomeric genes were synthesized by PCR amplification of oligonucleotides (Fig. 1b) which have to be designed to act simultaneously as primers and template. Their sequences should contain three functional regions (see Note 6). The 5′ terminal contains the sequences necessary for seamless cloning and/or multimerization and comprises the type IIS restriction enzymes (*Ear*I or *Sap*I) recognition site (see Note 7) along with the sequence of protruding DNA that governs the unidirectional ligation and codifies for the first and last codons of the ORF (in this case GTA for L-valine). The sequence of the codifying region was chosen taking into account the mRNA structure and the most preferred codons for the heterologous expression system (usually *E. coli*) (15, 16), searching for an equilibrium that avoids collapse of the bacterial translational system due to fabrication of such repetitive polypeptides. The 3′ terminal of every oligonucleotide couple is the complementary hybridization region that functions as primer. It is a codifying region that contains the most infrequent codons of the oligonucleotides in order to improve the specificity of the annealing zone (see Note 8). This is particularly important when the ELR monomer sequence is very repetitive (E and A blocks in our present case) but is not determinant when the ELR has a heterogeneous sequence in the middle of the monomer (CS5 peptide in REDV-ELR) used as the hybridization region. The melting temperature (Tm) of the complementary hybridization site should be higher than 60°C (Tm (°C) = 4·GC + 2·TA).

3.1.2. Preparative PCR

Despite all the precautions taken during primer design, the monotonous character of these genes makes this amplification a delicate step. It is therefore advisable to first perform several analytic PCRs with a minimum volume to test the effect of variable of dNTPs, $MgCl_2$, and primer concentrations or different annealing temperatures (Ta) (see Note 9). This procedure is described using construction of the REDV monomer gene (Fig. 1a) as an example:

1. Mix all the components of the PCR in a 0.2-mL, thin-walled tube on ice in the following order:
   - Nuclease-free water to final volume 50 μL.
   - 5 μL ×10 *Pfu* buffer.
   - 2.5 μL REDV for primer (stock 10 μM) (optimal final concentration range 0.5–1 μM).
   - 2.5 μL REDV rev primer (stock 10 μM) (optimal final concentration range 0.5–1 μM).

– 2 μL dNTPs Mix (stock 10 mM) (always after the buffer).
– Mix by pipetting.
– 2.5 U *Pfu* DNA polymerase (always last to be added).
– Spin down the mix and hold on ice (see Note 10).

2. Put the samples in the thermocycler once its temperature is stabilized at 95°C ("Good Start procedure") and run the following program:
   – First cycle: 5 min denaturalization at 95°C.
   – 20 cycles: 45 s denaturalization at 95°C.
     • 30 s annealing (see Note 11).
     • 45 s elongation at 72°C.
   – 15 min final elongation at 72°C.
3. Separate a 1/5 aliquot of PCR samples on a high resolution 2–3% agarose gel (MetaPhor) in TAE buffer (agarose and buffer of choice for small DNA fragment recovery). After staining with 1× SYBR green I solution visualize the bands by exposition to a UV lamp (Fig. 1b).

3.1.3. Cloning and Purification of ELR Monomeric Gene

In many cases, it is better to purify the PCR products using a preparative agarose gel instead of standard PCR clean-up kits before cloning due to the presence of nonspecific PCR fragments.

1. Load the rest of the PCR sample onto a preparative agarose gel. This is similar to the previous one with small specific characteristics. The comb has to be scaled to ensure separation of the whole sample. The gel should be run slowly (50 V for 3 h) to allow correct separation of the bands and, after staining, it should be visualized for a shorter time under a low power UV lamp to avoid DNA mutations.
2. Excise the slice of agarose containing the band from the gel with the help of a scalpel. A minimum quantity of agarose should be removed during band extraction.
3. Purify the monomer genes from the agarose with the "QIAquick Gel Extraction Kit" system, following the manufacturer's instructions. Elute the DNA in 20 μL of warm elution buffer.
4. Clone the purified band with the "Blunt-Ended PCR Cloning Kit" system and transform the competent *E. coli* cells. Plate cells on LB-agar culture dishes supplemented with antibiotic and incubate overnight at 37°C.
5. Select individual clones by colony screening PCR and agarose gel analysis (see Note 12). Prepare the PCR reaction mix in ice using external oligonucleotides (standard sequencing primers) to the polylinker of the cloning plasmid and the *Taq* DNA polymerase in a volume of 15 μL. Touch a single colony with a sterile toothpick (or yellow tip) in a flow chamber, and dip it into PCR tube and shake. Perform the PCR, adapting the

annealing and elongation times and temperatures to the size of the expected genes and type of polymerase. Run the samples in a 1.5–2% analytical agarose gel and select the positive clones on the basis of the length of the resulting bands.

6. Grow the selected transformants overnight in 10 mL LB medium supplemented with antibiotic (250 rpm, 37°C). Extract their plasmids using the BioRad miniprep kit and elute the DNA in 50 μL of warm elution buffer. The length of the insert can be tested by restriction mapping using *Ear*I or co-digestion with *Eco*RI and *Hin*dIII endonucleases and an analytical agarose gel.
7. Examine the selected plasmids (pMOS::REDVmon) by automated DNA sequence analysis to confirm the correctness of their inserts sequence.

3.2. Synthesis of ELR Polymeric Genes: Concatemerization Method

The first strategy to obtain the polymeric genes is the concatemerization method, a random, directional concatenation of the monomeric gene. This method allows multimeric genes with a defined distribution and discrete lengths to be obtained in a single cloning cycle. The first step involves obtaining a large and concentrated amount of monomer fragments by type IIS endonuclease digestion. REDVmon contains two *Ear*I restriction sites flanking its ORF. These sites are eliminated from the excised fragment after *Ear*I digestion and the nonpalindromic protruding bases generated (GTA/CAT), which leads to unidirectional seamless concatenation (Fig. 1d). The second step is the concatemerization reaction itself, which is performed in a controlled ligation reaction where fragment purity and concentration are crucial factors. Finally, the concatemerization mixture is directly cloned in an expression vector. It is often necessary to tailor commercial expression vectors to clone the concatemerization product. In our case, we customized the pET25 vector by consecutive site-directed mutagenesis to remove internal *Sap*I sites and modify its multiple cloning site region to obtain a suitable vector (see Note 13).

1. Digest 90 μg of plasmid pMOS::REDVmon with 40 U of EarI endonuclease (see Note 14) by overnight incubation at 37°C (see Note 15).
2. Purify the resulting gene fragment in a preparative 2–3% (Fig. 1c) agarose gel as described previously.
3. Perform a standard DNA precipitation to improve DNA fragment purity and concentration. Add 0.1 volumes of 3 M sodium acetate pH 5.2 (store at 4°C), 1 mL of 20 mg/mL glycogen (store at 4°C), and 2.5 volumes of cold 100% ethanol (store at –20°C). Leave overnight at –20°C for total recovery (see Note 16). Centrifuge the sample for 15 min at highest speed in a centrifuge at 4°C. Wash the DNA twice with cold 70% ethanol

(store at –20°C). Resuspend the pellet in the desired volume of type I water to achieve a concentrated solution (about 100 ng/μL). Store at –20°C if not required immediately.

4. Prepare the concatenation reaction on ice as follows: 50 ng/μL of REDVmon fragment, 400 U of T4 DNA ligase in 40 μL of 1× T4 DNA ligase reaction buffer. Incubate the mix overnight at 4°C.
5. Analyze a 1/10 aliquot of ligation product in a 1.5% agarose gel (Fig. 1e) to verify the multimer pattern produced (see Note 17).
6. Digest 10 μg of mutated pET25 (pET25^{mut}) with 10 U *Sap*I endonuclease in 100 μL of 1× reaction buffer overnight at 37°C.
7. Check vector cleavage by analyzing a 1/10 aliquot of the sample in a 1% agarose gel.
8. Add 5 U of SAP to the digestion mixture and incubate for 1 h at 37°C. Inactivate SAP at 70°C for 15 min and store at 4°C.
9. Purify the dephosphorylated linear pET25^{mut} in a preparative 1% agarose gel as described previously. After purification, check the concentration and purity by agarose electrophoresis.
10. Add 150 ng of purified pET25^{mut} to 500 ng of concatenation mixture in a minimum volume (10 μL) and incubate overnight at 4°C (see Note 18).
11. Transform competent *E. coli* cells with 1–5 μL of ligation product. Plate cells on LB-agar culture dishes supplemented with antibiotic and incubate overnight at 37°C.
12. Analyze by colony-screening PCR, restriction analysis (Fig. 1g), and examine by automated DNA sequence analysis of individual clones, as described previously.

This concatemerization method has allowed us to synthesize a library of REDV-ELR clones with different repetitions of a unique REDV monomer (Fig. 1a) in a single procedure.

Although this technique offers a powerful approach to obtain multimerizated genes, it nevertheless has some drawbacks. First of all, the impossibility of controlling the order or number of repetitions – if control of the monomer addition sequence is required then iterative–recursive methods must be used instead (see below). A second limitation concerns the presence of circular multimers of various lengths as by-products. In general, however, the amount of circular DNA in the concatenation mixture is not a major problem as, although it reduces the efficiency of the subsequent cloning step, it cannot be integrated.

3.3. Iterative–Recursive Method

The identity and sequence of the individual blocks within the polymer dictate the nature of the supramolecular assembly. To obtain reliable results it is therefore necessary to have absolute control

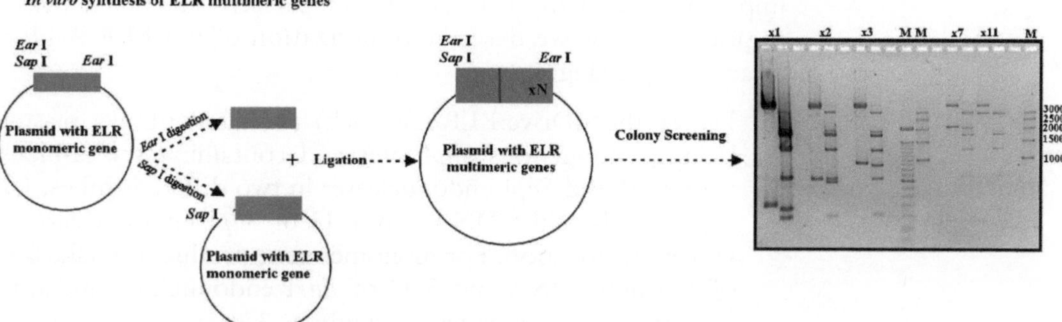

Fig. 2. Schematic representation of the iterative–recursive method for the synthesis of multimeric ELR genes.

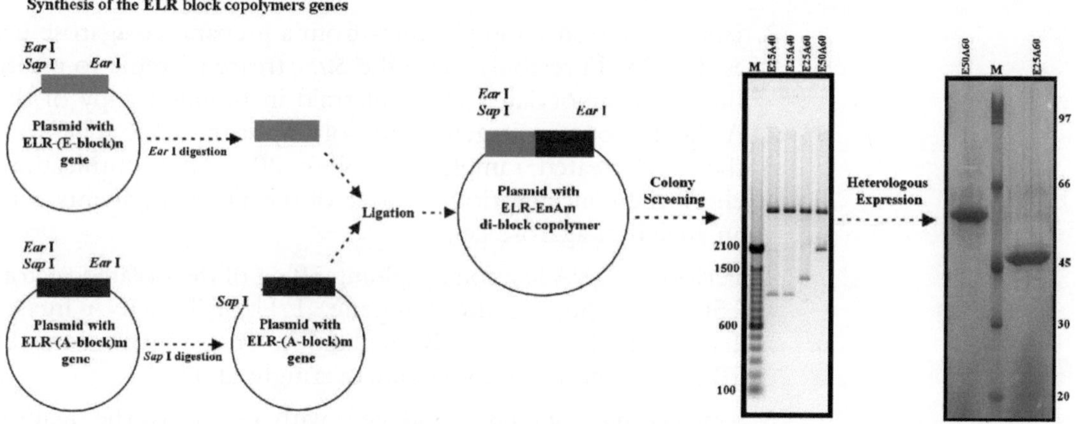

Fig. 3. Schematic representation of the iterative–recursive method for the synthesis of amphiphilic ELR block co-polymers.

over the polymer's sequence. The iterative–recursive method fits this need perfectly as it allows complete control of the size and distribution of the individual blocks that play central roles in formation of the supramolecular structures. This technique allows both polymeric genes (Fig. 2) and amphiphilic di- and tri-block copolymer genes to be synthesized (Fig. 3), although it has two important prerequisites. First of all, the ELR genes should be flanked by two *Ear*I restriction sites, in order to split the genes by action of this endonuclease, one of which must also be a *Sap*I restriction site (whose action will result in linearization of the receptor vector formed by the cloning plasmid and the ELR gene). Secondly, all the receptor vectors must lack endogenous *Sap*I restriction sites (see Note 19).

3.3.1. Synthesis of ELR Polymeric Genes

Construction of ELR homopolymeric genes can be performed randomly using the concatemerization method, as described above, or in a controlled manner using the recursive method. Although this step-by-step procedure is very time-consuming method, it allows tight control of the resulting products, which is often imperative when the construction of perfectly defined blocks is required. Below we describe dimerization of the ELR A-block as an example (Fig. 2).

1. Digest the pDrive:ELR-(A-block) (mutated pDrive plasmid lacking endogenous *Sap*I sites and containing the A-block) with *Ear*I and *Sap*I endonucleases in two different tubes. Use 1 μg of plasmidic DNA and 1 U of *Sap*I endonuclease for vector linearization. For monomer gene production, take 5 μg of plasmidic DNA and 5 U of *Ear*I endonuclease and leave both restriction enzymes overnight at 37°C.
2. Add 5 U of SAP phosphatase to the *Sap*I-containing tube and incubate for 1 h at 37°C, then transfer the sample to a 70°C water bath for 15 min and cool in ice.
3. Isolate the restriction products from a preparative agarose gel as described previously. Use the *Sap*I-treated sample to purify the dephosphorylated linear plasmid including a copy of the A-Block monomeric gene, and isolate just the ELR gene from the *Ear*I-treated sample (see Note 20). After purification, check the concentration and size of the DNA fragments with an analytical agarose gel.
4. Perform a DNA ligation by adding 50 ng of the isolated vector, 150 ng of pure monomer gene, 1 U of T4 DNA ligase, 1 μL of T4 DNA Ligase Buffer, and water to a final volume of 10 μL (see Note 21). Incubate overnight at 4°C.
5. Transform competent *E. coli* cells with 1–5 μL of the ligation mixture. Plate cells on LB-agar culture dishes supplemented with antibiotic and leave overnight at 37°C.
6. Analyze individual clones by colony-screening PCR, as described previously.
7. Grow the selected transformants in 10 mL of LB medium supplemented with antibiotic.
8. Extract the plasmids using the Quantum Prep Plasmid Mini Kit and elute the DNA in 50 μL of warm elution buffer. Prepare two reaction mixtures in ice, one with 0.2 U of *Ear*I endonuclease, 1 μL of buffer, and up to 10 μL of water, and the other with 0.2 U of *Eco*RI endonuclease, 1 μL of buffer, and up to 10 μL of water. Add 10 μL of reaction mixture and 5 μL of eluted plasmid DNA to cold Eppendorf tubes and leave to digest for 3 h at 37°C. Run the samples on a 1.5% agarose gel. Figure 2 shows the analysis of the plasmids with the polymeric ELR-(A-block) genes. Restrictions with *Eco*RI produce two bands, an upper band corresponding to the pDrive plasmid and a lower one corresponding to the multimeric gene. Digestion with *Ear*I leads to gene fragment liberation along with other bands from the pDrive vector.

Application of this recursive method therefore results in the synthesis of polymeric genes starting from a monomeric gene of a certain size. In this particular case, the monomer gene is the pentapeptide $(VPGAG)_{20}$ and all the resulting polymeric genes are repetitions of $20 \times n$ (where $n = 1, 2, 3, 7$, and 11).

3.3.2. Synthesis of Amphiphilic ELR Copolymer Genes

It has been reported that some of the physical properties of ELR copolymers are dependent on their architecture or block distribution and hence on the gene sequence (17). It is therefore necessary to determine which block in the di-block copolymer is going to be at the polypeptide N terminal before starting the experimental procedure. Although any possible distribution of the blocks can be obtained, in this example the hydrophilic block (ELR-E-block) is present at the N terminal and the hydrophobic block (ELR-A-block) at the C terminal of the polypeptide, i.e., E-A di-block or $E_{50}A_{60}$ considering the number of pentapeptides. The experimental protocol is similar to that described in Subheading 3.3.1.

1. Incubate the plasmid pDrive::ELR-$(\text{A-block})_3$ with *Sap*I endonuclease and the pDrive::ELR-$(\text{E-block})_{10}$ with *Ear*I. If the pDrive::ELR-$(\text{A-block})_3$ is linear, it must be dephosphorylated to avoid self-ligation in the next step.
2. Isolate the digestion products – the $(\text{E-block})_{10}$ fragment and the pDrive::ELR-$(\text{A-block})_3$ plasmid – as described above and prepare a DNA ligation reaction with a 1:1 vector:insert molar ratio.
3. Transform competent *E. coli* cells and perform colony screening by PCR or restriction mapping with the appropriate restriction enzymes for the isolated plasmids. Figure 3 shows the restriction analysis with *Eco*RI endonuclease on four ELR copolymer genes. The sizes of the lower bands show the existence of genes encoding the polymers $E_{25}A_{40}$, $E_{25}A_{60}$, and $E_{50}A_{60}$ (the first two constructs were prepared in a different experiment).
4. Subclone the $E_{50}A_{60}$ ELR copolymer gene into the $pET25^{mut}$ expression vector as described above.

This cloning strategy using a combination of *Sap*I and *Ear*I endonucleases allows the synthesis of several types of ELR di-block copolymer genes that can be used as receptor vectors or insert producers in new subcloning cycles to manufacture complex multiblock ELRs in just a few steps.

3.4. ELR Expression and Purification

ELR purification is based on a molecular transition of the polymer chains known as the inverse temperature transition (ITT), which is characteristic of this kind of genetically engineered protein polymer. In aqueous solution, and below a certain transition temperature (T_t), the free polymer chains remain disordered in the form of random coils that are fully hydrated, mainly by hydrophobic hydration. Above T_t, however, the chain loses essentially all of the ordered water structures and hydrophobically folds and assembles to form a phase-separated state in which the polymer chains adopt a regular structure known as a β-spiral, which is stabilized by hydrophobic contacts. This is known as the ITT. The hydrophobic association grows to form particles several hundred nanometers in diameter before settling into a visible phase-separated state. This folding is completely reversible on lowering the sample temperature below T_t, the value of which depends on the molecular mass, the mean polarity of the polymer, and the presence of other ions and molecules (see Note 22). The ELR therefore precipitates upon raising the temperature above T_t, whereupon it can be separated from the remaining bacterial proteins by centrifugation and re-solubilized by adding cold water and slow stirring. This protocol describes the expression and purification of the E–A diblocks, which is identical to REDV-ELR preparation.

1. Grow a single colony of *E. coli* BLR(DE3) harboring pET25^{mut}::ELRgene for 6 h at 37°C in an orbital shaker at 250 rpm in 5 mL of LB medium with antibiotic.
2. Reinoculate 2 mL of the resulting culture in 50 mL of fresh LB-amp medium and allow to grow overnight at 37°C at 250 rpm.
3. Inoculate 2 L of fresh LB-amp medium and allow to grow at 37°C in 2 L flasks (500 mL LB in each) to an optical density of 0.6–0.7 at 600 nm.
4. Induce the ELR expression with IPTG (0.5 mM) for 4 h at 37°C (see Note 23). Figure 1f (lanes 2–5) shows the SDS-PAGE analysis of the time-course induction of REDV-ELR (5).
5. Harvest the cells by centrifugation (6,500 × *g*, 15 min at 4°C) and discard the supernatant. Wash the cells twice by resuspension in TBS and further centrifugation.
6. Resuspend the bacteria in 1× SB buffer and disrupt using a 2-s sonication pulse every 5 s at 100 W. Keep the samples on ice to avoid protein denaturation and limit protease action (see Note 24). Centrifuge the suspension at 15,000 × *g* for 60 min at 4°C and keep the supernatant. The absence of ELR in the cellular debris (insoluble fraction) should be checked by SDS-PAGE (see Note 25).
7. Adjust the pH of the soluble fraction to 3.5 by the addition of dilute hydrochloric acid; this step must be done on ice and with vigorous shaking. Remove the denaturated material (acid proteins and DNA) by cold centrifugation at 15,000 × *g* for 20 min. Keep the supernatant and ensure that the pH remains acidic (see Note 26).
8. Heat the sample at 60°C for 2 h followed by a warm centrifugation for 20 min at 15,000 × *g*.

9. Resuspend the pellet in 2 mL of cold type I water. Neutral pH values help the polymer re-suspension process, therefore addition of dilute NaOH is often necessary. Shake for 2 h on ice to compensate the hysteresis effect of the A-block.
10. Repeat the procedure (steps 7–9) twice more while checking the pH values and the resuspension time on ice as these two factors dramatically influence the final yield of polymer.
11. Check the ELR purity by SDS-PAGE electrophoresis after staining with $CuCl_2$ (see Note 27). Isolated REDV-ELR is shown in Fig. 1f (lane 1), whereas purified polymers $E_{25}A_{60}$ and $E_{50}A_{60}$ are shown in Fig. 3 (see Note 28).

3.5. Biomedical and Nanotechnological Applications of ELRs

3.5.1. Construction of Bioactive Scaffolds with the REDV-ELR

ELRs have been used to produce different kinds of two- (18) and three-dimensional (19–21) tissue-engineering scaffolds. Thus, film- or layer-type supports facilitate the visualization and analysis of cell–material interactions, whereas the microstructure of ELR-based hydrogels, which present very high surface and internal porosity, simulates the structure of the ECM.

We have produced highly transparent 2D REDV-ELR films and 3D hydrogels that maintain the polymer's mechanical properties (20) (Fig. 4). Both REDV-ELR scaffolds are formed by covalent

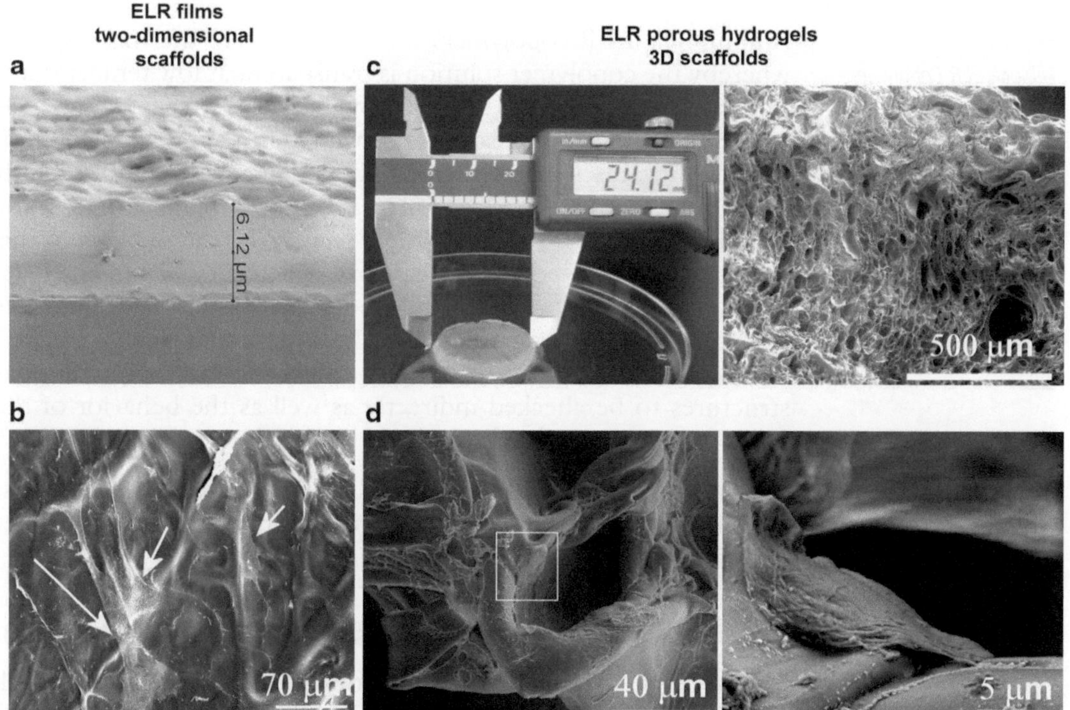

Fig. 4. 2D and 3D REDV-ELR scaffolds. Microscopy images of endothelial cells (HUVECs) seeded on chemically cross-linked ELR supports (20, 21). (**a**) Scanning electron micrographs of cross-linked films. (**b**) Cells seeded on REDV-ELR films. (**c**) Macroscopic pictures of swollen hydrogels in water and SEM micrograph of its cross-section. (**d**) Cross-sectional SEM micrographs and its magnified view shows the presence of cells within the internal pores.

chemical cross-linking between a homobifunctional HDI reagent and the ε-amine groups of the polymer's L-lysine moieties.

The film-manufacture protocol involves solvent casting deposition and subsequent cross-linking.

1. Deposit a 50-mg/mL aqueous REDV-ELR solution on a glass support and dry at 60°C for 8 h.
2. Dip the support in a 10% (v/v) HDI/acetone solution overnight at room temperature.
3. Wash the film thoroughly with type I water.
4. Sterilize by UV exposure overnight and store in 70% ethanol. Wash with sterile type I water and PBS before cell culture assay.

3D hydrogels were cast into molds by mixing a solution of REDV-ELR and cross-linker reagent in an organic solvent.

1. Mix 25 mg/mL HDI/DMF solution with REDV-ELR (80 mg/mL) dissolved in DMSO:DMF (80:20) (final polymer/crosslinker molar ratio of 1:3) and cast in cold Teflon molds for 3 h at room temperature.
2. Wash and sterilize as above.

On REDV-ELR films and hydrogels have cultured endothelial cells (HUVECs) with a well-spread morphology (Fig. 4) (20, 21).

3.5.2. Self-Assembly of Diblock ELR Copolymers

Amphiphilic ELR copolymers exhibit a two-step thermal response whereby the copolymer solution is transparent at low temperatures but the absorbance increases and remains well above baseline levels as the temperature increases. Upon further heating, the solution turbidity increases sharply before plateauing at a constant value. It is believed that the first small increase in absorbance is caused by the ITT of the ELR block with the lowest T_t (the A-block in this particular case). Upon heating above a determined temperature, the high-T_t ELR block (E-block) undergoes its ITT, which induces the formation of large aggregates, as indicated by the high OD values. The turbidity assays allow the formation of supramolecular structures to be checked indirectly as well as the behavior of the block copolymers when heated and cooled.

1. Prepare a 25-μM di-block ELR solution in cold water, adjust the pH to 7, and leave on ice for 30 min to ensure polymer solubilization (see Note 29).
2. Perform the OD measurements at 350 nm, starting at 5°C and increasing the temperature by 0.5°C every 5 min (see Note 30).

Figure 5 shows the turbidity profile for the $E_{50}A_{60}$ di-block copolymer (21). An increase in turbidity is a first sign of self-assembled supramolecular structure formation, which should be verified subsequently using specific techniques.

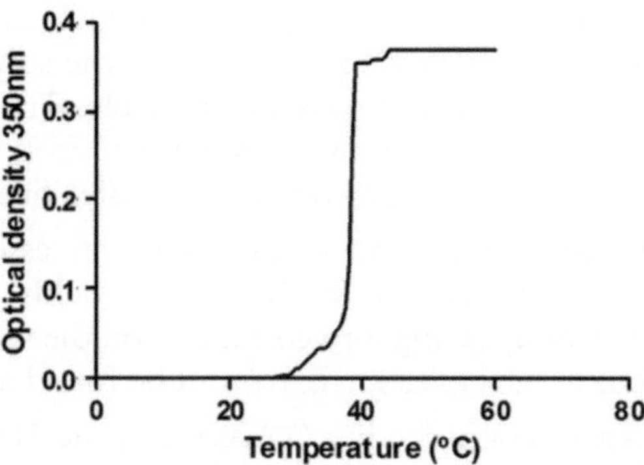

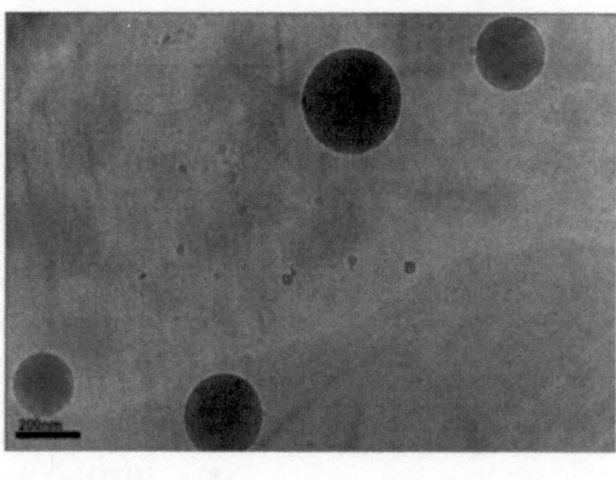

Fig. 5. Turbidity profile and cryo-TEM image of the $E_{50}A_{60}$ di-block ELR (21).

Cryo-TEM is a direct method that gives conclusive information regarding the type of self-assembled nanostructures obtained from multiblock copolymers at temperatures higher than the polymer's ITT.

1. Prepare a 1-mg/mL di-block copolymer solution in cold type I water (see Note 31). Stabilize the samples at 4°C and 50°C for 30 min prior to measurements (see Note 32).
2. Perform the cryo-TEM experiments. Figure 5 shows an image of the nanoparticles formed by the $E_{50}A_{60}$ ELR copolymer at 50°C; no recognizable structure was observed at 4°C.

An analysis of these images clearly shows the inherent ability of this recombinant amphiphilic ELR to form supramolecular structures (21). The diameter of this di-block copolymer (250 ± 15 nm) suggests the formation of vesicular structures when solutions of this polymer are heated above its temperature transition.

4. Notes

1. dNTP mix aliquot are stored at –20°C. Defrost on ice before use.
2. Leave the agarose to hydrate for at least 20 min before melting. Melt at low heating power and avoid boiling.
3. Defrost cells on ice before use.
4. Add ampicillin just before inoculation.
5. *E. coli* strain recommended for use with tandem repeat peptides.
6. In the last few years several companies (GenScript, Geneart, Sloning BioTechnology among others) have begun to offer the synthesis of any gene with no limits to its complexity, length, or scale. We have used some of these, although with contrasting results due to their repetitive structure.

7. Several type IIS restriction enzymes (*Pfl*MI, *Bgl*I *Bpi*I, *Eco31*I, *Bbs* I, *Bsm* BI, etc.), all of which can be used for the seamless cloning of ELRs with excellent results, are available. This particular enzyme pair was chosen on the basis of their recognition and restriction sequences and their widespread availability.
8. Although codon usage does not appear to be an essential parameter for high production levels, higher yields were obtained in some cases by taking into consideration the specific codon bias. Codon usage tables for *E. coli* can be found at:
   - http://www.kazusa.or.jp/codon/ (Accessed June 2010)
   - http://www.geneinfinity.org (Accessed June 2010)
   - pET System Manual – 11th Edition – Novagen
9. To achieve the expected amplification we suggest modifying the concentration of some PCR components in parallel (primers/template in the range 0.1–1 mM in increments of 0.2 mM; dNTPs in the range 100–250 μM) taking into account that higher specificity and lower yields are obtained at lower concentrations. $MgCl_2$ concentration also influences both yield and fidelity in inverse proportion. The standard concentration for this reagent is 2 mM, but the range 1–3 mM in 0.5 mM increments can be tested.
10. The preparation of two different negative-control PCR samples lacking one of each primer is recommended.
11. Hot Start or Touchdown techniques can often be used to improve the specificity of this amplification of extremely repetitive genes. There are several Hot Start amplification systems on the market. In the Touchdown technique, the program starts with an initial annealing temperature (T_a) of 72°C, 10–12°C above the expected melting temperature (T_m) of the hybridization zone of oligonucleotides. T_a is then progressively reduced every two PCR amplification cycles until the predetermined annealing temperature is reached, then an additional ten cycles are carried out at this temperature. The elongation time is chosen to be close to the minimum period required for sequence amplification.
12. PCR screening of the recombinant colony is a powerful method for positive clone identification. However, it can be inefficient for genes that encode ELRs as they are composed of very repetitive sequences with a high cytosine/guanine content, which means that the DNA melting temperature of the codifying sequence is higher than the annealing temperature for conventional primers. Consequently, part of the codifying sequence itself may act as a primer during PCR amplification in a nonspecific priming phenomena that significantly impedes PCR sensitivity and results in unspecific multiple products that smear on a gel. In our experience, the presence of a regular ladder of amplified products is a sign of the existence of repetitive genes.

13. pET25 vector mutations are generated by PCR using the "Quick Change Site-directed Mutagenesis Kit" and pairs of oligonucleotide primers designed with mismatching nucleotides at the middle of the primers. The *g* and *h* primers should be used to eliminate vector endogenous *Sap*I recognition sites without modification of the resulting ORFs and SitelinkerpETPhos primers (*i* and *j*) should be used to modify the multiple cloning site. We have obtained multiple mutations with a unique site-directed mutagenesis cycle: addition of seamless cloning recognition sites; inclusion of GTA codons by unidirectional ligation; insertion of the translation stop codon; deletion of the pelB leader sequence and substitution of some amino acid residues at the N terminal.
14. Many commercial cloning plasmids contain multiple type IIs endonuclease recognition sites and therefore result in extensive fragmentation. It is therefore often advisable to perform a simultaneous digestion with a suitable endonuclease (*Dpn*I, *Mse*I, etc.) which prevents the co-purification of unspecific fragments with a similar length without affecting the ELR gene.
15. To determine the starting amount of plasmid to digest, the amount of insert with respect to the plasmid size (in our case around 10%) and the efficiency of the separation and purification methods (in our experience both around 80–90%) should be calculated.
16. The addition of an inert carrier such as glycogen increases DNA recovery in dilute or short DNA samples.
17. The expected result is a ladder with defined bands corresponding to the size of the multimers (Fig. 1e). The band-size analysis shows that both linear and circular DNA multimers can coexist. When circular DNA limits the cloning efficiency, the linear type can be promoted by increasing the concentration and purity of the DNA starting fragments.
18. The insert:vector molar ratio is higher than for standard subcloning protocols as its molar ratio cannot be calculated exactly here because the inserts are a distribution of multimers. A successful cloning mix with a molar ratio of 60:1 was prepared by considering the inserts as monomers.
19. Site-directed mutagenesis of the vectors containing the individual blocks has to be performed in order to use the iterative–recursive method to generate polymeric genes. In our particular case, mutation is dependent on the presence of endogenous *Sap*I restriction sites on the cloning vector (pDrive) and was performed as described in Note 9.
20. It is advisable to purify the linearized plasmid from the gel because the presence of even a minute amount of undigested plasmid can complicate the following steps. An alternative

method for purifying extremely long linearized plasmids is the "Ultrafree-DA" (Millipore) filtration system. The filtered DNA should be precipitated as described above.

21. A vector–insert molar ratio of 1:5 should normally be used on account of their sizes. This ratio can be changed depending on the number of multimeric genes. A 1:1 molar ratio can be used when a simple duplication of the gene is required, although the resulting transformants will be fewer in number.

22. The presence of two residues (L-alanine and L-glutamic acid) with different properties in the $E_{50}A_{60}$ ELR affects the purification protocol. The hydrophobic A-block is pH-independent and shows hysteresis during cooling solubilization, whereas the hydrophilic E-block is pH-dependent since the γ-carboxylic group of the L-glutamic acid residue changes its state with the pH, thus affecting the T_t of the recombinamer.

23. Alternatively, ELR production can be performed with Circle Grow Media (Q Biogene) or the Overnight Expression System (Novagen).

24. In our experience, ELRs are usually very resistant to bacterial protease degradation, although the bioactive domains in some tissue engineering ELRs have been shown to be protease sensitive. In this case, we use protease inhibitors or reduce the induction temperature. Two common and inexpensive inhibitors are PMSF and EDTA.

25. The buffers used during purification have a pH close to neutral, where the γ-carboxylic group ($pK_a \approx 4.5$) of the L-glutamic acid residue is in its carboxylate form. This significantly increases the polymer's polarity and thus T_t, thereby affecting the final yield of the purification process.

26. The γ-carboxylic group of the L-glutamic acid residue is protonated under these conditions therefore the ITT of the E-block occurs at lower temperatures. The T_t of the E-block in $E_{50}A_{60}$ ranges from 33°C at pH 3.5 to >70°C at pH 7 (15).

27. The extremely hydrophobic amino acids of the elastomeric domains present in the ELR backbone are responsible for the failure of most of the conventional staining methods for protein detection in SDS-PAGE. The common choice for ELRs is copper chloride-based negative staining. Briefly, wash the gel with distilled water for 10 s and incubate for 5 min in a fresh 0.3 M $CuCl_2$ solution. The salt excess can be removed with distilled water. Another advantage of this fast technique is the fact that all the salt can be rapidly removed by incubation in an EDTA solution and the gel can be restained.

28. The observed rate of migration of ELRs in SDS-PAGE does not agree with their theoretical molecular weight: it is usually higher than the theoretical size, with a maximum increase in

band position of 20%. This unusually high electrophoretic mobility is typical for this type of hydrophobic polymer and is even more evident when the polypeptide sequence includes an L-alanine residue.

29. It is important to allow the samples to stabilize with agitation before starting to measure to ensure the absence of nonsolubilized polymer aggregates.
30. This 5 min period is key at higher temperatures as the supramolecular structures start to form at these temperatures and the elapsed time is essential for their stabilization.
31. Filter the copolymer solutions to remove nonresuspended polymer.
32. The maximum value is dependent on the copolymer's ITT, which should be at least 5°C higher than the polymer's transition temperature to ensure the stabilization of the supramolecular structures.

Acknowledgments

We are grateful for financial support from the MICINN (projects MAT 2007-66275-C02-01 and PSE-300100-2006-1), the JCyL (projects VA034A09, VA016B08, and VA030A08), the CIBER-BBN (project CB06-01-0003), the JCyL, and the Instituto de Salud Carlos III under the "Network Center of Regenerative Medicine and Cellular Therapy of Castilla and Leon."

References

1. Rodríguez-Cabello, J. C., Reguera, J., Girotti, A., Arias, F. J., and Alonso, M. (2006) Genetic Engineering of Protein-Based Polymers: The Example of Elastin-like Polymers, *Advances in Polymer Science* **200**, 119–167.
2. Arias, F. J., Reboto, V., Martin, S., Lopez, I., and Rodriguez-Cabello, J. C. (2006) Tailored recombinant elastin-like polymers for advanced biomedical and nano(bio)technological applications, *Biotechnol Lett* **28**, 687–695.
3. Rodríguez-Cabello, J. C., Reguera, J., Girotti, A., Alonso, M., and Testera, A. M. (2005) Developing functionality in elastin-like polymers by increasing their molecular complexity: the power of the genetic engineering approach, *Progress in Polymer Science* **30**, 1119–1145.
4. Rodriguez-Cabello, J. C., Prieto, S., Reguera, J., Arias, F. J., and Ribeiro, A. (2007) Biofunctional design of elastin-like polymers for advanced applications in nanobiotechnology, *J Biomater Sci Polym Ed* **18**, 269–286.
5. Girotti, A., Reguera, J., Rodríguez-Cabello, J. C., Arias, F. J., Alonso, M., and Testera, A. M. (2004) Design and bioproduction of a recombinant multi(bio)functional elastin-like protein polymer containing cell adhesion sequences for tissue engineering purposes, *Journal of Materials Science-Materials in Medicine* **15**, 479–484.
6. Mould, A. P., Wheldon, L. A., Komoriya, A., Wayner, E. A., Yamada, K. M., and Humphries, M. J. (1990) Affinity Chromatographic Isolation of the Melanoma Adhesion Receptor for the Iiics Region of Fibronectin and Its Identification as the Integrin-Alpha-4-Beta-1, *J. Biol. Chem.* **265**, 4020–4024.
7. Plouffe, B. D., Njoka, D. N., Harris, J., Liao, J., Horick, N. K., Radisic, M., and Murthy, S. K.

(2007) Peptide-mediated selective adhesion of smooth muscle and endothelial cells in microfluidic shear flow, *Langmuir* **23**, 5050–5055.

8. Lombard, C., Bouchu, D., Wallach, J., and Saulnier, J. (2005) Proteinase 3 hydrolysis of peptides derived from human elastin exon 24, *Amino Acids* **28**, 403–408.
9. Park, C., Yoon, J., and Thomas, E. L. (2003) Enabling nanotechnology with self assembled block copolymer patterns, *Polymer* **44**, 6725–6760.
10. Mortensen, K. (1998) Structural properties of self-assembled polymeric micelles, *Current Opinion in Colloid & Interface Science* **3**, 12–19.
11. Wright, E. R., and Conticello, V. P. (2002) Self-assembly of block copolymers derived from elastin-mimetic polypeptide sequences, *Advanced Drug Delivery Reviews* **54**, 1057–1073.
12. Mandal, B. B., and Kundu, S. C. (2009) Self-assembled silk sericin/poloxamer nanoparticles as nanocarriers of hydrophobic and hydrophilic drugs for targeted delivery, *Nanotechnology* **20**.
13. McMillan, R. A., Lee, T. A. T., and Conticello, V. P. (1999) Rapid assembly of synthetic genes encoding protein polymers, *Macromolecules* **32**, 3643–3648.
14. Won, J. I., and Barron, A. E. (2002) A new cloning method for the preparation of long repetitive polypeptides without a sequence requirement, *Macromolecules* **35**, 8281–8287.
15. Sorensen, M. A., Kurland, C. G., and Pedersen, S. (1989) Codon Usage Determines Translation Rate in Escherichia coli, *J Mol Biol* **207**, 365–377.
16. Tats, A., Tenson, T., and Remm, M. (2008) Preferred and avoided codon pairs in three domains of life, *BMC Genomics* **9**, 463.
17. Ribeiro, A., Arias, F. J., Reguera, J., Alonso, M., and Rodriguez-Cabello, J. C. (2009) Influence of the Amino-Acid Sequence on the Inverse Temperature Transition of Elastin-Like Polymers, *Biophysical Journal* **97**, 312–320.
18. Liu, J. C., and Tirrell, D. A. (2008) Cell response to RGD density in cross-linked artificial extracellular matrix protein films, *Biomacromolecules* **9**, 2984–2988.
19. McHale, M. K., Setton, L. A., and Chilkoti, A. (2005) Synthesis and in vitro evaluation of enzymatically cross-linked elastin-like polypeptide gels for cartilaginous tissue repair, *Tissue Engineering* **11**, 1768–1779.
20. Martin, L., Alonso, M., Girotti, A., Arias, F. J., and Rodriguez-Cabello, J. C. (2009) Synthesis and Characterization of Macroporous Thermosensitive Hydrogels from Recombinant Elastin-Like Polymers, *Biomacromolecules.* 10 (11), 3015–3022.
21. Rodríguez-Cabello, J. C., Martín, L., Alonso, M., Arias, F. J., and Testera, A. M. (2009) "Recombinamers" as advanced materials for the post-oil age, *Polymer* **50**, 5159–5169.

Chapter 3

Design of Biomolecules for Nanoengineered Biomaterials for Regenerative Medicine

Alvaro Mata, Liam Palmer, Esther Tejeda-Montes, and Samuel I. Stupp

Abstract

An important goal in the development of highly functional organic materials is to design self-assembling molecules that can reproducibly display chemical signals across length scales. Within the biomedical field, biomolecules are highly attractive candidates to serve as bioactive building blocks for the next generation of biomaterials. The peptide amphiphiles (PAs) developed by the Stupp Laboratory at Northwestern University generated a highly versatile self-assembly code to create well-defined bioactive nanofibers that have been proven to be very effective at signaling cells in vitro and in vivo. Here, we describe the basic steps necessary for synthesis and assembly of PA molecules into functional nanostructures.

Key words: Peptide amphiphiles, Peptide synthesis, Biomolecules, Regenerative medicine, Biomaterials

1. Introduction

Regenerative therapies are of enormous interest given their potential to increase quality of life in the world's population (1–4). Therefore, bioactive materials that can trigger or control biological processes could become important components of regenerative therapies. One promising approach is the development of bioactive molecules to form nanostructures that can interact specifically and reproducibly with cell receptors and proteins to control processes such as cell survival, cell proliferation, cell differentiation, and de-differentiation in the context of tissue and organ regeneration. These biomolecules can be designed not only to signal cells but also to self-assemble into well-defined supramolecular structures and materials. Furthermore, an attractive feature of these self-assembling

Melba Navarro and Josep A. Planell (eds.), *Nanotechnology in Regenerative Medicine: Methods and Protocols*, Methods in Molecular Biology, vol. 811, DOI 10.1007/978-1-61779-388-2_3, © Springer Science+Business Media, LLC 2012

systems is the possibility to deliver them through minimally invasive procedures, for example, by injecting them as dissolved molecules in aqueous media that rapidly form at the targeted tissues the signaling structures through programmed self-assembly. A number of research groups are currently taking advantage of the increased understanding of molecular self-assembly and nanoscience to develop bioactive, biomimetic, and multifunctional materials for regenerative medicine (5–13). The Stupp Laboratory at Northwestern developed a very broad class of peptide amphiphiles (PAs) capable of forming bioactive nanoscale filaments that mimic those in extracellular matrices. These filaments can display in tuneable densities peptide signals that promote regenerative processes (5, 6, 14–31). These molecules are composed of a peptide segment and a hydrophobic segment, typically an alkyl tail that provides a key driving force for self-assembly through hydrophobic collapse in water. Since an alkyl tail is always more hydrophobic than any peptide segment, it promotes the formation of nanostructures in water that display on their surfaces the peptide segments. The peptide segment contains at least two domains, one designed to create β-sheets and a terminal domain that is displayed on the nanostructure's surface containing the bioactive signal. The switch for self-assembly is electrolyte screening of charges that are introduced in the sequence if they are not present in the bioactive domain. The self-assembly process involves the formation of hydrogen bonding among PA molecules in the shape of long fibers and not spherical micelles as a result of β-sheet formation. The fibrous nanostructures are much more bioactive than nanospheres displaying a similar density of biological signals (32). In addition to bioactivity, the PA nanoscale fibers designed in the Stupp Laboratory are both biocompatible and biodegradable (33), and have been shown to have in vivo efficacy in neuronal regeneration (21), blood vessel regeneration (16), cartilage regeneration (29), bone regeneration (28), and bone marrow derived cell delivery (27). This chapter summarizes the key steps that are used to fabricate these PA molecules using Fmoc solid-phase peptide synthesis.

2. Materials

2.1. Method 1: PAs with the Alkyl Tail Located at the N Terminus

1. Preloaded Wang resin (Novabiochem, EMD Chemicals, Darmstadt, Germany).
2. *N,N*-Dimethylformamide (DMF).
3. Dichloromethane.
4. *N,N*-Diisopropylethylamine (DIEA), redistilled.
5. Fmoc-protected amino acids (Novabiochem, EMD Chemicals, Darmstadt, Germany).

6. 2-(1*H*-benzotriazole-1-yl) -1,1,3,3-tetramethyluronium hexafluorophosphate (HBTU).
7. Piperidine, redistilled.
8. Fluorenylmethoxycarbonyl-*O*-(benzylphospho)-serine (Fmoc-Ser(PO_3Bnzl)).
9. 1-(Mesitylene-2-sulfonyl)-3-nitro-1*H*-1,2,4-triazole.
10. Fatty acid (e.g., palmitic acid).
11. Ninhydrin (2,2-dihydroxyindane-1,3-dione).
12. Diethyl ether.
13. Dithiothreitol (DTT).
14. Triflouroacetic acid (TFA).
15. Triisopropylsilane (TIPS).
16. Ethanedithiol.
17. Water (deionized with a Millipore Milli-Q water purifier, resistance of 18 MΩ).
18. Automated peptide synthesizer (CS Bio Peptide Synthesizer, Menlo Park, CA).
19. Ultrasonicator (Branson 2510 bath sonicator, Danbury, CT).
20. Nuclear magnetic resonance (NMR) spectrometer (Varian Mercury 400 MHz, Palo Alto, CA).
21. Reverse-phase high-performance liquid chromatography (RP-HPLC) equipped with a Waters Atlantis C18 column (Agilent HP 1050 system, Santa Clara, CA).
22. Electrospray ionization (ESI) mass spectrometer (MSD, Agilent, Santa Clara, CA).
23. Matrix-assisted laser desorption/ionization (MALDI-TOF) (Apex III, Bruker, Bremen, Germany).
24. Rotary evaporator (RE-11 Rotovapor with a Welch 2025 dry vacuum pump, Büchi Laboretechnik, Postfach, Switzerland).
25. Lyophilizer (Labconco Freezone 6, Kansas City, MO).

2.2. Method 2: PAs with the Alkyl Tail Located at the C Terminus

1. *N*-Carbobenzyloxy-L-aspartic anhydride.
2. Dichloromethane.
3. Dodecylamine.
4. Triethylamine.
5. Ethanol.
6. Dioxane.
7. Hydrochloric acid.
8. Chloroform.
9. Magnesium sulfate.

10. Celite.
11. Fmoc-*O*-succinimide (Fmoc-OSu).
12. Rink resin.
13. Piperidine.
14. 2-(1*H*-benzotriazole-1-yl) -1,1,3,3-tetramethyluronium hexafluorophosphate (HBTU).

3. Methods

3.1. Method 1: PAs with the Alkyl Tail Located at the N Terminus

Figure 1 illustrates a PA molecule synthesized with the following protocol (24).

1. PAs are prepared on a 0.25-mmol scale by using standard fluorenylmethoxycarbonyl (Fmoc) chemistry using Wang resin preloaded with the first (C terminal) amino acid (see Note 1). The Wang resin should be swelled by soaking in dichloromethane for 30 min prior to use. The synthesis proceeds either by manual synthesis in a peptide synthesis vessel (e.g., ChemGlass Part #CG-1860) or on an automated peptide synthesizer.
2. The Fmoc group is cleaved with a 30% solution of piperidine in DMF for 10 min and the beads washed thoroughly with DMF. If using manual synthesis, several beads can be removed and the Kaiser (ninhydrin) test, which is used to detect ammonia or primary and secondary amines. The deprotection reaction is repeated until the ninhydrin test is positive.
3. The next amino acid is then activated by dissolving 4 equivalents of the amino acid, 3.95 equivalents 2-(1*H*-benzotriazole-1-yl)-1,1,3,3-tetramethyluronium hexafluorophosphate (HBTU), and 6 equivalents of DIEA in DMF (see Note 2). The activated ester solution is added to the reaction vessel and the reaction proceeds for 2 h. A Kaiser test should give a negative result when the coupling is complete.

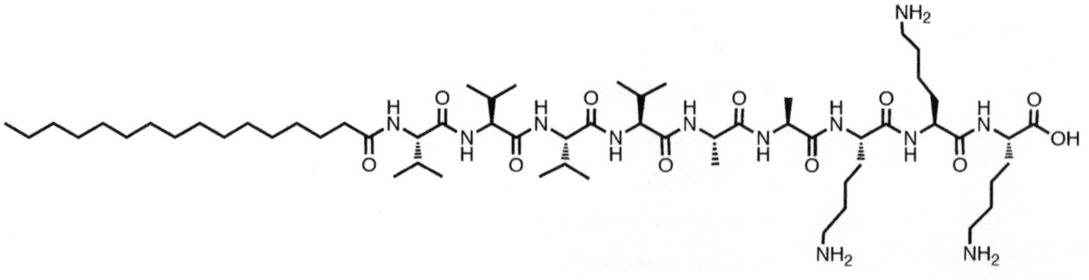

Fig. 1. Chemical structure of a PA molecule containing the alkyl tail on the N terminus. Reprinted with permission from ref. 34. Copyright 2005, American Chemical Society.

4. The deprotect–coupling procedure is repeated to form the full-length peptide sequences.
5. After the peptide portion of the molecule is prepared, the Wang resin is removed from the automated synthesizer and transferred to a peptide synthesis vessel.
6. The N terminus is capped with a fatty acid containing 6, 10, 16, or 22 carbon atoms. The alkylation reaction is accomplished by using 8 equivalents of the fatty acid, 8 equivalents HBTU, and 12 equivalents of DIEA in enough dichloromethane and DMF to fully dissolve the acid.
7. The acylation reaction is allowed to proceed for 6 h. Complete reaction is indicated by a negative Kaiser test.
8. Cleavage and deprotection of the PAs is achieved with a mixture of triflouroacetic acid (TFA), water, and triisopropylsilane in a ratio of 91:2.5:2.5 for 3 h at room temperature. For PAs containing cysteine, a mixture of TFA, water, triisopropylsilane, and ethanedithiol in a ratio of 91:3:3:3 is used.
9. The cleavage mixture and two subsequent dichloromethane washings are filtered into a round-bottom flask.
10. The solution is concentrated in vacuo by rotary evaporation to afford a viscous solution.
11. This solution is precipitated with cold diethyl ether.
12. The white precipitate is collected by filtration, washed with copious cold ether, and dried under vacuum.
13. The purity of the crude material is determined by analytical reverse-phase high-performance liquid chromatography (RP-HPLC) equipped with a Waters Atlantis C18 column (5 μm particle size, 150 × 4.6 mm or 250 × 4.6 mm) (see Note 3). The crude residue is then dissolved and purified by preparative reverse-phase HPLC with a Waters Atlantis C18 preparative column (5 μm particle size, 250 × 30.0 mm) (see Note 4).
14. The appropriate fractions are then collected, concentrated in vacuo by rotary evaporation to remove the organic solvent, followed by freezing with liquid nitrogen and lyophilization to give a fluffy white solid (see Notes 5 and 6). The identity of the purified material is confirmed by ESI and MALDI-TOF mass spectrometers (see Note 7).

3.2. Method 2: PAs with the Alkyl Tail Located at the C Terminus

Figure 2 illustrates a PA molecule synthesized with the following protocol (34).

1. The first step consists of synthesizing the molecule illustrated in Fig. 3

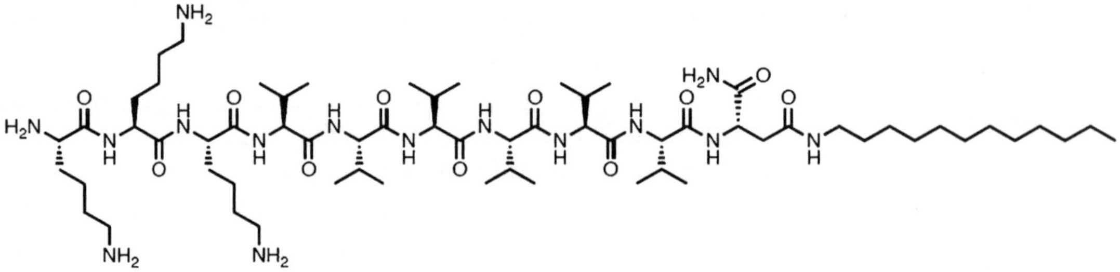

Fig. 2. Chemical structure of a PA molecule containing the alkyl tail on the C terminus. Reprinted with permission from ref. 34. Copyright 2005, American Chemical Society.

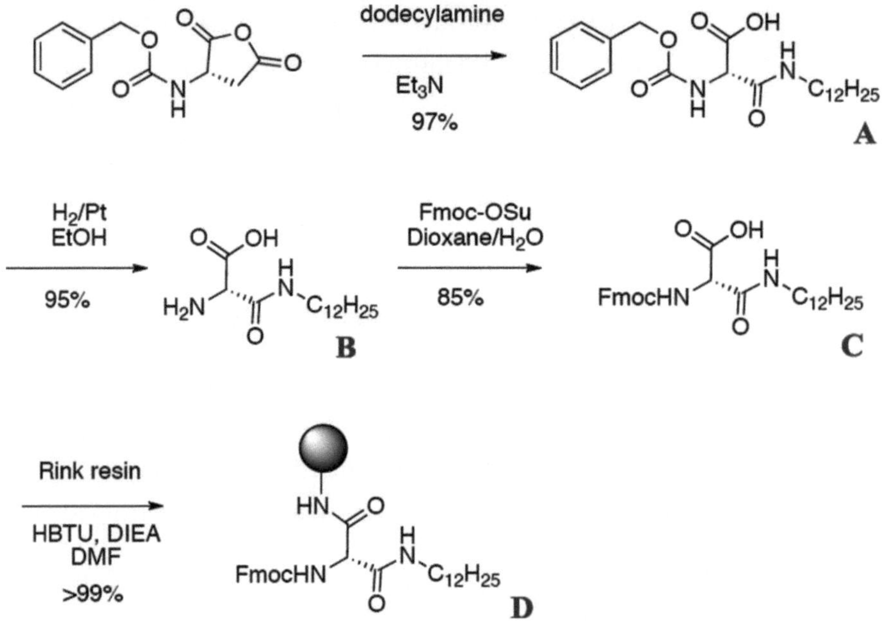

Fig. 3. Steps to synthesize the fatty acid amino acid D, which serves as the first amino acid in the synthesis of PAs with the alkyl tail at the N terminus. Reprinted with permission from ref. 34. Copyright 2005, American Chemical Society.

Product A (*N*-Dodecyl-2-carbobenzyloxyaminosuccinamic acid):

(a) *N*-Carbobenzyloxy-L-aspartic anhydride (1 mmol) is dissolved in 50 mL of dichloromethane.

(b) Addition of 1.05 equivalents of dodecylamine and 1.1 equivalents of triethylamine.

(c) The reaction is sealed with a tight cap to prevent evaporation, and stirred for 12 h until no trace of starting material is detected by thin-layer chromatography (TLC) (5% MeOH in CH_2Cl_2).

(d) The reaction is quenched with 20 mL of 1 M hydrochloric acid followed by extraction with chloroform (5×).

(e) The organic layer is dried over magnesium sulfate, and product A is obtained as a white solid (yield 97%).

Product B (2-Amino-*N*-dodecylsuccinamic acid):

(a) In 100 mL of ethanol, 100 mmol of product A are dissolved and transferred to a reaction vessel containing palladium on carbon (10 wt%).

(b) The vessel is then placed under hydrogen (35 Torr) for 3 h. The reaction mixture is filtered over Celite, and the product is obtained as a white solid after evaporation to dryness under reduced pressure (yield 95%).

Product C (*N*-Dodecyl-2-Fmoc-aminosuccinamic acid):

(a) In 200 mL of a water/dioxane (1:1 v/v) mixture, 6.6 mmol of product B are dissolved with 1.3 mL (1.5 equivalents) of triethylamine and 1 equivalent of Fmoc-*O*-succinimide (Fmoc-OSu).

(b) The reaction is monitored by TLC (10% MeOH in CH_2Cl_2) and after 2–3 h all of the Fmoc-OSu is consumed.

(c) The reaction is quenched with acid, resulting in a white precipitate (product C) that is collected by filtration (yield 85%).

Product D:

(a) The standard Rink resin is placed in a reaction vessel, deprotected three times with 30% piperidine in NMP.

(b) The resin is then coupled with 2 equivalents of product C overnight using HBTU as a coupling reagent.

(c) The coupling is repeated until a ninhydrin test is negative.

2. The amino acids are added to product D using standard Fmoc solid-phase techniques on the automated synthesizer to give the full-length peptide.

3.3. Method 3: Self-Assembly of Peptide Amphiphiles

These molecules self-assemble into nanofibers with no detectable critical micelle concentration (cmc). Electrostatic screening of the charged residues with salt or by changing the pH results in greater bundling of the nanofibers and is observed macroscopically as gelation (Fig. 4) (see Note 8).

1. Gelation of peptide amphiphiles with basic residues

(a) The solid PA is dissolved in water to the desired concentration (typically about 1 wt%). Ultrasonication (Branson 2510 bath sonicator, Danbury, CT) or gentle heating may be required for complete dissolution (see Note 9).

Fig. 4. Electrostatic screening of the PA molecules leads to self-assembly into well-defined nanofibers and subsequent gel formation (6).

   (b) Gelation is induced by lowering the pH by introducing aqueous HCl or by exposing the solution to HCl vapors in a sealed container for approximately 10 min (see Note 10). Gelation is indicated in samples that are self-supporting upon inversion of the container.

2. Gelation of peptide amphiphiles with acidic residues

   (a) The solid PA is dissolved in water to the desired concentration (typically on the order of 1 wt%). Sonication or gentle heating may be required for complete dissolution.

   (b) Gelation is induced by raising the pH by introducing aqueous NaOH or by exposing the solution to ammonia vapors in a sealed container for approximately 10 min (see Notes 10 and 11).

4. Notes

1. All reactions described in Subheadings 3.1–3.3 are conducted at room temperature.
2. Using an excess of HBTU can result in capping of the N terminus (34).
3. These peptide amphiphile molecules can be purified by HPLC using either water–acetonitrile or water–methanol solvent gradients. However, the self-assembly behavior after purification is most consistent using water–acetonitrile.
4. In some cases, the peptide amphiphile can be difficult to solubilize, adding a small amount of a co-solvent like methanol or hexafluoroisopropanol (HFIP) can be used to aid solubility. Furthermore, using a horn sonicator can also be quite useful.
5. The solid material after lyophilization is very low density. It is recommended that static electricity be minimized to avoid loss of material or weighing errors. An antistatic gun is very useful for this purpose.
6. The lyophilized material can be stored in a freezer at −20°C for many months without any significant degradation.
7. The preferred matrix for MALDI of these compounds is α-cyano-4-hydroxycinnamic acid.
8. If the samples will be used for cell studies, then the trifluoracetate counterions should be exchanged for chloride. Typically, 600 mg of PA is dissolved in 600 mL of 0.01 M HCl and the sample is immediately frozen immediately and then lyophilized. Cold packs placed beside the samples can help to minimize melting. The concentration of residual TFA can be determined by fluorine-19 NMR (Varian Mercury 400 MHz, Palo Alto, CA) in methanol-*d4* with trifluoroethanol, hexafluorobenzene, or 4-trifluoromethylacetanilide as an external standard.
9. Lower PA concentrations also result in nanofiber formation, but gelation may not be observed macroscopically. The nanofiber formation can be observed by transmission electron microscopy or atomic force microscopy.
10. Gelation can be induced by adding aqueous NaOH 1 μL at a time (for positively charged PAs) or adding aqueous HCl 1 μL at a time (for negatively charged PAs) until a self-supporting material is observed.
11. For sequences with acidic residues, gelation can also be induced by addition of $CaCl_2$ (0.5 molar equivalents of Ca^{2+} per carboxylic acid) (35).

References

1. Huebsch, N., Mooney, D. (2009) Inspiration and application in the evolution of biomaterials *Nature* **462**, 426–432.
2. Stevens, M.M., George, J.H. (2005) Exploring and engineering the cell surface interface *Science* **310**, 1135–1138.
3. Furth, M. E., Atala, A. (2008) Current and future perspectives of regenerative medicine. Principles of Regenerative Medicine. Burlington MA: Elsevier.
4. Stupp, S. I. (2005) Biomaterials for regenerative medicine *MRS Bulletin* **30**, 546–553.
5. Hartgerink, J.D., Beniash, E., Stupp, S. I. (2001) Self-assembly and mineralization of peptide-amphiphile nanofibers *Science* **294**, 1684–1688.
6. Hartgerink, J.D., Beniash, E., Stupp, S.I. (2002) Peptide-amphiphile nanofibers: A versatile scaffold for the preparation of self-assembling materials *Proceedings of the National Academy of Sciences of the United States of America* **99**, 5133–5138.
7. Paramanov, S.E., Gauba, V., Hartgerink, J.D. (2005) Synthesis of collagen-like peptide polymers by native chemical ligation *Macromolecules* **38**, 7555–7561.
8. Betre, H., Setton, L.A., Meyer, D.E., Chilkoti, A. (2002) Characterization of a genetically engineered elastin-like polypeptide for cartilaginous tissue repair *Biomacromolecules* **3**, 910–916.
9. Zhao, X., Zhang, S. (2007) Designer self-assembling peptide materials *Macromolecular Biosciences* 7, 13–22.
10. Aggeli, A., Bell, M., Carrick, L.M., Fishwick, C.W.G., Harding, R., Mawer, P.J., Radford, S.E., Strong, A.E., Boden, N. (2003) pH as a trigger of peptide β-sheet self-assembly and reversible switching between nematic and isotropic phases *Journal of the American Chemical Society* **125**, 9619–9628.
11. Williams, R.J., Smith, A.M., Collins, R., Hodson, N., Das, A.K., Ulijn, R.V. (2009) Enzyme-assisted self-assembly under thermodynamic control *Nature Nanotechnology* **4**, 19–24.
12. Schneider, J.P., Pochan, D.J., Ozbas, B., Rajagopal, K., Pakstis, L., Kretsinger, J. (2002) Responsive hydrogels from the intramolecular folding and self-assembly of a designed peptide *Journal of the American Chemical Society* **124**, 15030–15037.
13. Zhang, S., Holmes, T., Lockshin, C., Rich, A. (1993) Spontaneous assembly of a self-complementary oligopeptide to form a stable macroscopic membrane *Proceeding of the National Academy of Sciences of the United States of America* **90**, 3334–3338.
14. Silva, G.A., Czeisler, C., Niece, K.L., Beniash, E., Harrington, D.A., Kessler, J. A., Stupp, S. I. (2004) Selective differentiation of neural progenitor cells by high-epitope density nanofibers *Science* **303(5662)**, 1352–1355.
15. Guler, M.O., Soukasene, S., Hulvat, J.F., Stupp, S.I. (2005) Presentation and recognition of biotin on nanofibers formed by branched peptide amphiphiles. *Nano Letters* **5(2)**, 249–252.
16. Rajangam, K., Behanna, H.A., Hui, M.J., Han, X., Hulvat, J.F., Lomasney, J.W., Stupp, S.I. (2006) Heparin binding nanostructures to promote growth of blood vessels *Nano Letters* **6**, 2086–2090.
17. Storrie, H., Guler, M.O., Abu-Amara, S.N., Volberg, T., Rao, M., Geiger, B., Stupp, S.I. (2007) Supramolecular crafting of cell adhesion *Biomaterials* **28**, 4608–4618.
18. Stendahl, J.C., Want, L.J., Chow, L.W., Kaufman, D.B., Stupp, S.I. (2008) Growth factor delivery from self-assembling nanofibers to facilitate islet transplantation *Transplantation* **86(3)**, 478–481.
19. Capito, R., Azevedo, H., Gelichko, Y., Mata, A., Stupp, S.I. (2008) Self-assembly of large and small molecules into hierarchically ordered sacs and membranes *Science* **319**, 1812–1816.
20. Kapadia, M. R., Chow, L.W., Tsihlis, N.D., Ahanchi, S.S., Eng, J.W., Murar, J., Martinez, J., Popowich, D.A., Jiang, Q., Hrabie, J.A., Saavedra, J.E., Keefer, L.K., Hulvat, J.F., Stupp, S.I., Kibbe, M.R. (2008) Nitric oxide and nanotechnology: A novel approach to inhibit neointimal hyperplasia *Journal of Vascular Surgery* **47(1)**, 173–182.
21. Tysseling-Mattiace, V.M., Sahni, V., Niece, K.L., Birch, D., Czeisler, C., Fehlings, M.G., Stupp, S.I., Kessler, J.A. (2008) Self-assembling nanofibers inhibit glial scar formation and promote axon elongation after spinal cord injury *The Journal of Neuroscience* **28(14)**, 3814–3823.
22. Rajangam, K., Arnold, M.S., Rocco, M.A., Stupp, S.I. (2008) Peptide amphiphile nanostructure-heparin interactions and their relationship to bioactivity *Biomaterials* **29**, 293298–293305.
23. Sargeant, T.D., Guler, M.O., Oppenheimer, S.M., Mata, A., Satcher, R.L., Dunand, D.C., Stupp, S.I. (2008) Hybrid bone implants: Self-assembly of peptide amphiphile nanofibers within porous titanium *Biomaterials* **29(2)**, 161–171.

24. Sargeant, T. D., Rao, M.S., Koh, C.Y., Stupp, S.I. (2008) Covalent functionalization of NiTi surfaces with bioactive peptide amphiphile nanofibers *Biomaterials* **29(8),** 1085–1098.
25. Spoerke, E.D., Anthony, S.G., Stupp, S.I. (2009) Enzyme Directed Templating of Artificial Bone Mineral *Advanced Materials* **21(4)**, 425–430.
26. Mata, A., Hsu, L., Capito, R., Aparicio, C., Henrikson, K., Stupp, S.I. (2009) Micropatterning of bioactive self-assembling gels *Soft Matter* **5**, 1228–1236.
27. Webber, M.J., Tongers, J., Renault, M.A., Roncalli, J.G., Losordo, D.W., Stupp, S.I. (2009) Development of bioactive peptide amphiphiles for therapeutic cell delivery *Acta Biomateriala* **6**, 3–11.
28. Mata, A., Geng, Y., Henrikson, K., Aparicio, C., Stock, S., Satcher, R., Stupp, S.I. (2010) Bone regeneration mediated by biomimetic mineralization of a nanofiber matrix *Biomaterials* **31**, 6004–6012.
29. Shah, R.N., Shah, N.A., Del Rosario Lim, M.M., Hsieh, C., Nuber, G., Stupp, S.I. (2010) Supramolecular design of self-assembling nanofibers for cartilage regeneration *Proceedings of the National Academy of Sciences of the United States of America* **107(8)**, 3293–3298.
30. Chow, L.W., Wang, L.J., Kaufman, D.B., Stupp, S.I. (2010) Self-assembling nanostructures to deliver angiogenic factors to pancreatic islets *Biomaterials* **31(24)**, 6154–6161.
31. Standley, S.M., Toft, D.T., Cheng, H., Soukasene, S., Chen, J., Raja, S.M., Band, V., Band, H., Cryns, V.L, Stupp, S.I. Induction of Cancer Cell Death by Self-assembling Nanostructures Incorporating a Cytotoxic Peptide *Cancer Research.* (available online Res. 0: 0008–5472.CAN-09-3267v1).
32. Muraoka, T., Koh, C.Y., Cui, H., Stupp, S.I. (2009) Light-triggered bioactivity in three-dimensions *Angewante Chemie International Edition* **48**, 5946–5949.
33. Ghanaati, S., Webber, M.J., Unger, R.E., Orth, C., Hulvat, J.F., Kiehna, S.E., Barbeck, M., Rasic, A., Stupp, S.I. (2009) Dynamic in vivo biocompatibility of angiogenic peptide amphiphile nanofibers *Biomaterials* **30**, 6202–6212.
34. Behanna, H.A., Donners, J.J.J.M, Gordon, A.C., Stupp, S.I. (2005) Coassembly of amphiphiles with opposite polarities into nanofibers *Journal of the American Chemical Society* **127**, 1193–1200.
35. Greenfield, M.A., Hoffman, J.R., de la Cruz, M.O., Stupp, S.I. (2010) Tunable Mechanics of Peptide Nanofiber Gels *Langmuir* **26(5)**, 3641–3647.

Chapter 4

Stimuli Responsive Polymers for Nanoengineering of Biointerfaces

Szczepan Zapotoczny

Abstract

There is an increasing demand on the development of "smart" switchable interfaces since controlling surface topography and chemical functionality on a nanometer scale is crucial for numerous biomedical applications. Those surfaces, which are based on stimuli responsive polymers (SRPs), are able to modify their interactions with cells, biomolecules responding to different physical (e.g., temperature) or chemical (e.g., pH) stimuli. Such behavior may partially mimic complex dynamic properties of natural systems that are regulated by many biological stimuli. This paper reviews major studies and applications of SRPs as biointerfaces in a form of thin polymeric films (gels) and surface tethered polymers (brushes).

Key words: Stimuli responsive polymers, Polymer brush, Polymer film, Cell adhesion, Poly(*N*-isopropylacrylamide), Elastin-like polypeptides

1. Introduction

Polymeric biomaterials used to support cell and tissue growth have originally been designed to fulfill requirements mainly on their bulk mechanical and structural properties while the properties of interfaces between biological systems and man-made devices (biointerfaces) also appear to be crucial. For example, wettability is an important parameter related to surface relevant phenomena as adhesion and can be controlled by varying surface chemical functions and topography on a nanometer scale. Many of the commonly used biodegradable polymers are hydrophobic so in order to enable adhesion of bio-objects (proteins, cells) from aqueous environment, while preserving their functions, the surface of these polymers should be modified to some extent. Plasma treatment

Melba Navarro and Josep A. Planell (eds.), *Nanotechnology in Regenerative Medicine: Methods and Protocols*, Methods in Molecular Biology, vol. 811, DOI 10.1007/978-1-61779-388-2_4, © Springer Science+Business Media, LLC 2012

with argon, oxygen and other gases (1, 2), or other harsh chemical etching methods (3) are able to introduce charged groups to the hydrophobic surfaces of polymeric biomaterials. Unfortunately, those techniques lead to some degradation of the polymers, what is more important, and the obtained effects may be nonpermanent and difficult to control due to spontaneous surface reconstruction.

Biologically motivated surface engineering (4) has been developing as an answer to rising demands for active and switchable biointerfaces. Recent approaches to surface modification target the spatially controlled design of interfaces as well as the development of surfaces that respond to external stimuli (5, 6). These "smart" surfaces provide a dynamic control of their mechanical properties and chemical functionalities to direct cell function or biomolecule adhesion. They may also be designed to accommodate active biomolecules (growth factors, drugs) and release them on demand upon application of a certain stimulus.

Stimuli responsive polymers (SRP) form a group of macromolecular materials that undergo significant changes of their structure and properties upon change of the environmental parameters like temperature (7, 8), pH (9), ionic strength (10) electric field (11), and chemical and biological analytes (12) in a relatively small parameter range. Polymers responding to irradiation at specific wavelength (13) and changes of the solvent quality (14) are also considered to be stimuli-responsive ones. The transition mechanism involves mainly conformational changes of the polymeric chains, sometimes associated with breaking and formation of chemical bonds.

There are a number of synthetic and natural SRPs (15). The most studied temperature-responsive polymer is poly(*N*-isopropylacrylamide) (PNIPAM) that exhibits the so-called lower critical solution temperature (LCST) at 32°C (16). PNIPAM is well soluble in water at temperatures below LCST but precipitates after rising the temperature even slightly above 32°C. This phase transition is related to the formation of intermolecular hydrogen bonds with water molecules forcing the chains to adopt conformation that prevents the hydrophobic isopropyl groups to be exposed to the aqueous environment. Those bonds break above LCST leading to exclusion of water from the chains and formation of intramolecular hydrogen bonds. As a result, the polymer becomes insoluble, forming aggregates and precipitating from the solution. Importantly, the process may easily be reversed just by lowering the temperature below LCST and its kinetics depend on the mobility of the chains.

Upon application of a stimulus, SRPs undergo dramatic changes on molecular scale that are manifested also as macroscopic changes. This "smart" behavior of SRP has already been widely explored in solution, but their more advanced applications are related to functional biointerfaces that may be engineered on nanoscale. Development of substrates that dynamically regulate

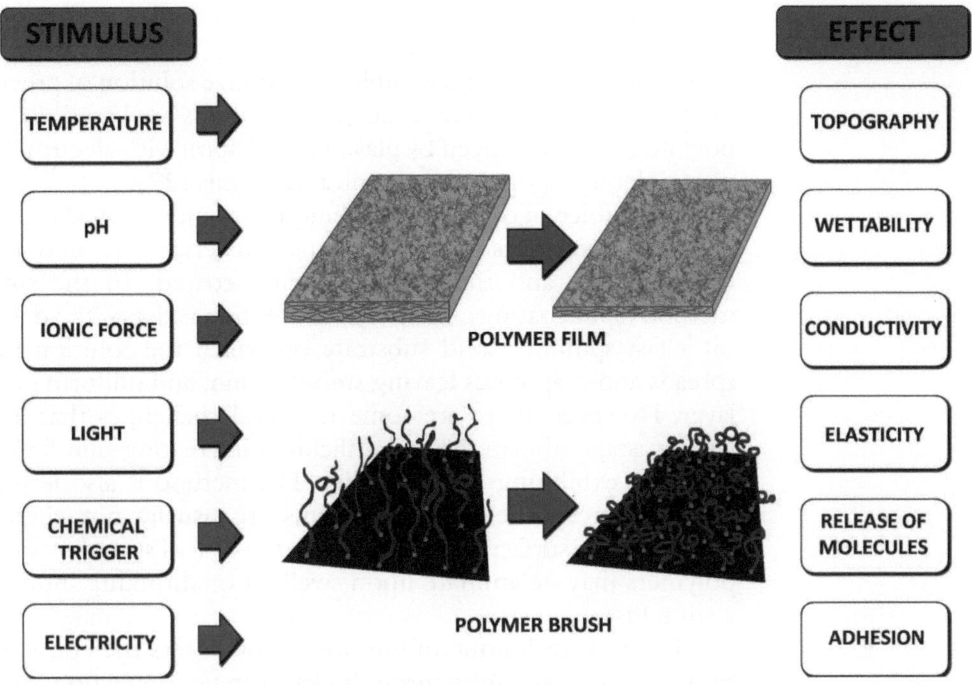

Fig. 1. Schematic structural changes of stimuli responsive polymer films and brushes upon application of different stimuli and the implied changes of surface properties.

biological functions in response to applied stimuli, thereby mimicking the dynamic properties of biological systems is an important research area in the field of nanomedicine. Surfaces that can modulate biomolecule activity (17), protein immobilization (18), and cell adhesion or migration (19) at the liquid–solid contact can be useful in diverse biomedical applications. SRPs have already found numerous applications in regenerative medicine, minimally invasive surgery, controlled drug delivery, diagnostics, and as tools for the study of basic cell biology.

Progress in solution applications of SRPs like drug delivery systems, affinity separation of biomolecules, and enzyme recovery using proteins conjugated with SRPs has been reported and reviewed (20, 21). Also, bulk materials based on SRPs like shape-memory biomaterials (22) were fabricated.

This paper reviews major studies and applications of SRP as smart biointerfaces (23, 24) in a form of (A) thin polymeric films/gels and (B) surface-tethered polymers (brushes) (see Fig. 1). Examples of surfaces responding to different stimuli are presented indicating important features/advantages of the approaches used. The most important fabrication methods and applications are also described.

1.1. Thin Polymer Films and Gels

Polymer films having thickness ranging from several nanometers to micrometers and more can be prepared on substrate surfaces using several deposition techniques with different complexity and applicability. Methods, such as spin coating (25) and layer-by-layer (LbL) electrostatic self-assembly (26), utilize solution of previously prepared polymers while some other methods are based on in situ polymerizations induced by plasma, irradiation with electron beam, ultraviolet light, or electrochemical reactions (27).

The choice of deposition technique depends upon the physicochemical properties of the polymer material, the film quality requirements, and the substrate being coated. In the simplest method (spin coating), the polymer solution is deposited dropwise on a fast spinning solid substrate on which the solution rapidly spreads and evaporates leaving smooth, thin, and uniform polymer layer. However, there are some technical challenges that include large wastage of material and difficulties in creating thin films with polymers exhibiting low solubility. The method is also limited to flat substrates. The so-formed films are usually not chemically anchored to a surface and especially in the case of stimuli-responsive polymers may delaminate upon swelling or shrinking induced by a stimulus.

The LbL technique of film formation allows the construction of multilayer structures through electrostatic interaction between oppositely charged polyelectrolytes, which are alternatingly deposited on a charged substrate. In such a way, multilayer films of desired thickness may be prepared by repeating the procedure. LbL is the method which allows controlling film fabrication on the nanometer scale thus nanoengineering of film structure and chemical composition in a relatively easy way (28, 29). Stimuli responsive LbL films have also been fabricated and applied as smart coatings for biomedical applications (30). They may be assembled on substrates of complex geometry like tissue scaffolds or implantable devices (31, 32). There are a number of reports on LbL capsules employed as smart microcontainers, for example in drug delivery (33) or biocompatible coatings (34), but this topic is beyond the scope of this paper.

Other deposition methods usually require chemical reactions to take place. Typically, monomers from vapor phase or solution are left to adsorb on a surface, which is subjected to further treatment in a plasma reactor, electrochemical cells, or using an electron beam (e-beam) or laser irradiation in order to run the radical polymerization. Special initiator that efficiently produces radicals upon the mentioned treatments is usually introduced into the mixture. So-prepared thin films often form a network of crosslinked macromolecules thanks to recombination of radicals generated on neighboring chains. Such gels may be formed in a more controlled way by employing bifunctional crosslinkers that may bind covalently at two sites of the chains bridging them. Ultraviolet (UV) light may also be used to cross-link the copolymer containing, e.g., cinnamoyl groups that undergo photodimerization (35). In biomedical applications, the most common are hydrogels that are highly hydrated gels composed of hydrophilic polymers (36).

The protocol for e-beam-assisted film formation is simple and involves three steps: spreading a dilute monomer solution on a surface, e-beam irradiation, and subsequent washing. The thickness of the grafted polymer film can be controlled by monomer concentration and radiation energy. A typical thickness obtained for grafted PNIPAM is about 100 nm.

The protocol for plasma polymerization (37) is similar to the one mentioned above but the necessary apparatus is more easily available in laboratories compared to the e-beam technology. Nevertheless, both methods may induce chemical changes to the deposited polymers as well as produce gels with too high level of cross-linking that do not support cell adhesion (38).

One of widely used method for film deposition is electropolymerization (39). It is definitely the choice for fabrication of films of conductive polymers, but its application is limited only to conductive support so it is not suitable for majority of polymeric biomaterials that are electrical insulators.

1.1.1. Temperature-Responsive Films

As mentioned above, PNIPAM is one of the most intensively studied temperature-responsive polymer. The reversible phase transition of surface-grafted PNIPAM has been utilized to develop thermo-responsive culture dishes for cells (40). Okano et al. (41) (see Fig. 2) demonstrated that tissue culture polystyrene (TCPS) dishes covered with e-beam-grafted PNIPAM film allowed cells to adhere, spread and proliferate above the LCST of the polymer. A decrease in culture temperature below the LCST resulted in detach-

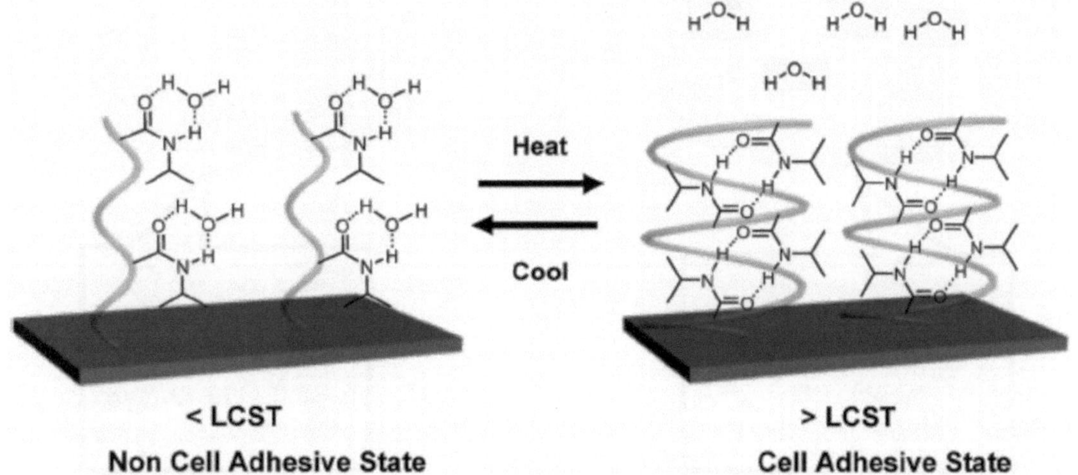

Fig. 2. Diagram illustrating the temperature-induced switching of a PNIPAM-modified surface. PNIPAM chains are shown forming intermolecular hydrogen bonds with water molecules at temperatures below the LCST (*left*) and forming intramolecular hydrogen bonds between C=O and N–H groups at temperatures above the LCST (*right*). Reprinted with permission from ref. 41.

ment of cells. Further investigation into the mechanism of cell attachment revealed that strong interactions between hydrophobic PNIPAM and fibronectin, an ECM (extracellular matrix) protein, mediated cell attachment above the LCST. Decreasing the temperature below the LCST makes the surface hydrophilic and resulted in desorption of the ECM along with the cultured cells.

Temperature-triggered cell desorption from PNIPAM surfaces provides a gentler alternative to traditional cell removal methods such as mechanical dissociation and enzymatic digestion (42). The removal of cells grown on TCPS by mechanical dissociation and enzymatic digestion has been shown to be damaging to both cells and ECM (42). Cell sheets released from surfaces grafted with PNIPAM using a modest temperature drop were detached with an intact extracellular matrix (43) and preserved tight junctions between cells (44). Such cell sheet engineering (45, 46) (see Fig. 3) in comparison to the cell detachment by mechanical scraping or trypsin digestion, causes less damage to the cells and to the extracellular matrix as determined by immunofluorescence microscopy (42). The released cells maintained their substrate adhesivity,

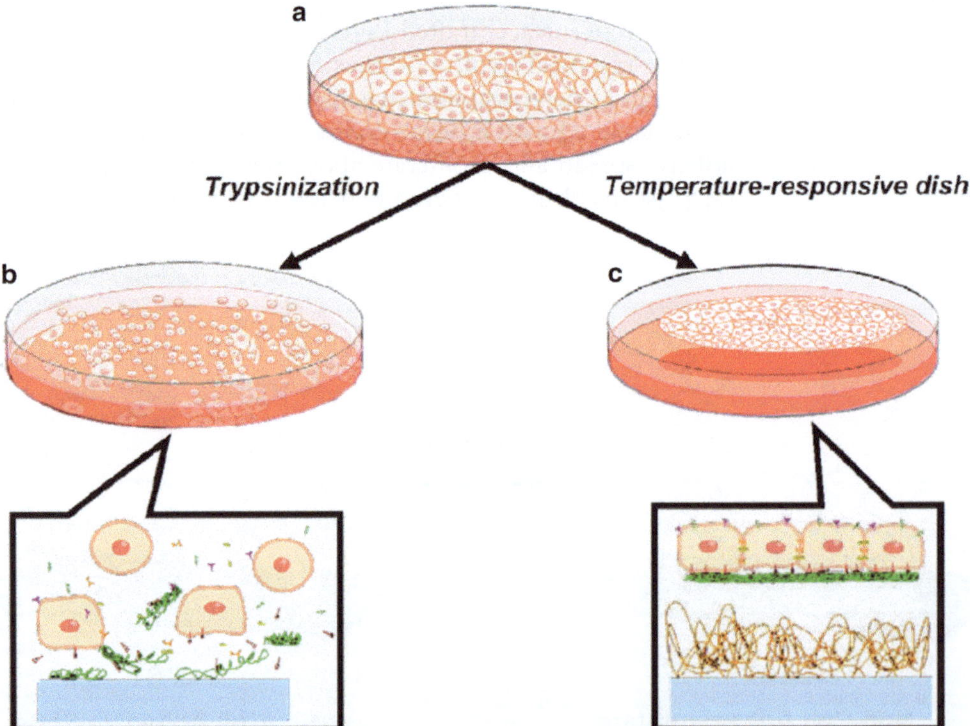

Fig. 3. Temperature-responsive culture dishes. (**a**) During cell culture, cells deposit extracellular matrix (ECM) molecules and form cell-to-cell junctions. (**b**) With typical proteolytic harvest by trypsinization, both ECM and cell-to-cell junction proteins are degraded during cell recovery. (**c**) In contrast, cells harvested from temperature-responsive dishes are recovered as intact sheets along with their deposited ECM, by simple temperature reduction. Reprinted with permission from ref. 46.

growth and secretion activities (42). In addition, it was shown that thickness of the PNIPAM film in the nanometer range plays a role in cell adhesion and detachment (47). This noninvasive cell recovery method has promising applications in tissue engineering (48).

Cellular interactions with PNIPAM-grafted surfaces can be regulated vertically by controlling the thickness of the grafted polymer layers in nanometer-scale range. 15–20 nm thick layers of PNIPAM exhibit temperature dependent cell adhesion/detachment, while surfaces with layers thicker than 30 nm do not support cell adhesion. These changes in cell adhesion are explained by the limited mobility of the surface grafted polymer chains as a function of grafting, hydration, and temperature (49).

Lateral regulation of the cell adhesion on the smart surface may be achieved by micro or nanopatterning of surface polymer coatings (50, 51). With respect to tissue engineering, cell patterning techniques can be used for controlled cell growth, differentiation as well as for the development of cell-based diagnostics (52, 53).

As an example the patterns of PNIPAM were grafted on tissue culture polystyrene (TCPS) using e-beam irradiation and masking (54). In the first step, PNIPAM film was formed on TCPS. Then, the film was covered with the metal mask exposing only selected areas to e-beam irradiation. In the presence of BMA monomer, hydrophobic PBMA polymer was co-grafted in the exposed areas (see Fig. 4). On such chemically patterned surface, site-selective adhesion and growth of rat primary hepatocytes and bovine carotid endothelial cells allowed for patterned coculture at 27°C. Seeded hepatocytes adhered exclusively onto hydrophobic, dehydrated poly(*N*-isopropylacrylamide-*co*-*n*-butyl methacrylate) (PNIPAM-PBMA)co-grafted domains, but not onto neighboring hydrated PNIPAM domains. Sequentially seeded endothelial cells then adhered exclusively to PNIPAM domains which become hydrophobic upon increasing temperature to 37°C, achieving patterned

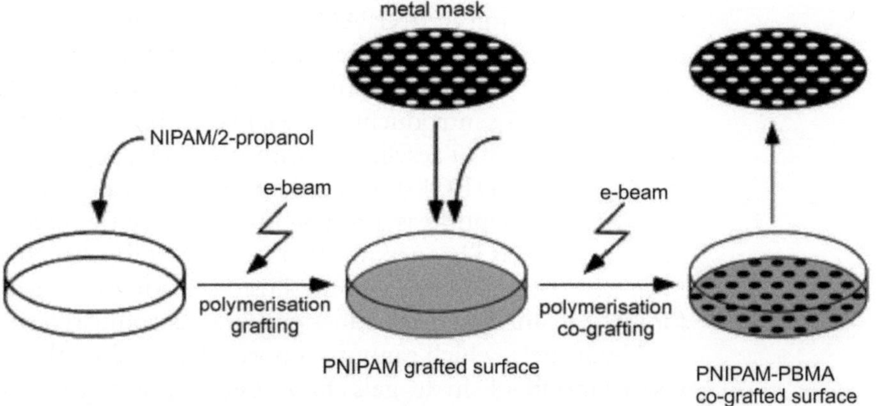

Fig. 4. Preparation of patterned dual thermoresponsive polymer-grafted culture dishes. Adapted with permission from ref. 54.

cocultures. Reducing culture temperature to 20°C promoted hydration of both polymer-grafted domains, and release of the cocultured, patterned cells as continuous cell sheets. A coculture of three or more cell types might also be realized since the transition temperature of grafted thermosensitive copolymer can readily be varied.

The transition temperatures of PNIPAM may also be tuned, e.g., to reach appropriate physiological range. Copolymerization of NIPAM with hydrophobic butyl methacrylate decreases the LCST of the aqueous copolymer solution and copolymerization with hydrophilic comonomers, such as acrylic acid or hydroxyethyl methacrylate, results in an increase in LCST (55).

Thermo-responsive polymers containing redox groups have also been employed as dynamic enzyme supports (56). The poly(*N*-isopropylacrylamide-*co*-vinylferrocene) thermo-shrinking redox gel was used to control the loading and unloading of both glucose oxidase (GOx) and lactate oxidase (LOx) enzymes from electrode surfaces. The copolymer-coated indium tin oxide (ITO) electrode was loaded with either GOx or LOx enzyme at a temperature below 10°C, followed by subsequent elevation of the temperature to 35°C. This increase in temperature caused the gel film to shrink and entrap the enzyme. In order to unload the enzyme from the redox gel-coated electrode, the temperature was decreased to 10°C and a fresh or different enzyme could be reloaded.

One of the major challenges of temperature-induced switching is the localized application of temperature gradients. Recent advances in microfabrication have enabled the use of miniaturized components, such as microheaters, (57) in combination with switchable surfaces. Cheng et al. reported the use of addressable microheaters in combination with a plasma-deposited PNIPAM coating for cell patterning (58).

To tailor the properties of PNIPAM to the specific needs of a given application, while maintaining its thermoresponsiveness, structural derivatives of PNIPAM have been prepared. Copolymerization of PNIPAM with functional comonomers and/or postmodification of grafted chains allow for the control of the rate of cell adhesion and spreading or alternatively the rate of cell detachment. By introducing a reactive carboxylic acid group via copolymerization of acrylic acid with PNIPAM, peptides containing the Arg-Gly-Asp (RGD) cell binding sequence can be immobilized. The RGD tripeptide is present in several cell-contacting ECM proteins, such as fibronectin, vitronectin, type I collagen, and it has been widely studied as an immobilized cell adhesion ligand specific for integrin-mediated cell adhesion (59). As compared to nonmodified hydrogels, which cannot support cell adhesion, the mentioned above modified hydrogels have been shown to increase the spreading and proliferation of, e.g., rat osteoblasts (60). The surfaces of the copolymer modified with RGD sequence can mediate cell adhesion in the absence of serum and retain cell detachment properties in response to temperature change (61).

Another class of frequently studied thermoresponsive molecules is based on elastin-like polypeptides (ELPs) which are composed of repeated Val-Pro-Gly-X-Gly sequence, where X may be any amino acid other than Pro (62, 63). These polypeptides are soluble in water below their LCST and become reversibly desolvated and form aggregates above LCST. At a molecular level, the LCST transition of an ELP is accompanied by a conformational change in the polymer chain from a disordered, random hydrophilic coil to a more ordered, collapsed hydrophobic globule. The value of LCST for ELPs may easily be varied by changing pH and ionic strength (64) of a medium which brings additional opportunities for these systems. What is more important, their chain length and composition can be controlled through genetic synthesis and ELPs can be produced with tailored LCST ranging from 0°C to 100°C (65).

Similarly to PNIPAM, ELPs have been explored for protein purification processes. Fusion proteins tagged with ELPs were expressed in the *E. coli* expression system and purified simply by inverse transition cycling, through the application of a slight change in temperature to alter the state of the ELPs (66). Owing to the presence of the ELPs, the target protein became thermoresponsive, allowing it to be separated through ELP aggregation. In a recent advancement, this system was modified for an array of immobilized, micropatterned ELPs, where the ELP fusion protein was captured by the immobilized ELPs via hydrophobic interactions upon LCST transition (67). Moreover, the capture and release of target proteins were also made possible (see Fig. 5) (65). Using a similar approach, the ELPs were combined with antibody-binding proteins to specifically precipitate antibodies from solutions containing a mixture of proteins (68).

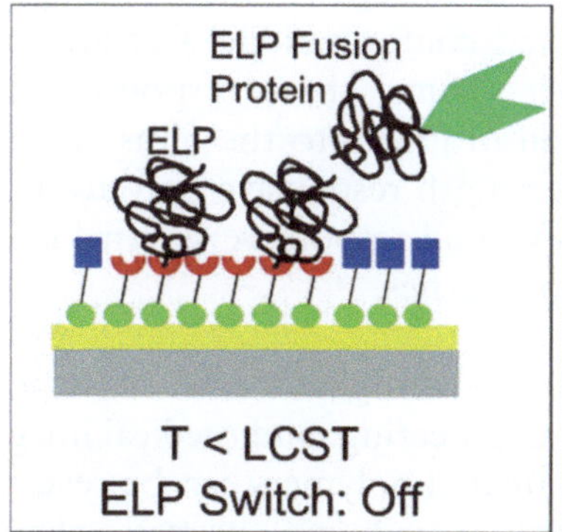

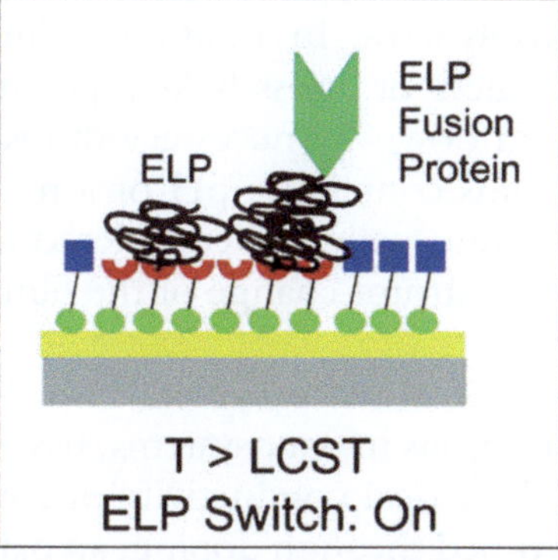

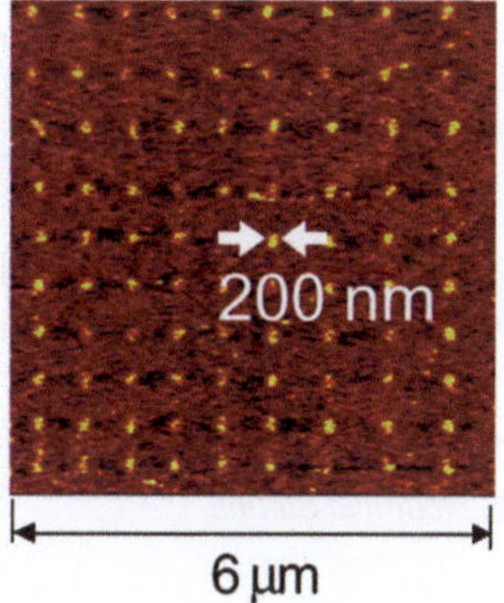

Fig. 5. Schematic of temperature-based protein capture using an ELP and atomic force micrograph of ELP dot array. Reprinted with permission from ref. 65.

Serpe et al. were the first who applied the LbL technique to self-assemble microgel particles at surfaces. Temperature-responsive poly(*N*-isopropylacrylamide)-*co*-acrylic acid microgels and polyallylamine were alternately deposited at a surface via electrostatic self-assembly (69). The fact that microgel particles retained their thermoresponsivity within the multilayer allowed the use of these films for pulsed thermally triggered release of insulin (70) and doxorubicin (71). The approach based on microgel self-assembly is promising as it blends the advantages known for microgels, such as high loading capacity and fast response with the facile LbL technique of surface modification.

There are natural polymers, like some derivatives of cellulose that also show temperature-responsive behavior. For example, hydroxypropylcellulose (72) and methylcellulose (73) exhibiting LCST have been used for the synthesis of temperature-responsive hydrogels (74). Such polymers have very promising applications, but they have not been explored so far as extensively as PNIPAM.

1.1.2. Ionic Strength and pH Responsive Films

Common stimuli used to regulate intermolecular interaction in electrostatically assembled systems include pH and ionic strength. The pH-induced excess charge can be utilized, for example, to bind and release low molecular weight compounds, such as drugs in LbL films consisting of weak polyelectrolytes (75). As Hiller and Rubner showed (76) poly(allylamine hydrochloride) and poly(styrenesulfonic acid) (PAH/PSS) can be incorporated into a multilayer film with specifically designed molecular architectures (by changing pH conditions during LbL deposition) that make the coating either virtually insensitive or highly responsive to small changes in post-assembly pH. Moreover, the elastic moduli of these films were dependent on the pH at which the polyelectrolyte multilayers were assembled. Thus, the attachment and proliferation of human microvascular endothelial cells were regulated through changes in the compliance of these films (77). However, stimulating changes of polymer chains conformation in already prepared films/gels by varying pH or ionic strength of the environment is usually relatively slow. In addition to limited conformational freedom of entangled or cross-linked polymer chains, these responses are slowed down by necessary diffusion of ions into the films. Thus, the films containing pH or ionic strength responsive polyelectrolytes are usually not considered as stimuli-responsive systems that require abrupt change of the film properties.

1.1.3. Films Responsive to Electrical Stimuli

Electrosensitive polymers have been investigated extensively for applications in nanosystems, tissue engineering, and medical imaging. Electrical conductivity of conjugated polymers can be reversibly tuned through doping, an oxidation/reduction process, where charge carriers are introduced to the polymeric backbone either chemically or electrochemically. Mainly due to the reversibility of doping, conducting polymers are of considerable interest for a variety of biomedical applications (78). A particular application of conductive polymers is drug delivery, where control of the current allows one to control the amount of ionic drug that can be released from a polymer film on an electrode. On the basis of this process, a variety of anions including biotin (79), and ATP (80) have been electrostatically entrapped into conducting polymer films and released by electrical potential stimulus in a controlled way. Furthermore, conductive polymer films may act as vehicles for delivery of positively charged drugs (81). As an example, a conducting, composite polymer, poly(*N*-methylpyrrolylium)/poly (styrenesulfonate) (PMP/PSS) has been employed for controlled delivery of the neurotransmitter dopamine (81). The PMP/PSS films were prepared on glassy carbon disks by the anodic polymerization of *N*-methylpyrrole from an aqueous solution of PSS. In its reduced state, the PMP/PSS film was able to bind dopamine cations, which could be released by oxidizing the polymer film.

Polypyrrole is an interesting material for biomedical applications, due to its chemical and thermal stability, and low cytotoxicity. Polypyrrole may be especially useful as substrates for cell cultures since it provides a noninvasive way to regulate cell form and function. By reversibly changing their oxidation state and, consequently, their properties and surface binding characteristics, polypyrrole polymer films on ITO-coated glass substrates have been shown (82) to act as dynamic cell culture substrates. In vitro studies demonstrated that extracellular matrix molecules, such as fibronectin, adsorb efficiently onto the oxidized (polycation) polypyrrole thin films, and support cell attachment under serum-free conditions. On the other hand, electrochemical reduction of the oxidized polypyrrole film to its neutral state by applying an electrical potential resulted in inhibition of both cell spreading and DNA synthesis, but without adversely affecting cell viability.

Other potential biomedical application of the polypyrroles includes highly localized stimulation of neurite outgrowth and guidance for neural tissue regeneration. Langer et al. (83) electrochemically synthesized polypyrrole films on ITO-conductive borosilicate glass and evaluated their application as a substrate for enhancing nerve cell interactions in culture. Application of an external electrical stimulus through the polymer film resulted in enhanced neurite outgrowth.

1.1.4. Light-Responsive Films

Polymer films comprising photo-responsive molecules, such as spiropyran, (84, 85) also represent attractive candidates for controlling protein and cell adhesion on surfaces. Spiropyran isomerizes under illumination with UV light from the more hydrophobic spiro conformation to the polar, hydrophilic zwitterionic merocyanine conformation, while reverse isomerization can be triggered by irradiation with visible light. This change from hydrophobic to hydrophilic state upon isomerization has been applied to demonstrate UV light-induced detachment of fibrinogen, platelets, and mesenchymal stem cells from poly(spiropyran-*co*-methyl methacrylate)-coated glass plates (84). The other approach by Edahiro et al. (85) involved the use of a polymer material composed of PNIPAM having spiropyran chromophores as side chains to develop a reversible photo-responsive culture surface. UV irradiation with a wavelength of 365 nm was shown to induce an enhancement to adhesion of living cells, which could be reversed by visible light irradiation with a wavelength of 400–440 nm followed by thermal annealing at 37 °C (85). The effect of UV irradiation on cells is dependent on the intensity and wavelength of the radiation. Also, different cell types and lines have somewhat different sensitivity toward UV radiation. In these cell cultivation studies based on spiropyran-derived polymer films, the cells were apparently not significantly affected by the UV irradiation and exhibited sufficient viability after exposure.

A further means of controlling biomolecule activity of surfaces by photochemical stimulus relied on the *trans–cis* isomerization of the azobenzene molecule. The azo chromophore isomerizes by illumination with UV light (λ_{irr} = 300–400 nm) from the stable *trans* form to the *cis* state while reverse isomerization can be triggered by irradiation with visible light (λ_{irr} = 425–500 nm). Isomerization of azobenzene is accompanied by an appreciable shape change as the *trans* isomer adopts a more linear conformation than the *cis* isomer. Hayashi et al. (86) have shown that, when azobenzene moieties were incorporated into a peptide immobilized on a carboxymethylated dextran-coated gold surface, the structural changes in the azobenzene could lead to the photoregulation of peptide binding to its RNA aptamer.

1.1.5. Films Sensitive to (Bio)Chemical Triggers

Enzyme-triggered activation of surfaces shows great promise as a method for altering biological properties of surfaces in a controlled manner (87). Ulijn et al. (88) developed a strategy that enzymatically switched polymer surfaces from a state that prevented cell adhesion to another state in which cell adhesion and spreading were promoted. This strategy was based on the fabrication of poly(ethylene glycol) (PEG) acrylamide films on epoxy-coated glass slides by spin coating and UV curing, followed by the functionalization of the polymer film with the RGD peptide, which was initially protected with a bulky blocking group. Surface-tethered RGD sequences were shown to be inactive in promoting cell adhesion when capped with the blocking group, which was selectively removed upon exposure to the serine protease chymotrypsin, thereby activating the RGD and triggering cell attachment in situ.

To use signals derived from living cells to control their adhesion and detachment, a polyethylene membrane grafted with a copolymer of NIPAM and benzo-18-*crown*-6-acrylamide (BCAm) has been developed (89). This system makes use of the shift in the LCST of the copolymer that occurs when BCAm acts as a receptor to capture potassium ions at a constant temperature. Cells cultured on this membrane can be released with the addition of exogenous potassium ions. In a recent development, this system was shown to specifically detach dead cells upon UV irradiation, as dead cells release potassium, whereas live cells do not. Cell release was confirmed to be mediated by potassium released from the necrotic cells, as the addition of free poly-BCAm used to capture released potassium prevented the dead cells from detaching (see Fig. 6).

1.2. Surface-Tethered Polymer Brushes

Polymer brushes are polymer coatings in which polymer chains are tethered on one end to a surface at sufficiently high grafting density such that steric repulsion between the chains causes chain stretching (90). The stretched chain conformation in brushes significantly differs from the random-walk conformation of free polymer chains in solution or in conventional solution cast or spin-coated polymer films. The thickness of those surface-tethered coatings is dictated by both the polymer chain length and surface graft density, and ranges typically from a few nanometers to several hundred nanometers. At the smaller densities of the tethered chains, they adopt rather "pancake" or "mushroom" than "brush" conformations (see Fig. 7). Polymer brushes comprise the advantages of both self-

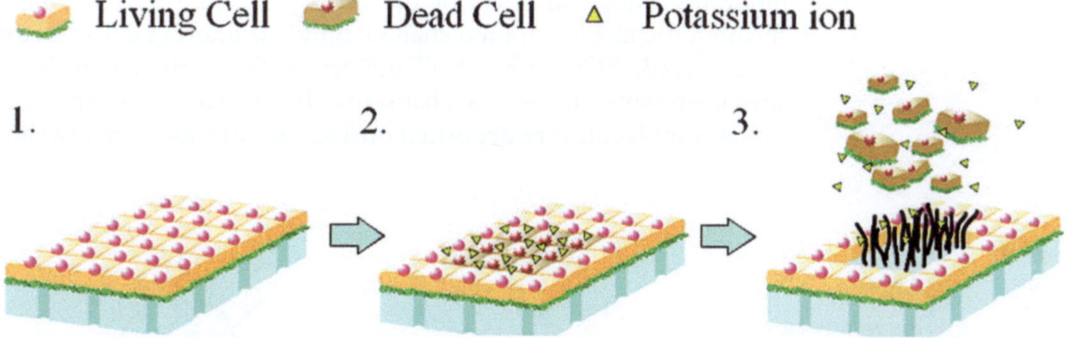

Fig. 6. Cell culture on potassium cation sensitive substrate that selectively releases dead cells. The polymer substrate changes its surface hydrophilic properties in the presence of K^+ ions. When cells die, potassium ions are released from the dead cells, the polymer recognizes the ion signal, and it becomes hydrated. As a result, dead cells are selectively removed from the surface. Reprinted with permission from ref. 89.

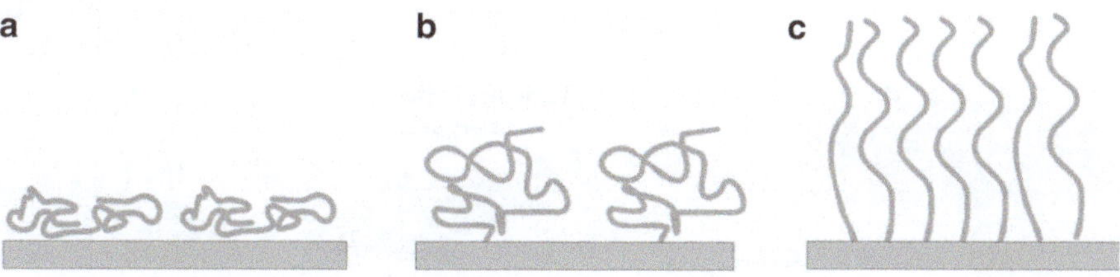

Fig. 7. Scheme of the conformations of the surface-tethered polymer chains: (**a**) "pancake", (**b**) "mushroom", and (**c**) "brush".

assembled monolayers (SAMs) and classical polymer films used as functional coatings.

Polymer brushes can be formed on a variety of solid substrates, including metal, semiconductors, and polymeric supports (91). Other attractive features that characterize these coatings are represented by their structural intrinsic properties (dense assemblies of grafted molecules), which enable a very high interfacial density of chemical functional groups and large capacity for binding, e.g., biomolecules. Polymer brushes, due to lack of entanglements, are characterized by higher chain mobility compared to conventional polymer films or surface-tethered gels and respond usually much faster and to a greater extent to external stimuli. What is more, chemical functionality can be incorporated in specific positions from the top surface that allows precise embedding of desired molecules in the brush layer (92). Postfunctionalization of brush films with biomolecules or nanoobjects offers additional smart characteristics to these coatings.

The surface-tethered SRPs have been formed using physical adsorption (93), as well as covalent anchoring in "grafting to"(94), and "grafting from" (95–97) approach (see Fig. 8). Covalent attachment is often preferred due to the inherent resistance of the coating to degradation by temperature and solvents. In "grafting to" approach, end-functional macromolecules are attached to a surface via physical adsorption or chemical reactions. Due to steric hindrance (slow diffusion of the macromolecules to the surface through the already grafted chains), only low grafting densities can be achieved. "Grafting from" utilizes surface-tethered initiating sites from which polymeric chains may be grown. In the first step, initiator molecules are deposited on a surface in a form of SAM and

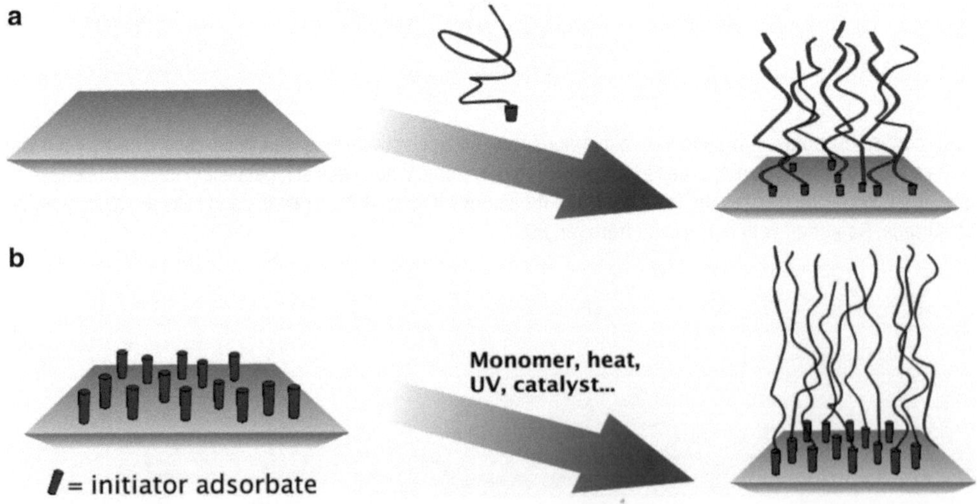

Fig. 8. Scheme of (**a**) "grafting to" and (**b**) "grafting from" techniques for fabricating polymer brushes.

later polymers are grown. This method allows controlling the surface concentration of the active sites and leads to synthesis of high density polymer brushes. In addition, patterned polymeric brushes may easily be formed exploiting such soft lithographic techniques as microcontact printing (μCP) (98, 99) or tip-assisted nanolithography (100, 101) in order to deposit the initiators from which the polymer chains may eventually be grown. The surface covered with the brushes may be further engineered on molecular scale by, e.g., formation of copolymer chains and fabrication of mixed brushes consisting of two types of surface-tethered chains.

Polymer brushes resulting from the "grafting from" technique have been prepared via a number of surface-initiated polymerizations, including free-radical, cationic, and anionic. Particularly, useful for this technique are controlled radical polymerization methods: atom transfer radical polymerization (ATRP), reversible addition–fragmentation transfer polymerization (RAFT), and nitroxide-mediated radical polymerization (NMRP). Detailed description of these surface-initiated polymerization techniques can be found in a textbook (102) or a review paper (103). Surface-initiated polymerization techniques are now routinely used with a very wide range of monomers to generate functional polymer brushes with a high degree of control over the thickness, composition, chain architecture, and grafting density of the brush. While SAMs offer ease of preparation and versatile surface chemistry, polymer brushes can be produced by surface-initiated polymerization techniques with similar control over surface coverage and composition but in addition they offer control in 3D by tunable thickness and level of loading of the functional groups.

Surface-tethered polymers possessing brush morphologies have been employed so far as very effective lubricating coatings for various materials (104, 105), stabilizing coatings for colloids, platforms for the micro/nanoelectronics and to an increasing extent as functional biointerface of biomedical materials (106, 107) or protective layers against marine biofouling (108). Stimuli-responsive brushes have a great application potential thanks to their exceptional dynamic response compared to classical polymer films or gels with entangled or crosslinked chains. Mechanical actuators based on expansion and shrinking of responsive polymer brushes were fabricated (109, 110). Microfluidic channels were also coated with responsive mixed brushes in order to control the passage of liquids through the channels (111), but the most promising technical advances in using polymer brushes have been associated with their biointerfacial applications. Polymer brushes provide exciting platforms for the design of precision molecular interfaces for biological applications. They can dynamically control the presentation of regulatory signals to modulate biomolecule activity cell and protein adhesion or control drug permeation through nanoporous membranes.

1.2.1. Temperature-Responsive Brushes

PNIPAM as a prototype "smart" polymer was mainly used for fabrication of biologically relevant temperature-responsive brushes. Huber et al. used in situ-free radical polymerization of NIPAM on functionalized SAM substrates with high grafting densities (112). The grafting method has been used to integrate a 4 nm thick coating into a microfluidic hot plate device. The authors were able to fabricate a rapidly switching microfluidic device that can adsorb proteins as myoglobin, BSA, hemoglobin, and cytochrome *C*, from solution at specific sites in less than 1 s, hold them without major denaturation, and release on demand.

Functional polymer brushes were shown to be a versatile platform for the immobilization of peptides which influence the adhesion of cells. By grafting copolymer brushes, it is possible to make surfaces that can reversibly switch wettability, thereby cell adhesion as well, even more dramatically.

RGD moiety was immobilized to temperature-responsive poly(*N*-isopropylacrylamide-*co*-2-carboxyisopropylacrylamide) (PNIPAM-co-CIPAM) copolymer brushes. These surfaces promoted cell adhesion and spreading under serum-free conditions at 37°C (above the LCST of the copolymer) (113). At this temperature, the copolymer chains collapsed and formed a compact structure, allowing the integrin receptors on the cell membrane to recognize the conjugated RGD sequences to promote cell adhesion. By lowering culture temperature below the LCST, the copolymer chains were swollen, shielding immobilized RGD peptides from integrin access, limiting cell–surface attachment. These studies showed that specific interactions between cell integrins and immobilized RGD moieties can be noninvasively thermally regulated for cell attachment/detachment under serum-free conditions. The use of flexible polymer brushes present a main advantage over immobilization of adhesive motifs on SAMs. Cell adhesion depends not only on the integrin/RGD interaction, but also on the receptors (integrins) clustering to form cell focal points. When the peptide sequence is coupled to brushes, the flexibility and mobility of the brushes enables an easier formation of clusters.

PNIPAM polymers have also been exploited as bacterial (114), and protein (115), adhesion mediators. As with cell adhesion, changes in surface properties governed by polymer transitions from hydrophilic to hydrophobic states mediated bioadhesion. Such regulatory control over adhesion of different bioactive analyte classes can be applied to develop novel chromatographic matrixes.

SRP brushes may also be used to control the activity of enzymes. PNIPAM–ferrocene polymer bearing oxirane side groups has been covalently immobilized by the "grafting-to" approach to an amino-terminated SAM on gold. The grafted polymer could act as a mediator for electron transfer between the cofactor pyrroloquinoline quinone (PQQ) of soluble glucose dehydrogenase (sGDH) and the gold electrode (116). In the swollen state, the

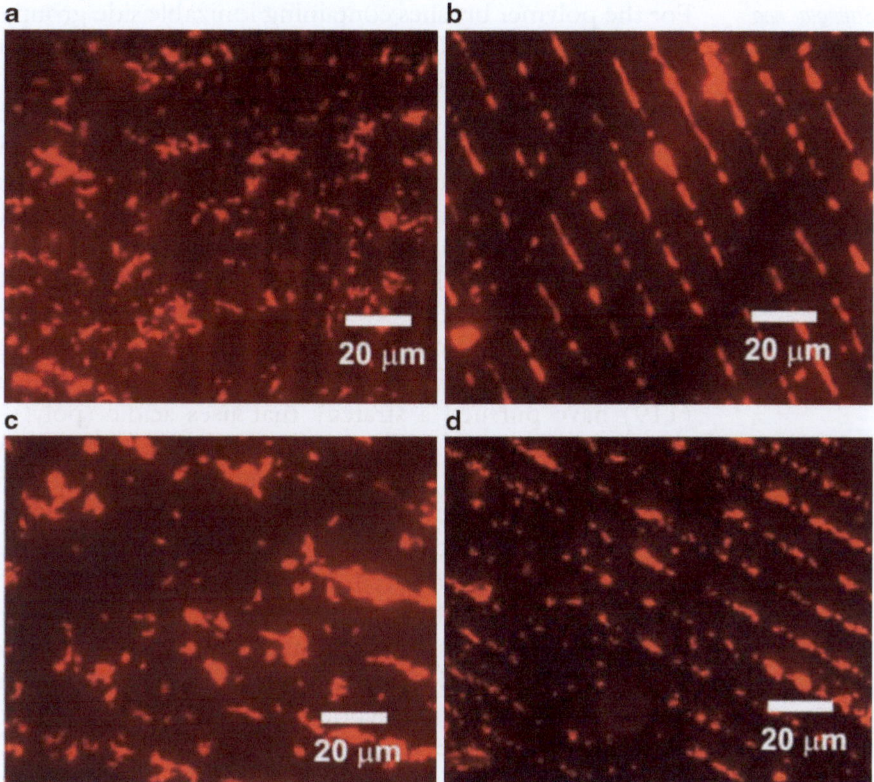

Fig. 9. Fluorescence microscopy of micropatterned surfaces of PNIPAM brushes incubated in suspensions of *Streptococcus mutans*. Images were recorded over two reversibility cycles. Images (**a–d**): initial incubation at 37°C for 1 h (**a**), followed by incubation at 4°C for 1 h (**b**), transfer to culture at 37°C for 1 h (**c**), and a second incubation at 4°C for 1 h (**d**). Reprinted with permission from ref. 117.

PNIPAM–ferrocene polymer formed brush-like structures, offering optimal conditions for mediated electron transfer between the enzyme and electrode surface while at higher temperatures the brushes collapsed limiting the penetration of the polymer layer by the enzyme molecules that led to less efficient electron transfer.

The lateral distribution of the switchable adhesive sites may easily be obtained using patterned PNIPAM brushes (117). As shown in Fig. 9, the alignment of the cells along the micropatterned PNIAM brushes was obtained only upon incubation at 4°C indicating preferential adsorption of the cells to the hydrophilic brushes. The cells while incubated at 37°C showed their nonselective adsorption across the surface. Subsequent changes of temperature of incubation lead to alignment (at 4°C) and loss of their alignment at temperature above LCST. The LCST-triggered adsorption and desorption of proteins has also been demonstrated on ELP nanopatterned surfaces (65).

1.2.2. Ionic Strength- and pH-Responsive Brushes

For the polymer brushes containing ionizable side groups, changes in the ionization (due to changes in salt concentration or, for weakly acidic or basic functional groups, due to changes in pH) result in alternation of the counterion concentration around the chains. This transition results in the swelling and expansion of the thin brush film due to large osmotic pressure. An increase in the grafting density of chains for brushes results in decreased swelling of the coating. This phase transition behavior has been used to dynamically control the immobilization of proteins on solid surfaces (118).

Efforts have also been directed toward the development of pH-responsive switchable surface-tethered brushes. Higashi et al. (119) have pursued a strategy that uses acidic (poly(L-glutamic acid), PLGA) and basic (poly(L-lysine), PLL) block polypeptides immobilized on gold surfaces to undergo a pH-induced phase transition between peptide surfaces with different charge distributions. Two types of diblock polypeptides, PLGA-*block*-PLL and PLL-*block*-PLGA, were prepared on gold substrates via the "grafting-from" method. The surface charge distribution on both polypeptide brushes has been found to be switchable by changing pH and to be strongly dependent on the conformation and ionization degree of the outer peptide block.

Poly(methacrylic acid) (PMAA), a weak polyelectrolyte with inherent pH-responsive properties (120), is a polymer of special interest in molecular biology and biomaterials related to applications in the synthesis of platforms for protein adsorption, in cell adhesion studies, and in developing biosensors (121, 122). Furthermore, PMAA has been used to improve surface hydrophilicity in tissue engineering scaffolds as well as in drug delivery devices, among other applications (123). Also, by the manipulation of ionic strength, polyelectrolyte brush nanoparticles immobilized on glass substrates have been used to function as nanocontainers for the controlled uptake and release of proteins while preserving their structural integrity (124).

The polyelectrolyte brush nanoparticles consisted of poly(styrene) core particles of 110 nm diameter onto which long chains of poly(styrenesulfonate) were grafted. The individual immobilized nanoparticles could be loaded with up to 30,000 green fluorescent protein molecules in a buffer solution of low ionic strength. The bound protein was fully released upon increasing the ionic strength of the buffer solution. Their high storage density, good retention, and controlled uptake and release of the proteins make these polyelectrolyte brush nanoparticles promising candidates for applications in drug delivery systems.

ELPs, which were tethered onto either glass (67) or gold (65) surfaces by the "grafting-to" approach, have been shown to undergo a switchable and reversible, hydrophilic–hydrophobic phase transition. The LCST transition resulted in the reversible capture of an ELP fusion protein, thioredoxin (Trx), onto the ELP surfaces directly from cell lysate. The steric accessibility of Trx was confirmed by its binding to specific anti-thioredoxin monoclonal antibody.

In a similar system adsorption of bovine serum albumin (BSA) on spherical polyelectrolyte brushes was studied. Those systems consist of a solid polystyrene core of 100 nm diameter onto which linear polyelectrolyte chains (poly(acrylic acid)) were grafted. The adsorption of BSA is strongest at low salt concentration and decreases drastically with increasing amounts of added salt. Virtually, no adsorption takes place at salt concentration of 0.1 M (125).

1.2.3. Light-Responsive Brushes

There are several ways to control surface properties by the application of a photochemical stimulus. Photochemical reactions that are accompanied by the change of shape/size of a molecule are usually used for that purpose. The most commonly used is *trans–cis* isomerization of the azobenzene molecule. Taking advantage of the change in azobenzene molecular dimensions, Kessler et al. (19) reported the control of cell adhesion properties on RGD-functionalized surfaces. The azobenzene derivative was incorporated into the RGD peptide and tethered to a poly(methyl methacrylate) surface by UV irradiation. The photoswitchable RGD peptide-coated surfaces exhibited enhanced cell adhesion in the *trans*-azobenzene configuration. On the other hand, the surfaces that had previously been UV irradiated at 366 nm showed a reduced cell plating efficiency as a result of shortening the distances of the RGD peptides to the surface due to the *trans–cis* isomerization of the azobenzene derivative. The azobenzene molecule has also been incorporated into an enzyme inhibitor to provide a means of modulating the binding of α-chymotrypsin to a surface (see Fig. 10) (126). Photoregulation was achieved by using a phenylalanine-based trifluoromethyl ketone inhibitor containing a photoisomerizable azobenzene group. The surface-bound azobenzene inhibitor in the *trans* state exhibited a reduced binding of the α-chymotrypsin enzyme. However, irradiation of the functionalized surface with UV light (360 nm) induced isomerization from the *trans* to the *cis*-azobenzene configuration, which was accompanied by a significant increase in enzyme binding to the surface.

1.3. Outlook

Whereas important progress on static biological surfaces has been already made, much research is now focusing on the development of stimuli-responsive biointerfaces. A number of switchable biological surfaces based on SAMs, polymer films and, only recently, on polymer brushes, have been reported. The smart behavior modulates their interactions with cells or biomolecules like proteins, DNA, and change the response of cells and tissues that come into contact with these surfaces.

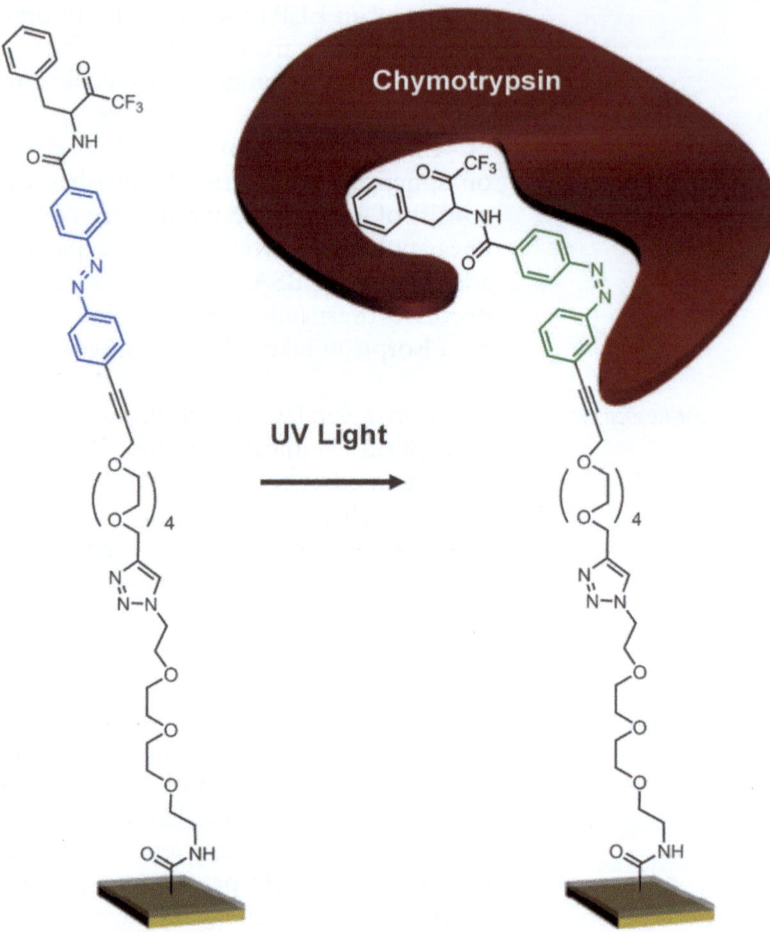

Fig. 10. Photoswitchable surface wherein the photo-induced structural changes in the azobenzene moiety give rise to changes in the binding of chymotrypsin to the surface. Reprinted with permission from ref. 126.

Different methodologies were applied to achieve smart surfaces for variety of applications, including preventing biofouling, chromatography, and bioanalytical devices. Both synthetic and natural polymers were used for that purpose. An important goal in developing switchable surfaces is to increase their biological relevance. Thus, a single stimulus-responsive surface has limited practical application and multistimuli-responsive surfaces will have to be developed in order to efficiently work in challenging and complex biological environments. These surfaces should more closely mimic the properties of the natural ECM, which is responsive to different cell–matrix and cell–cell interactions that are also spatially varied. Micro/nanofabrications of such biointerfaces and investigation of novel alternative stimuli (biological ones) also seem to be very important. These developments will be crucial for the future of tissue engineering. Still a lot of collaborative efforts of chemists, physicists, medical researchers, biologists, and others is required to improve the responsive properties of the smart surfaces and mimic complex interactions at biointerfaces so that more biological and medical applications of such devices could become successful.

1.4. Synthesis of Temperature Responsive Polymer Brushes on Gold Surface

Controlled surface-initiated photopolymerization of NIPAM using the so-called iniferter-bearing SAMs on gold is described in this protocol (99). Iniferter groups play a role of initiator, chain transfer agent, and terminator of the growing polymeric chain (127). This type of controlled radical polymerization allows the growth of very thick films during short polymerization time (*it proceeds very fast because is carried out in water and for the particular monomer which is highly reactive the same grafting process for MMA in bulk or solution does not result in such thick films*) at room temperature and enables precise control of the film thickness. It is especially appropriate for grafting polymer brushes from gold surfaces since the adsorbed monolayers of thiols or disulfides on gold cannot sustain high temperatures usually required for radical polymerizations. What is more, the polymerization can easily be stopped and restarted simply by switching the light on and off. PNIPAM brushes obtained in this procedure are thermosensitive with clear and abrupt changes in the topography and adhesion properties of the surface around 32°C – physiologically relevant temperature.

2. Materials

2.1. Gold Substrate

1. The glass plates (1 cm^2 area, Ssens BN, Hengelo, The Netherlands) covered with 200 nm layer of sputtered gold are cleaned using "piranha" solution made of perhydrol (30% H_2O_2) and H_2SO_4 (99%) mixed in 3:7 ratio (see Note 1).

2.2. Disulfide-Based Adsorbates with Photoiniferter Functionality and Formation of SAMs

1. 11-Mercaptoundecan-1-ol.
2. 4-(Chloromethyl) phenyl isocyanate.
3. Dibutyltin dilaurate.
4. Diethylammonium salt of diethyldithiocarbamic acid.
5. Octadecylthiol.
6. Equimolar aqueous solution of iodine and potassium iodide.
7. Organic solvents: chloroform, acetone, ethanol, methanol, toluene of p.a. quality.
8. Deionized water (resistivity of 18.2 MΩ).

2.3. Monomer

N-Isopropylacrylamide, (NIPAM) is purified by reprecipitation from hexane and subsequently dried under vacuum.

2.4. Photoreactor

Photoreactor equipped with six UV-B lamps (15 W G15T8E, Ushio Japan) is used for the photopolymerization. The chamber is constantly purged with nitrogen during irradiation in order to avoid diffusion of oxygen into the reaction system.

3. Methods

3.1. Synthesis of Disulfide-Based Adsorbates with Photoiniferter Functionality

Disulfide-based photoiniferter (DTCA) (see Fig. 11) is synthesized in a three-step procedure.

1. 11-Mercaptoundecan-1-ol is oxidized to the corresponding disulfide, 11,11′-dithiodiundecan-1-ol, using an equimolar aqueous solution of iodine and potassium iodide at room temperature.
2. The obtained disulfide is then reacted with 4-(chloromethyl) phenyl isocyanate in chloroform with dibutyltin dilaurate as a catalyst, giving dithiodiundecane-11,1-diylbis{[4-(chloromethyl)phenyl] carbamate} as a product.
3. The latter compound is reacted with diethylammonium salt of diethyldithiocarbamic acid in acetone at 40°C for 12 h. The solvent is later evaporated and the resulting oil is redissolved in toluene and washed several times with water. The final product dithiodiundecane-11,1-diylbis[4({([diethylamino)carbonothioyl]thioethyl)phenyl]carbamate} (DTCA) is obtained with 98% purity so no further purification is necessary.

3.2. Preparation of SAM on Gold

Freshly cleaned gold substrate is placed in 1 mM solution of octadecylthiol (ODT) (40 mol%) mixed with DTCA (60 mol%), with respect to the iniferter groups in ethanol at room temperature overnight. After withdrawal, the sample is extensively rinsed with chloroform, ethanol, and finally dried under the stream of nitrogen.

3.3. Surface-Initiated Photopolymerization

1. The sample bearing DTCA-SAMs is placed in a quartz flask equipped with a 280 nm optical cut-off filter. Alternatively, any glassware that transmits UV light at least from ca. 280 nm may be used (see Note 2).

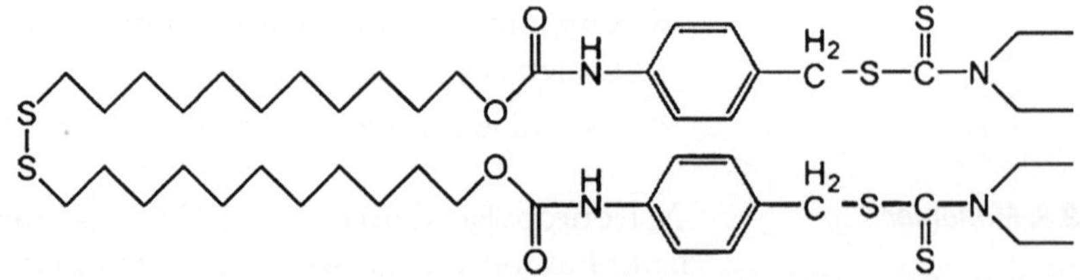

Fig. 11. The adsorbate molecule (DTCA) containing disulfide and photoiniferter groups.

2. The appropriate glassware with the sample is filled with 5% aqueous solution of purified NIPAM monomer and sealed with a septum (see Note 3). The monomer solution is purged with argon for 30 min before UV-irradiation in order to remove oxygen.
3. The deoxygenated and sealed flask is placed in the photoreactor at a distance of 20 cm from the lamps and irradiated for the necessary time (see Note 4). At the given conditions, the PNIPAM brushes grow roughly 10 nm high after 5 min of irradiation.
4. After the photopolymerization, the substrate is quickly transferred to water, subsequently extensively rinsed with water and methanol and finally dried under stream of nitrogen. Such prepared PNIPAM brushes on gold surface are stored at room temperature in nitrogen atmosphere prior the usage.

4. Notes

1. Mixtures of perhydrol and sulfuric acid in other ratios also called "piranha" solution have been reported in literature. Thus, the reported here 3:7 ration should not be treated very strictly. The role of "piranha" is to oxidize all organic pollutants adsorbed on gold surface.
2. The UV light shorter than 280 nm cannot be used since it may initiate polymerization of the NIPAM monomer in solution and more importantly it may trigger oxidation of sulfur-Au linkages and thus desorption of the initiating SAMs. If the source of monochromatic light (in the range 280–320 nm) is used to perform photopolymerization, there is no need of cut-off filter and simply quartz glassware should be used.
3. The monomer solution should cover the sample by at least several millimeters but should not reach the septa in order to avoid any pollution that might be extracted to the reaction medium.
4. Other lamps of similar spectral characteristics may be used. The distance from the lamps may be varied in order to change the polymerization rate.

Acknowledgments

The author acknowledges the financial support from the project operated within the Foundation for Polish Science Team Programme cofinanced by the EU European Regional Development Fund, PolyMed, TEAM/2008-2/6. Dr. Edmondo M. Benetti (ETH Zürich) is kindly acknowledged for his comments to the protocol part.

References

1. Ikada, Y. (1994) Surface modification of polymers for medical applications. *Biomaterials* **15**, 725–736.
2. Chu, P. K., Chen, J. Y., Wang, L. P., Huang, N. (2002) Plasma-surface modification of biomaterials *Mater. Sci. Eng. R-Reports* **36**, 143–206.
3. Park, G. E., Pattison, M. A., Park, K., Webster, T. J. (2005) Accelerated chondrocyte functions on NaOH-treated PLGA scaffolds, *Biomaterials* **26**, 3075–3082.
4. Chilkoti, A. Hubbell, J. A. (2005) Biointerface science. *MRS Bulletin* **30**, 175–176.
5. Lahann, J., Langer, R. (2005) Smart materials with dynamically controllable surfaces. *MRS Bulletin* **30**, 185–188.
6. Mrksich, M. (2005) Dynamic substrates for cell biology. *MRS Bulletin* **30**, 180–184.
7. Heskins, M. Guillet, J. E. (1968) Solution properties of poly(N-isopropylacrylamide). *J. Macromol. Sci., Chem.* **A2**, 1441–1455.
8. Winnik, F. M. (1990) Fluorescence studies of aqueous solutions of poly(N- isopropylacrylamide) below and above their LCST. *Macromolecules* **23**, 233–242.
9. Siegel, R.A., Firestone, B. A. (1988) pH-dependent equilibrium swelling properties of hydrophobic polyelectrolyte copolymer gels. *Macromolecules* **21**, 3254.
10. Kontturi, K., Mafé, S., Manzanares, J. A., Svarfvar, B. L., Viinikka, P. Modeling of the Salt and pH Effects on the Permeability of Grafted Porous Membranes. (1996) *Macromolecules* **29**, 5740.
11. Kwon, I. C., Bae, Z. H., Kim, S. W. (1991) Electrically erodible polymer gel for controlled release of drugs. *Nature* **354**, 291–293.
12. Miyata, T., Asami, N., Uragami, T. (1999) A reversibly antigen-responsive hydrogel. *Nature* **399**, 766.
13. Dai, S., Ravi, P., Tam, K.C. (2009) Thermo- and photo-responsive polymeric systems. *Soft Matter* **5**, 2513–2533.
14. Kumar, A., Srivastava, A., Galaev, I.Y., Mattiasson, B. (2007) Smart polymers: Physical forms and bioengineering applications. *Prog. Polym. Sci. (Oxford)* **32**, 1205–1237.
15. Alexander, C., Shakesheff, K.M. (2006) Responsive polymers at the biology/materials science interface.*Adv. Mat.* **18**, 3321–3328.
16. Schild, H.G. (1992) Poly (N-isopropylacrylamide): experiment, theory and application. *Prog. Polym. Sci.* **17**, 163–249.
17. Yeo, W. S., Yousaf, M. N., Mrksich, M. (2003) Dynamic Interfaces between Cells and Surfaces: Electroactive Substrates that Sequentially Release and Attach Cells. *J. Am. Chem. Soc.* **125**, 14994.
18. Mendes, P. M., Christman, K. L., Parthasarathy, P., Schopf, E., Ouyang, J., Yang, Y., Preece, J. A., Maynard, H. D., Chen, Y., Stoddart, J. F. (2007) Electrochemically controllable conjugation of proteins on surfaces. *Bioconjugate Chem.* **18**, 1919.
19. Auernheimer, J., Dahmen, C., Hersel, U., Bausch, A., Kessler, H. (2005) Photoswitched Cell Adhesion on Surfaces with RGD Peptides. *J. Am. Chem. Soc.* **127**, 16107.
20. Kost, J., Langer, R. (2001) Responsive polymeric delivery systems. *Adv. Drug Deliv. Rev.* **46**, 125–148.
21. Chilkoti, A., Dreher, M. R., Meyer, D. E., Raucher, D. (2002) Targeted drug delivery by thermallyresponsive polymers. *Adv. Drug Deliv. Rev.* **54**, 613–630.
22. Langer, R., Tirrell, D. A. (2004) Designing materials for biology and medicine. *Nature* **428**, 487–492.
23. Cole, M.A., Voelcker, N.H., Thissen, H., Griesser, H.J. (2009) Stimuli-responsive interfaces and systems for the control of protein-surface and cell-surface interactions. *Biomaterials* **30,** 1827–1850.
24. Alexander, C., Shakesheff, K. M. (2006) Responsive Polymers at the Bilology/Materials Science Interface *Adv. Mat.* **18**, 3321–3328.
25. Norrman, K., Ghanbari-Siahkali, A., Larsen, N. B. (2005) Studies of spin-coated polymer films. *Annu. Rep. Prog. Chem., Sect. C: Phys. Chem.* **101**, 174.
26. Decher, G. (1997) Fuzzy Nanoassemblies: Toward Layered Polymeric Multicomposites. *Science* **277**, 1232.
27. Mittal, K. L., Lee, K.-W. (1997) Polymer Surfaces and Interfaces: Characterization, Modification and Applications, VSP, Utrecht.
28. Bertrand, P., Jonas, A., Laschewsky, A., Legras, R. (2000) Ultrathin polymer coatings by complexation of polyelectrolytes at interfaces: Suitable materials, structure and properties. *Macromol. Rapid Commun.* **21**, 319.
29. Boudou, T., Crouzier, T., Ren, K., Blin, G., Picart, C. (2009) Multiple Functionalities of Polyelectrolyte Multilayer Films: New Biomedical Applications. *Adv. Mat.* **21**, 1–27.
30. Sukhishvili, S. A. (2005) Responsive polymer films and capsules via layer-by-layer assembly. *Curr. Opinion Coll. Interface Sci.* **10**, 37–44.

31. Tang, Z., Wang, Y., Podsiadlo, P., Kotov, N.A. (2006) Biomedical applications of layer-by-layer assembly: From biomimetics to tissue engineering. *Adv. Mat.* **18**, 3203–3224.

32. Picart, C. (2008) Polyelectrolyte multilayer film: From physico-chemical properties to the control of cellular processes. *Curr. Medicinal Chem.* **15,** 685–697.

33. De Geest, B. G., De Koker, S., Sukhorukov, G. B., Kreft, O., Parak, W. J., Skirtach, A. G., Demeester, J., De Smedt, S. C., Henninka, W. E. (2009) Polyelectrolyte microcapsules for biomedical applications. *Soft Matter* **5**, 282.

34. Bulwan, M, Zapotoczny, S., Nowakowska, M. (2009) Robust one-component chitosan-based ultrathin films fabricated using layer-by-layer technique. *Soft Matter* **5**, 4726 – 4732.

35. von Recum, H.A., Kim, S.W., Kikuchi, A., Okuhara, M., Sakurai, Y., Okano, T. (1998) Novel thermally reversible hydrogel as detachable cell culture substrate. *J. Biomed. Mater. Res.* **40**, 631–639.

36. Jeonga, B., Kim, S.W., Bae, Y. H. (2002) Thermosensitive sol–gel reversible hydrogels. *Adv. Drug Deliv. Rev.* **54**, 37–51.

37. Pan, Y.V., Wesley, R.A., Luginbuhl, R., Denton, D.D., Ratner, B.D. (2001) Plasma polymerized N-isopropylacrylamide: Synthesis and characterization of a smart thermally responsive coating. *Biomacromolecules* **2**, 32–36.

38. da Silva, R. M. P., Mano, J. F., Reis, R. L. (2007) Smart thermoresponsive coatings and surfaces for tissue engineering: switching cell-material boundaries. *Trends Biotechnology* **25**, 577–583.

39. Skotheim, T. A., Reynolds, J. R. (2007) *Handbook of Conductive Polymers*, CRC Press, Boca Raton.

40. Ahn, Y. H., Mironov, V. A., Gutowska, A. (2001) Reversible gelling culture media for in-vitro cell culture in three-dimensional matrices. US patent, US6103528.

41. Yamada, N., Okano, T., Sakai, H., Karikusa, F., Sawasaki Y., Sakurai, Y. (1990) Thermo-responsive polymeric surfaces; control of attachment and detachment of cultured cells. *Makromol. Chem., Rapid Commun.,* **11**, 571.

42. Canavan, H. E., Cheng, X. H., Graham, D. J., Ratner B. D., Castner, D. G. (2005) *J. Biomed. Mater. Res. Part A* **75A**, 1–13.

43. Kushida, A., Yamato, M., Konno, C., Kikuchi, A., Sakurai, Y., Okano, T. (1999) Decrease in culture temperature releases monolayer endothelial cell sheets together with deposited fibronectin matrix from temperature-responsive culture surfaces. *J. Biomed. Mater. Res. Part A* **45**, 355–362.

44. Ide, T., Nishida, K., Yamato, M., Sumide, T., Utsumi, M., Nozaki, T., Kikuchi, A., Okano, T., Tano, Y. (2006) Structural characterization of bioengineered human corneal endothelial cell sheets fabricated on temperature-responsive culture dishes. *Biomaterials* **27**, 607–614.

45. Yamato, M., Okano, T. (2004) Cell sheet engineering. *Mater Today* 7, 42–47.

46. Yang, J., Yamato, M., Konno, C., Nishimoto, A., Sekine, H., Fukai, F., Okano, T. (2005) Cell sheetengineering: Recreating tissues without biodegradable scaffolds. *Biomaterials* **26**, 6415–6422.

47. Akiyama, Y., Kikuchi, A., Yamato, M., Okano, T. (2004) Ultrathin poly(N-isopropylacrylamide) grafted layer on polystyrene surfaces for cell adhesion/detachment control. *Langmuir* **20**, 5506–5511.

48. Yang, J., Yamato, M., Shimizu, T., Sekine, H., Ohashi, K., Kanzaki, M., Ohki, T., Nishida, K., Okano, T. (2007) Reconstruction of functional tissues with cell sheet engineering. *Biomaterials* **28**, 5033–5043.

49. Kikuchi, A., Okano, T. (2005) Nanostructured designs of biomedical materials: applications of cell sheet engineering to functional regenerative tissues and organs. *J. Control Release* **101**,69–84.

50. Falconnet, D., Csucs, G., Michelle Grandin, H., Textor, M. (2006) Surface engineering approaches to micropattern surfaces for cell-based assays. *Biomaterials* **27**, 3044–3063.

51. Engel, E., Michiardi, A., Navarro, M., Lacroix, D., Planell, J.A. (2008) Nanotechnology in regenerative medicine: the materials side. *Trends Biotechnology* **26**, 39–47.

52. Zinger, O., Zhao, G., Schwartz, Z., Simpson, J., Wieland, M., Landolt, D., Boyan, B., (2005) Differential regulation of osteoblasts by substrate microstructural features. *Biomaterials* **26**, 1837–1847.

53. Teixeira, A.I. McKie, G. A., Foley, J. D., Bertics, P. J., Nealey, P. F., Murphy, C. J. (2006) The effect of environmental factors on the response of human corneal epithelial cells to nanoscale substrate topography. *Biomaterials* **27**, 3945–3954.

54. Tsuda Y, Kikuchi A, Yamato M, Nakao A, Sakurai Y, Umezu M, Okano, T. (2005) The use of patterned dual thermoresponsive surfaces for the collective recovery as co-cultured cell sheets. *Biomaterials* **26**, 1885–1893.

55. Kujawa, P., Winnik, F. M. (2001) Volumetric studies of aqueous polymer solutions using pressure perturbation calorimetry: a new look at the temperature-induced phase transition of poly(N-isopropylacrylamide) in water and D_2O. *Macromolecules* **43**, 4130–4135.
56. Tatsuma, T., Saito, K., Oyama, N. (1994) Enzyme-exchangeable enzyme electrodes employing a thermoshrinking redox gel. *J. Chem. Soc., Chem. Commun.* 1853.
57. Yang, H., Choi, C. A., Chung, K. H., Jun, C.-H., Kim, Y. T. (2004) An independent, temperature controllablemicroelectrode array. *Anal. Chem.* **76**, 1537–1543.
58. Cheng, X., Wang, Y., Hanein, Y., Bohringer, K., Ratner, B. D. (2004) Novel cell patterning using microheater-controlled thermoresponsive plasma films. *J. Biomed. Mater. Res. Part A* **70**, 159–168.
59. Ruoslahti, E., Pierschbacher, M. D. (1987) New perspectives in cell adhesion: RGD and integrins. *Science* **238**, 491.
60. Stile, R. A., Healy, K. E. (2001) Thermo-responsive peptide-modified hydrogels for tissue regeneration.*Biomacromolecules* **2**, 185–194.
61. Ebara, M., Yamato, M., Aoyagi, T., Kikuchi, A., Sakai, K., Okano, T. (2004) Immobilization of cell adhesivepeptides to temperature-responsive surfaces facilitates both serum-free cell adhesion and noninvasive cell harvest. *Tissue Eng.* **10**, 1125–1135.
62. Urry, D. W., Luan, C. H., Parker, T. M., Gowda, D. C., Prasad, K. U., Reid, M. C., Safavy, A. (1991) Temperature of polypeptide inverse temperature transition depends on mean residue hydrophobicity. *J. Am. Chem. Soc.* **113**, 4346–4348.
63. Rodríguez-Cabello, J. C., Prieto, S., Reguera, J., Arias, F. J., Ribeiro, A. (2007) Biofunctional design of elastin-like polymers for advanced applications in nanobiotechnology. *J. Biomater. Sci. Polym. Ed.* **18,** 269–286.
64. Urry, D. W. (1997) Physical chemistry of biological free energy transduction as demonstrated by elastic protein-based polymers. *J. Phys. Chem. B* **101**, 11007–11028.
65. Hyun, J., Lee, W.-K., Nath, N., Chilkoti, A., Zauscher, S. (2004) Capture and release of proteins on the nanoscale by stimuli-responsive elastin-like polypeptide "switches". *J. Am. Chem. Soc.* **126**, 7330–7335.
66. Meyer, D. E., Chilkoti, A. (1999) Purification of recombinant proteins by fusion with thermally responsivePolypeptides. *Nat. Biotechnol.* **17**, 1112–1115.
67. Nath, N., Chilkoti, A. (2003) Fabrication of a reversible protein array directly from cell lysate using a stimuli-responsive polypeptide. *Anal. Chem.* **75**, 709–715.
68. Kim, J. Y., Mulchandani, A. Chen, W. (2005) Temperature-triggered purification of antibodies. *Biotechnol. Bioeng.* **90**, 373–379.
69. Serpe, M. J., Jones, C. D., Lyon, L. A. (2003) Layer-by-layer deposition of thermoresponsive microgel thin films. *Langmuir* **19**, 8759.
70. Nolan, Ch. M., Serpe, M. J., Lyon, L. A. (2004) Thermally modulated insulin release from microgel thin films. *Biomacromolecules* **5**, 1940.
71. Serpe, M. J., Yarmey, K. A., Nolan, Ch. M., Lyon, L. A.(2005) Doxorubicin uptake and release from microgel thin films. *Biomacromolecules* **6**, 408.
72. Gao, J., Haidar, G., Lu, X., Hu, Z. (2001) Self-association of hydroxypropylcellulose in water. *Macromolecules* **34,** 2242–2247.
73. Li, L., Thangamathesvaran, P. M., Yue, C. Y., Tam, K. C., Hu, X., Lam, Y. C. (2001) Gel Network Structure of Methylcellulose in Water. *Langmuir* **17**, 8062–8068.
74. Heitfeld, K.A., Guo, T., Yang, G., Schaefer, D.W. (2008) Temperature responsive hydroxypropyl cellulose for encapsulation. *Mater. Sci. Eng. C* **28**, 374–379.
75. Burke, S.E., Barrett, Ch. J. (2004) pH-dependent loading and release behavior of small hydrophilic molecules in weak polyelectrolyte multilayer films. *Macromolecules* **37,** 5375.
76. Hiller, J., Rubner, M. F. (2003) Reversible molecular memory and pH-switchable swelling transitions in polyelectrolyte multilayers. *Macromolecules* **36**, 4078.
77. Thompson, M. T., Berg, M. C., Tobias, I. S., Rubner M. F., Van Vliet, K. J. (2005) Tuning compliance of nanoscale polyelectrolyte multilayers to modulate cell adhesion. *Biomaterials* **26**, 6836–6845.
78. Gerard, M., Chaubey, A., Malhotra, B. D. (2002) Application of polyaniline as enzyme based biosensor *Biosens. Bioelectron.* **17**, 345.
79. George, P. M., La Van, D. A., Burdick, J. A., Chen, C. Y., Liang E., Langer, R. (2006) Electrically Controlled Drug Delivery from Biotin-Doped Conductive Polypyrrole. *Adv. Mater.* **18**, 577.
80. Pyo, M., Maeder, G., Kennedy, R. T., Reynolds, J. R. (1994) Controlled release of biological molecules from conducting polymer modified electrodes. The potential dependent release of adenosine 5′-triphosphate from poly(pyrrole adenosine 5′-triphosphate) films. *J. Electroanal. Chem.* **368**, 329.
81. Miller, L. L., Zhou, Q. X. (1987) Poly(N-methylpyrrolylium) poly(styrenesulfonate) - a

conductive, electrically switchable cation exchanger that cathodically binds and anodically releases dopamine. *Macromolecules* **20**, 1594–1597.

82. Wong, J. Y., Langer, R., Ingber, D. E. (1994) Electrically conducting polymers can noninvasively control theshape and growth of mammalian cells. *Proc. Natl. Acad. Sci. USA.* **91**, 3201.

83. Schmidt, C.E., Shastri, V. R., Vacanti, J. P. , Langer, R. (1997) Stimulation of neurite outgrowth using an electrically conducting polymer. *Proc. Natl. Acad. Sci. USA.* **94**, 8948–8953.

84. Higuchi, A., Hamamura, A., Shindo, Y., Kitamura, H., Yoon, B. O., Mori, T., Uyama T., Umezawa, A. (2004) Photon-Modulated Changes of Cell Attachments on Poly(spiropyran-co-methyl methacrylate) Membranes. *Biomacromolecules* **5**, 1770.

85. Edahiro, J., Sumaru, K., Tada, Y., Ohi, K., Takagi, T., Kameda, M., Shinbo, T., Kanamori T., Yoshimi, Y. (2005) In Situ Control of Cell Adhesion Using Photoresponsive Culture Surface. *Biomacromolecules* **6**, 970–974.

86. Hayashi, G., Hagihara, M., Dohno, C., Nakatani, K. (2007) Photoregulation of a Peptide – RNA Interaction on a Gold Surface. *J. Am. Chem. Soc.* **129**, 8678–8679.

87. Ulijn R. V. (2006) Enzyme-responsive materials: a new class of smart biomaterials. *J. Mater. Chem.* **16**, 2217–2225.

88. Todd, S. J., Farrar, D., Gough, J. E., Ulijn R. V. (2007) Enzyme-Triggered Cell Attachment to Hydrogel Surfaces. *Soft Matter* **3**, 547–550.

89. Okajima, S., Sakai, Y., Yamaguchi, T. (2005) Development of a regenerable cell culture system that senses and releases dead cells. *Langmuir* **21**, 4043–4049.

90. Milner, S. T. (1991) Polymer Brushes. *Science* **251**, 905.

91. Barbey, R., Lavanant, L., Paripovic, D., Schüwer, N., Sugnaux, C., Tugulu, S., Klok H.-A. (2009) Polymer brushes via surface-initiated controlled radical polymerization: synthesis, characterization, properties, and applications. *Chem. Rev.* **109**, 5437–5527.

92. Navarro, M., Benetti, E. M., Zapotoczny, S., Planell J. A., Vancso, G. J. (2008) Buried, Covalently Attached RGD Peptide Motifs in Poly(methacrylic acid) Brush Layers: The Effect of Brush Structure on Cell Adhesion. *Langmuir* **24**, 10996–11002.

93. Tanahashi, T., Kawaguchi, M., Honda, T., Takahashi, A. (1994) Adsorption of poly(N-isopropylacrylamide) on silica surfaces. *Macromolecules* **27**, 606–607.

94. Cho, E. C., Kim, Y. D., Cho, K. (2004) Thermally responsive poly(N-isopropylacrylamide) monolayer on gold: synthesis, surface characterization, and protein interaction/adsorption studies. *Polymer* **45**, 3195–3204.

95. Zhao, B., Brittain, W. J. (2000) Polymer brushes: surface-immobilized macromolecules. *Prog. Polym. Sci.* **25**, 677–710.

96. Edmondson, S., Osborne, V.L., Huck, W. T. S. (2004) Polymer brushes via surface-initiated polymerizationsChem. Soc. Rev. **33**, 14–22.

97. Jordan, R. (2006) Surface-Initiated Polymerization. Springer-Verlag, New York, 1st edn.

98. Jones, D. M., Smith, J. R., Huck, W. T. S., Alexander, C. (2002) Variable Adhesion of Micropatterned Thermoresponsive Polymer Brushes: AFM Investigations of Poly(N-isopropylacrylamide) Brushes Prepared by Surface-Initiated Polymerizations. *Adv. Mater.* **14**, 1130–1134.

99. Benetti, E. M., Zapotoczny, S., Vancso, G. J. (2007) Tunable thermoresponsive polymeric platforms on gold by photoiniferter-based surface grafting. *Adv. Mater.* **19**, 268–271.

100. Zapotoczny, S., Benetti, E. M., Vancso, G. J. (2007) Preparation and characterization of macromolecular "hedge" brushes grafted from Au nanowires. *J. Mater. Chem.* **17**, 3293–3296.

101. Benetti, E. M., Chung, H. J., Vancso, G. J. (2009) pH Responsive Polymeric Brush Nanostructures: Preparation and Characterization by Scanning Probe Oxidation and Surface Initiated Polymerization. *Macromol. Rapid Commun.* **30**, 411–417.

102. Advincula, R. C., Brittain, W. J., Caster, K.C., Rühe, J., (2004) Polymer Brushes: Synthesis, Characterization, Applications. Wiley-VCH Verlag GmbH &Co. KGaA: Weinheim, Germany.

103. Luzinov, I., Minko, S., Tsukruk, V.V. (2004) Adaptive and responsive surfaces through controlled reorganization of interfacial polymer layers. *Progr. Polym. Sci. (Oxford)* **29**, 635–698.

104. Raviv, U., Giasson, S., Kampf, N., Gohy, J.-F., Jérôme, R., Klein, J. (2003) Lubrication by charged polymers. *Nature* **425**, 163–165.

105. Chen, M., Briscoe, W. H., Armes, S. P., Klein, J. (2009) Lubrication at Physiological Pressures by Polyzwitterionic Brushes. *Science* **323**, 1698–1701.

106. Kaholek, M., Lee, W., Ahn, S., Ma, H., Caster, K. C., LaMattina, B., Zauscher, S. (2004) Stimulus-responsive poly(N-isopropylacrylamide) brushes and nanopatterns prepared by

surface-initiated polymerization. *Chem. Mater.* **16**, 3688–3696.

107. Tugulu, S., Silacci, P., Stergiopulos, N., Klok, H. A. (2007) RGD–functionalized polymer brushes as substrates for the integrin specific adhesion of human umbilical vein endothelial cells. *Biomaterials* **28**, 2536–2546.
108. Arifuzzaman, S., Özçam, A. E., Efimenko, K., Fischer, D. A., Genzer, J. (2009) Formation of surface-grafted polymeric amphiphilic coatings comprising ethylene glycol and fluorinated groups and their response to protein adsorption. *Biointerphases* **4**, FA33–FA44.
109. Ryan, A. J., Crook, C. J., Howse, J. R., Topham, P., Geoghegan, M., Martin, S. J., Parnell, A. J., Ruiz-Perez, L. and Jones, R. A. L. (2005) Mechanical actuation by responsive polyelectrolyte brushes and triblock gels. *J. Macromol. Sci. Phys.* **B44**, 1103–1121.
110. Ryan, A. J., Crook, C. J., Howse, J. R., Topham, P., Jones, R. A. L., Geoghegan, M., Parnell, A. J., Ruiz-Perez, L., Martin, S. J., Cadby, A., Menelle, A., Webster, J. R. P., Gleeson, A. J. and Bras, W. (2005) Responsive brushes and gels as components of soft nanotechnology. *Faraday Discuss.* **128**, 55–74.
111. Ionov, L., Houbenov, N., Sidorenko, A., Stamm, M. and Minko, S. (2006) Smart microfluidic channels. *Adv. Funct. Mater.* **16**, 1153–1160.
112. Huber, D.; Manginell, R.; Samara, M.; Kim, B.-I.; Bunker, B. C. (2003) Programmed adsorption and release of proteins in a microfluidic device. *Science* **301**, 352–354.
113. Ebara, M., Yamato, M., Aoyagi, T., Kikuchi, A., Sakai, K., Okano, T. (2004) Temperature-Responsive Cell Culture Surfaces Enable "On - Off" Affinity Control between Cell Integrins and RGDS Ligands *Biomacromolecules* **5**, 505.
114. Cunliffe, D., Alarcon, C. D., Peters, V., Smith, J. R., Alexander, C. (2003) Thermoresponsive Surface-Grafted Poly(N-isopropylacrylamide) Copolymers: Effect of Phase Transitions on Protein and Bacterial Attachment. *Langmuir* **19**, 2888–2899.
115. Huber, D. L., Manginell, R. P., Samara, M. A., Kim B. I., Bunker, B. C. (2003) Programmed Adsorption and Release of Proteins in a Microfluidic Device. *Science* **301**, 352.
116. Nagel, B., Warsinke, A., Katterle, M. (2007) Enzyme Activity Control by Responsive Redox Polymers. *Langmuir* **23**, 6807–6811.
117. de las Heras Alarcón, C., Farhan, T., Osborne, V. L., Huck, W. T. S., Alexander, C. (2005) Bioadhesion at micro-patterned stimuli-responsive polymer brushes. *J. Mat. Chem.* **15**, 2089.
118. Frey, W., Meyer, D. E., Chilkoti, A. (2003) Thermodynamically Reversible Addressing of a Stimuli Responsive Fusion Protein onto a Patterned Surface Template. *Langmuir* **19**, 1641–1653.
119. Koga, T., Nagaoka, A., Higashi, N. (2006) Fabrication of a Switchable Nano-surface Composed of Acidic and Basic Block Polypeptides. *Colloids Surf. A* **284**, 521–527.
120. Israels, R.; Gersappe, D.; Fasolka, M.; Roberts, V. A.; Balazs, A. C. (1994) pH-Controlled Gating in Polymer Brushes. *Macromolecules* **27**, 6679–6682.
121. Mei, Y.; Wu, T.; Xu, C.; Langenbach, K. J.; Elliott, J. T.; Vogt, B. D.; Beers, K. L.; Amis, E. J.; Washburn, N. R. (2005) Tuning Cell Adhesion on Gradient Poly(2-hydroxyethyl methacrylate)-Grafted Surfaces. *Langmuir* **21**, 12309–12314.
122. Harris, B. P., Kutty, J. K., Fritz, E. W., Webb, C. K., Burg, K. J. L., Metters, A. T. (2006) Photopatterned Polymer Brushes Promoting Cell Adhesion Gradients. *Langmuir* **22**, 4467–4471.
123. Zhu, Y. B., Gao, C. Y., Guan, J. J., Shen, J. C. (2003) Engineering porous polyurethane scaffolds by photografting polymerization of methacrylic acid for improved endothelial cell compatibility. *J. Biomed. Mater. Res. A* **67A**, 1367–1373.
124. Anikin, K., Röcker, C., Wittemann, A., Wiedenmann, J., Ballauff, M., Nienhaus, G. U. (2005) Polyelectrolyte-Mediated Protein Adsorption: Fluorescent Protein Binding to Individual Polyelectrolyte Nanospheres. *J. Phys. Chem. B* **109**, 5418.
125. Wittemann, A., Haupt, B., Ballauff, M. (2003) Adsorption of proteins on spherical polyelectrolyte brushes in aqueous solution. *Phys. Chem. Chem. Phys.* **5**, 1671–1677.
126. Pearson, D., Downard, A. J., Muscroft-Taylor, A., Abell, A. D. (2007) Reversible Photoregulation of Binding of α-Chymotrypsin to a Gold Surface. *J. Am. Chem. Soc.* **129**, 14862–14863.
127. Otsu, T., Matsumoto, A. (1998) Controlled Synthesis of Polymers Using the Iniferter Technique: Developments in Living Radical Polymerization. *Adv. Polym. Sci.* **136**, 75.

Chapter 5

Micro/Nanopatterning of Proteins Using a Nanoimprint-Based Contact Printing Technique

Elena Martínez, Mateu Pla-Roca, and Josep Samitier

Abstract

Micro and nanoscale protein patterning based on microcontact printing technique on large substrates have often resolution problems due to roof collapse of the poly(dimethylsiloxane) (PDMS) stamps used. Here, we describe a technique that overcomes these issues by using instead a stamp made of poly(methyl methacrylate) (PMMA), a much more rigid polymer that do not collapse even using stamps with very high aspect ratios (up to 300:1). Conformal contact between the stamp and the substrate is achieved because of the homogeneous pressure applied via the nanoimprint lithography instrument, and it has allowed us to print lines of protein 150 nm wide, at a 400 nm period. This technique, therefore, provides an excellent method for the direct printing of high-density submicrometer scale patterns, or, alternatively, micro/nanopatterns spaced at large distances.

Key words: Microcontact printing, Protein, Poly(methyl methacrylate), Nanoimprint lithography

1. Introduction

Increasingly, there is an interest in the biomedical field for the controlled patterning of proteins on surfaces as a key factor for cell-surface interaction, cell differentiation, or tissue culturing experiments (1, 2). Microcontact printing (μCP) has been developed for rapid patterning of molecules over large areas onto a variety of surfaces (e.g., gold surfaces with thiols, glass with silanes) (3). This technique, that uses a PDMS stamp, is parallel in nature, involves process conditions which are nondetrimental to the (bio)-molecules, requires no special equipment, and can be undertaken by a nonskilled scientist at the bench top. Unfortunately, conventional μCP suffers from problems that limit pattern resolution (4), such as diffusion of the molecules to be printed or deformation of the stamp under the applied pressure (5). These problems

Melba Navarro and Josep A. Planell (eds.), *Nanotechnology in Regenerative Medicine: Methods and Protocols*, Methods in Molecular Biology, vol. 811, DOI 10.1007/978-1-61779-388-2_5,

have been limiting the available aspect ratios for printing to a maximum of 10:1 (6), thus making the printing of nanopatterns difficult. We have recently shown that the problem of structural collapse can be alleviated if the μCP is carried out in a liquid medium that supports the stamp during printing (7). Renault and coworkers have shown that it is also possible to circumvent this problem, thus making contact printing of proteins possible at nanodimensions, by using domed structural features in their stamps (8). Similarly, sub-100 nm patterns have been obtained using V-shaped stamps of h-PDMS, a high Young's modulus PDMS silicone (9). The large Young's modulus value provides increased polymer stiffness and minimizes roof collapse, but in this case, the separation of pattern features is about 2.5 μm. Here, we describe the use of a new contact printing technique for patterning proteins on surfaces at the micro/nanoscale by using stamps with aspect ratios much higher than those used in conventional μCP. To achieve this, we use poly-(methyl methacrylate) (PMMA) stamps and pressure applied by nanoimprint lithography apparatus.

2. Materials

2.1. Oxidized Silicon-Based Master Fabrication

1. Commercial microstructured oxidized silicon wafer piece (800 nm of thermal silicon oxide grown on the surface), 1 cm^2 of area (Centro Nacional de Microelectrónica, CNM-CSIC, Barcelona, Spain). Masters were ordered to have both positive (where the features are higher than the master surface) and negative (where the features are below the master surface) superficial microstructures. Microstructures had square shape, 5 × 5 μm in dimension, 5 μm pitch separation and were 500 nm depth.
2. Oxidized silicon wafer piece, 1 cm^2 of area (Centro Nacional de Microelectrónica, CNM-CSIC, Barcelona, Spain).
3. Reagent quality H_2O_2 and H_2SO_4 products (Sigma-Aldrich, Spain) for Piranha preparation (3 mL H_2O_2 and 7 mL of H_2SO_4). *Caution: Piranha is an extremely strong oxidant and should be handled very carefully.*
4. Milli-Q ultrapure water (Millipore, Spain).
5. Absolute ethanol and *n*-heptane (Sigma-Aldrich, Spain).
6. Dry nitrogen (2 bar, Abelló, Spain).
7. (Heptadecafluoro-1,1-2,2-tetrahydrodecyl)trimethoxysilane (United Chemical Technologies, USA).

2.2. Poly(Methyl Methacrylate) Stamp Fabrication

1. Poly(methyl methacrylate) (PMMA) polymer sheets 125 μm thick (Goodfellow, UK).

2. Round aluminum foils 2 in. in diameter, 100 μm thick (Obducat AB, Sweden).
3. Tweezers and scissors.

2.3. Silicon-Based Master and PMMA Stamp Characterization

1. Atomic Force Microscopy (AFM) cantilevers to work in tapping mode, resonant frequency of 300 kHz and spring constant of 40 N/m (Mikromask, Estonia).
2. Tweezers.

2.4. Inking of the PMMA Stamp

1. Inking protein solutions: 0.1 mg/mL Streptavidin-Texas Red® (Molecular Probes, US) and 0.1 mg/mL neutravidin (Molecular Probes, US) in PBS buffer (Aldrich, Spain).
2. Parafilm, tweezers.
3. Dry nitrogen (2 bar, Abelló, Spain).

2.5. Micro/Nanopatterning of Proteins with PMMA Stamps

1. Inked PMMA stamp.
2. Microscope slides (Deltalab, Spain).
3. Round aluminum foils 2 in. in diameter, 100 μm thick (Obducat AB, Sweden).
4. Tweezers.

2.6. Patterning Characterization by Fluorescence Microscopy

1. Staining solution: 0.25 mg/mL of biotin-4-fluorescein (Molecular Probes, USA) in PBS containing 2.5% of BSA (Sigma-Aldrich, Spain).
2. Deionized water.
3. Image L software (Image L 1.33 U, National Institutes of Health, USA).

3. Methods

3.1. Preparation of the Oxidized Silicon-Based Master

1. Microstructured masters were cleaned by immersion in fresh Piranha solution during 5 min., then rinsing with MilliQ water and absolute ethanol. Finally, they were dried by blowing nitrogen (2 bar). After this step, they were ready for silanization (explained in what follows).
2. Nanostructured masters were produced by Focused Ion Beam (FIB) Lithography technique (Strata DB235; FEI Co., Netherlands). FIB lithography uses an energetic ion beam (Ga ions for example), which is accelerated and focused on the sample, thus producing collisions with the atoms of the surface and resulting in the etching of these atoms. The beam can be controlled to scan the sample in the shape of the desired patterns with a maximum lateral resolution of about 20 nm.

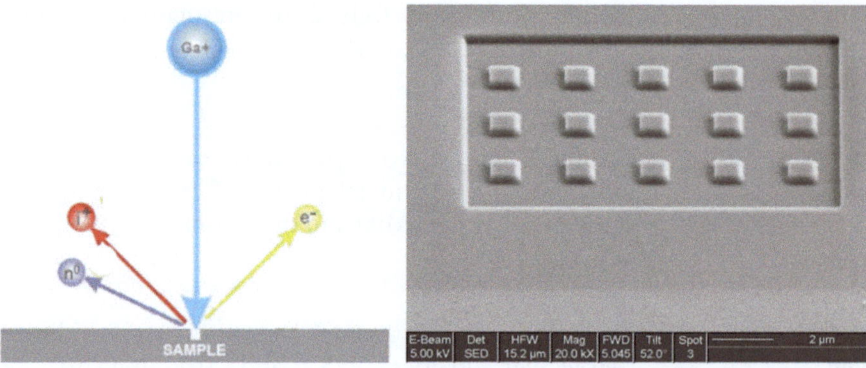

Fig. 1. (**a**) Schematics of the working principle of a focused ion beam (FIB), where a highly energetic Ga beam is focused on the sample surface to produce localized material etching. (**b**) SEM image of a milled pattern performed on a silicon wafer.

The equipment used is a double beam apparatus that combines the etching capabilities of an FIB with the image capabilities of a high-resolution scanning electron microscope (SEM) (Fig. 1). Oxidized silicon masters with line structures 200 nm wide, 100 nm depth, and 400 nm periodicity were fabricated with this instrument for subsequent use.

3. In order to avoid master-replica sticking while doing the nanoimprinting procedure, masters were silanized at this point. For this purpose, Piranha cleaned micro or nanostructured masters were put inside an exsiccator over a glass surface (glass Petri dish). Two or three drops of the (heptadecafluoro-1,1-2,2-tetrahydrodecyl)trimethoxysilane were poured on a vial and the vial was introduced on the exsiccator close to the master. The vacuum exsiccator was then closed and vacuum was produced during 1 h.

4. After removing the vacuum carefully, the master is put inside an oven at 110°C during 1 h. Then, it is sonicated with heptane for 15 min and well dried under an N_2 stream.

3.2. Fabrication of the PMMA Stamps

1. PMMA polymer sheets were used as supplied (see Note 1) in order to fabricate the PMMA stamps. For this purpose, they were micro and nanostructured using a nanoimprint lithography apparatus (Obducat AB, Sweden). The hard (oxidized silicon) master is place on the base of the operating head of the apparatus while the PMMA sheet (cut to the same dimension as the master) sits on top of the master (Fig. 2). The sandwich master-PMMA is then covered by a thick lamella of aluminum and vacuum is performed to keep all the pieces in place.

2. The nanoimprinting machine applies a cycle of pressure and heat to the master-PMMA sandwich in order to soften the

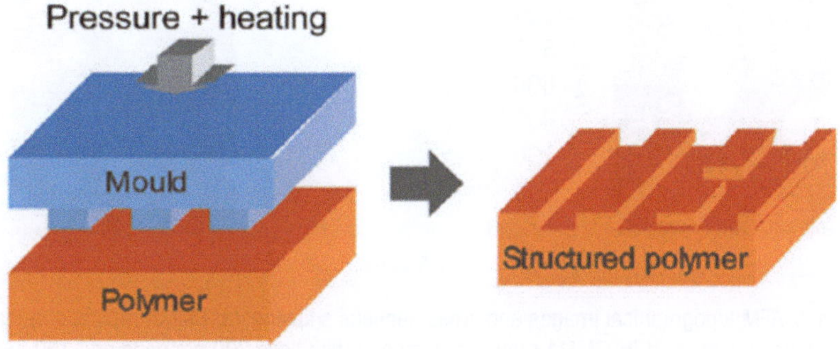

Fig. 2. A schematic diagram of the nanoembossing technique performed to obtain micro and nanostructured PMMA replicas by a nanoimprinting apparatus.

polymer by heating it above its glass transition temperature. The imprinting parameters used to get PMMA replicas are as following:

(a) Step 1. Heating the sandwich at 80°C for 5 min at a pressure of 5 bar.

(b) Step 2. Heating the sandwich at 130°C for 10 min at a pressure of 5 bar.

(c) Step 3. Increase the pressure to 30 bar while keeping the temperature at 130°C for 20 min.

(d) Step 4. Cooling down to 80°C while keeping the pressure at 30 bar for 10 min.

(e) Step 5. Cooling down to room temperature and releasing the pressure.

3. Demolding the PMMA replica by simply peeling it off the silicon-based mold.

3.3. Silicon-Based Master and PMMA Stamp Characterization

1. Place the master or the PMMA replicas on the sample holder of the Atomic Force Microscope, a Nanoman (Veeco, USA) in our case. Fix the sample with the vacuum provided on the holder of the instrument.
2. Perform AFM measurements on tapping mode and scanning fields appropriated for the micro and nanostructured samples. If the imprinting process has been successful, the topography of the replica should be the reverse of the master topography (Fig. 3).

3.4. Inking of the PMMA Stamp

1. Wash the PMMA stamps with ultrapure water by sonication (15′) and dry well with N_2 (2 bar). Use tweezers during the drying process.
2. Prepare the inking protein solution (0.1 mg/mL in PBS). Fresh protein solutions are recommended in order to avoid protein aggregates.

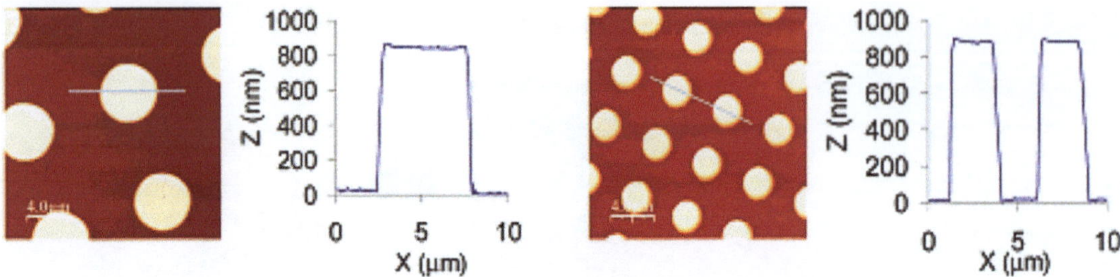

Fig. 3. AFM topographical images and cross-sections showing: (**a**) PMMA replica of a master with holes 5 μm in size and 800 nm in depth and (**b**) PMMA replica of a master with lines 200 nm wide and 100 nm deep (adapted from ref. 10 with permission from Elsevier Science).

3. Place the stamp (features facing up) on a plastic Petri dish.
4. Pour a drop (200 μL is the stamp is 1 cm × 1 cm) over the structured PMMA stamp and wait 60 min to "ink" properly the stamp. In order to avoid buffer evaporation and PBS crystallization, close the Petri, and seal the Petri with parafilm.
5. Dry the stamp surface (2 min) using N_2 (2 bar). It is important to blow the entire drop in one single blow (flow directionality not important) (see Note 2).

3.5. Micro/Nanopatterning of Proteins with PMMA Stamps

1. Clean a microscope glass slide (the substrate to be patterned) with acetone, ethanol and ultrapure MilliQ water. Blow dry well the slide using the N_2 gun (2 bar).
2. Place the glass substrate on the sample holder of the nanoimprinting machine and carefully place the protein-inked PMMA stamp on top of it with the inked structures to be printed over the substrate surface. At this point, no protein will be transferred from the PMMA stamp to substrate due to both surfaces are not in conformal contact.
3. Run the nanoimprinter with the following parameters:
   (a) Step 1. Apply 7.5 bar of pressure at room temperature for 600 s (see Note 3).
   (b) Step 2. Release the pressure and demold the sandwich.

3.6. Patterning Characterization by Fluorescence Microscopy

1. In order to fluorescently label the neutravidin patterns place onto the patterned areas a drop (200 μL) of PBS solution containing 0.25 mg/mL of biotin-4-fluorescein and 2.5% of BSA (for blocking) for 10 min.
2. Rinse 3–4 times with deionized water.
3. Store within a fridge in a light-protected box if required.
4. Fluorescence pictures of the Streptavidin-Texas Red® patterns and fluorescently stained neutravidin patterns are performed

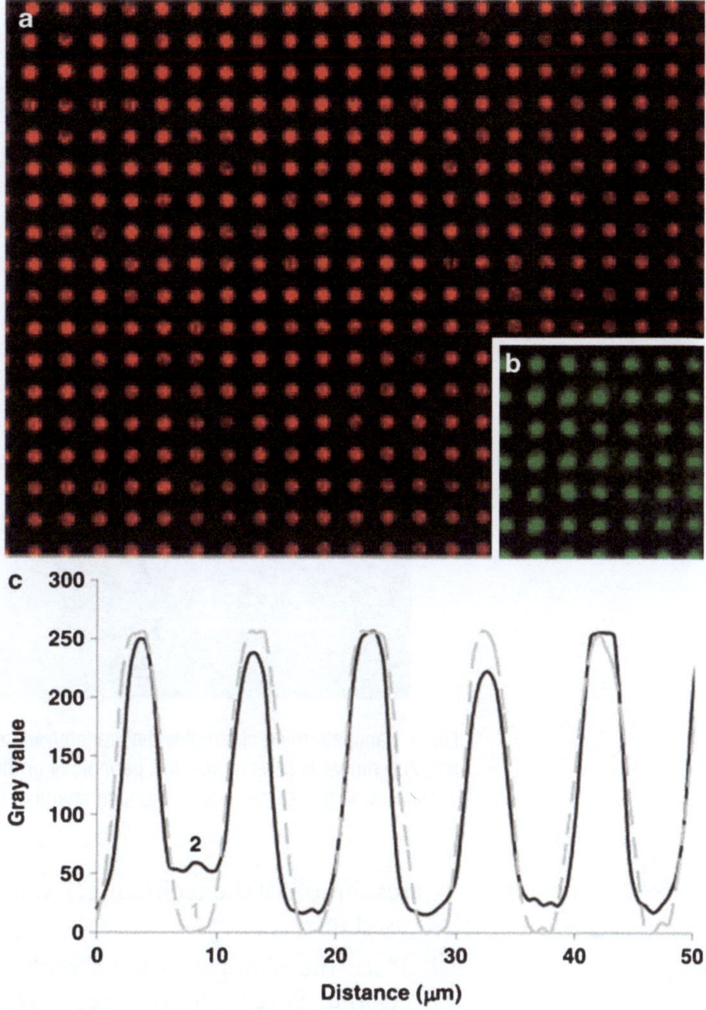

Fig. 4. (**a**) Fluorescence microscopy picture showing the successful transfer of Streptavidin-Texas Red (seen in *red*) by the nanoimprinting assisted technique using PMMA stamps. (**b**) In order to check for the protein functionality, neutravidin patterns were conjugated with biotin-4-fluorescein (seen in *green*), showing the successful protein recognition. (**c**) Line profiles of fluorescence intensity after biotin-4-fluorescein staining of Neutravidin transferred by using a PMMA stamp (*gray dotted line*) and by using a conventional PDMS stamp (*black solid line*) (adapted from ref. 10 with permission from Elsevier Science). The similar fluorescence values suggest that the amount of active protein transferred with the new technique is comparable to that obtained using the well-established microcontact printing methodology.

by a fluorescence microscope with EPI illumination and the appropriated filters for the fluorophores used. Fluorescence intensity profiles were obtained using Image L software (Image L 1.33 U, National Institutes of Health, USA) (Fig. 4).

3.7. Patterning Characterization by AFM

1. Nanostructured PMMA stamps allowed for the patterning of proteins with nanometric resolution but no direct imaging of the patterning through fluorescence was possible due to the

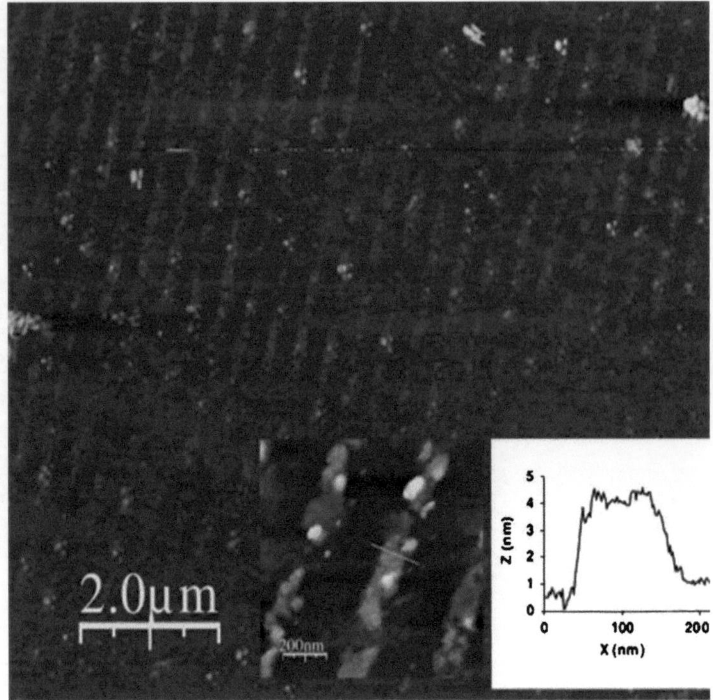

Fig. 5. Nanopatterning of streptavidin, transferred using PMMA stamps containing 100 nm tall, 200 nm wide lines of 400 nm periodicity on flat glass. The printed protein lines are 150 nm wide due to the slightly rounded shape of the structures in the PMMA stamp.

resolution of the technique, so an atomic force microscope was used instead.

2. Place the nanopatterned sample on the sample holder of the Atomic Force Microscope, a Nanoman (Veeco, US) in our case. Fix the sample with the vacuum provided on the holder of the instrument.
3. Perform AFM measurements on tapping mode and scanning fields appropriated for nanostructured samples. If the patterning process has been successful, the printed protein patterns (lines on the example presented) should be seen on the sample (Fig. 5).

4. Notes

1. As supplied, the PMMA sheets are protected both sides by transparent and blue adhesive layers. These layers have to be removed immediately before using the polymer sheet in the nanoimprinting apparatus. This way the polymer will remain clean without the need of using solvents or other cleaning methods.

2. Excess of protein solution is removed by blow drying with a nitrogen. Washing steps for the stamps were omitted for fear of washing off the protein from the stamp.
3. The pressure applied by the nanoimprinter should be high enough to keep a conformal contact between the stamp and the substrate but it should be low enough to avoid protein denaturation.

Acknowledgments

E.M. is grateful to the Spanish Ministry of Science and Education for the provision of grants through the I3 system. We acknowledge financial support from the Spanish Ministerio de Ciencia e Innovación through the project entitled "Regenerative Stem Cell Therapies for Heart Failure" and from the Science Support Program of the Fundación Marcelino Botín.

References

1. Whitesides, G.M., Ostuni, E., Takayama, S., Jiang, X.Y., Ingber, D.E. (2001) Soft lithography in biology and biochemistry. *Annu Rev Biomed Eng.* **3**, 335–373.
2. Kane, R. S., Takayama, S., Ostuni, E., Ingber, D. E., Whitesides, G. M. (1999) Patterning proteins and cells using soft lithography. *Biomaterials* **20**, 2363–2376.
3. Xia, Y., Whitesides, G.M. (1998) Soft lithography. *Angew Chem Int Ed* **37**, 550–575.
4. Hui, C. Y., Jagota, A., Lin, Y. Y., Kramer, E. J. (2002) Constraints on microcontact printing imposed by stamp deformation. *Langmuir* **18**, 1394–1407.
5. Lauer, L., Klein C., Offenhäusser, A. (2001) Spot compliant neuronal networks by structure optimized micro-contact printing. *Biomaterials* **22**, 1925–1932.
6. Delamarche, E., Schmid, H., Biebuyck, H. A., Michel, B. (1997) Stability of molded polydimethylsiloxane microstructures. *Adv. Mater.* **9**, 741–746.
7. Bessueille, F., Pla-Roca, M., Mills, C. A., Martinez, E., Samitier, J., Errachid A. (2005) Submerged microcontact printing (SμCP): an unconventional printing technique of thiols using high aspect ratio, elastomeric stamps. *Langmuir* **21**, 12060–12063.
8. Renault, J.P., Bernard, A., Bietsch, A., Michel, B., Bosshard, H.R., Delamarche, E., Kreiter, M., Hecht, B., Wild, U.P. (2003) Fabricating arrays of single protein molecules on glass using microcontact printing, *J. Phys. Chem. B* **107**, 703–711.
9. Hong-Wei, L., Beinn, V. O., Muir, G. F., Huck, W. T. S. (2003) Nanocontact printing: a route to sub-50-nm-scale chemical and biological patterning. *Langmuir* **19**, 1963–65.
10. Pla-Roca, M., Fernandez, J. G., Mills, C. A., Martínez, E., Samitier J. (2007) Micro/nanopatterning of proteins via contact printing using high aspect ratio PMMA stamps and nanoimprint apparatus. *Langmuir* **23**, 8614–8618.

Chapter 6

Functionalization of Surfaces with Synthetic Oligonucleotides

Brendan Manning and Ramon Eritja

Abstract

There is a large interest in the use of nucleic acids covalently bound to surfaces for a variety of biomedical uses: biosensors, microarrays, drug delivery, lab-on-chip devices, and gene therapy, etc. Most of these applications require the covalent attachment of oligonucleotides via specific reactive groups on both modified oligonucleotide and/or surface. The purpose of this chapter is to provide experimental protocols for the synthesis of oligonucleotides and for the immobilization of these synthetic oligonucleotides onto surfaces such as gold and silicon oxide.

Key words: DNA, Nucleic acids, Biosensors, Surface analysis, Oligonucleotides, Microarrays

1. Introduction

The repair and maintenance of living tissues and organs in vivo is a complex task; regenerative medicine aims to achieve this goal through an amalgamation of synthetic and natural means. The covalent coupling and patterning of biomolecules is expected to play an important role in the development of new biomedical devices (1). High-throughput parallel screening of biomolecules by "lab-on-chip" devices offers the possibility of rapid diagnosis and analysis (2–5). Furthermore, it is well known that the physicochemical properties of a biomaterial affect cellular function, and recently it has been shown that the nanotopography greatly affects cell behavior (6–9). Also of interest is the development of targeted delivery systems based on nanoparticles or liposomes (10, 11). A sophisticated delivery system could provide multiple complementary therapies in a single dose while also sending feedback via reporter molecules (12).

Melba Navarro and Josep A. Planell (eds.), *Nanotechnology in Regenerative Medicine: Methods and Protocols*, Methods in Molecular Biology, vol. 811, DOI 10.1007/978-1-61779-388-2_6, © Springer Science+Business Media, LLC 2012

To fabricate devices requiring the attachment of a biomolecule, specifically an oligonucleotide, to a substrate a specific and robust coupling procedure is necessary (13). There are two primary substrates identified, gold and silicon oxide; however, the coupling methods described are transposable to other suitable surfaces. Covalent coupling to each substrate requires different modifications to the synthetic oligonucleotide. This can be achieved with the use of specialized reagents that are compatible to DNA synthesis conditions and that are able to generate the appropriate functional groups at specific sites on a synthetic oligonucleotide. For a gold surface, the modification of choice is the addition of a thiol moiety to the 3′ or 5′ end of the oligonucleotide. The strong affinity of a thiol group for gold results in the chemisorption of the oligonucleotide followed by oxidation of the gold to form a covalent bond (14, 15). When grafting to a silicon oxide substrate there is a greater variety of methods to choose from. Of the current methods, it has been shown that use of an amine layer followed by the coupling with a homobifunctional linker, 1,4-phenylenediisothiocyanate, followed by grafting of an amine modified oligonucleotide provides a useful and robust method for the covalent attachment of oligonucleotides (see Note 1) (16–18).

2. Materials

2.1. Synthesis of Oligonucleotides Carrying Amino or Thiol Groups at the 5′- or 3′-Ends

1. Oligonucleotides sequences were prepared using solid-phase methodology and 2-cyanoethyl phosphoramidites as monomers. The phosphoramidites of the natural nucleosides are commercially available (see Note 2).
2. Ancillary reagents used during oligonucleotide synthesis are: 0.4 M 1*H*-tetrazol in ACN (activation); 3% trichloroacetic acid in DCM (detritylation), acetic anhydride/pyridine/tetrahydrofurane (1:1:8) (capping A), 10 % *N*-methylimidazole in tetrahydrofurane (capping B), 0.01 M iodine in tetrahydrofurane/pyridine/water (7: 2: 1) (oxidation). These solutions can be obtained from the same companies that provide phosphoramidites (see Note 2).
3. Reagent used for the introduction of an amino group at the 3′-end: 3′-amino-modifier C7 CPG (see Fig. 1 and Note 2).
4. Reagent used for the introduction of an amino group at the 5′-end: MMT-amino modifier C6 phosphoramidite (see Fig. 1 and Note 2).
5. Reagent used for the introduction of a thiol group at the 3′-end: 3′-thiol-modifier C3S-S CPG (see Fig. 1 and Note 2).

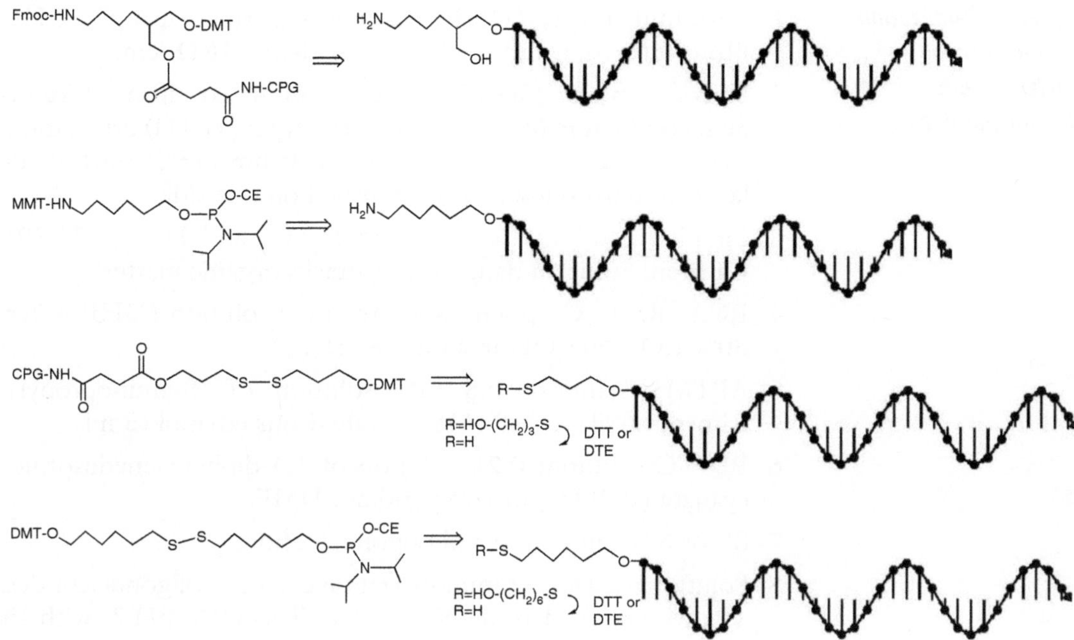

Fig. 1. Phosphoramidites and solid supports used in the preparation of oligonucleotides carrying amino or thiol groups at their 3′- or 5′-ends.

6. Reagent used for the introduction of a thiol group at the 5′-end: 5′-thiol modifier C6S-S phosphoramidite (see Fig. 1 and Note 2).
7. HPLC solutions are as follows. Solvent A: 5% acetonitrile in 100 mM triethylammonium acetate (pH 6.5) and solvent B: 70% acetonitrile in 100 mM triethylammonium acetate pH 6.5.
8. Rhodamine derivative carrying a maleimido group: Tetramethylrhodamine 5-maleimide (Fluka).

2.2. Functionalization of Gold Surfaces with Oligonucleotides

1. Ultrahigh purity (UHP) water was purified using a UHQ filtration system and used with resistivity >18 Ω/cm.
2. Substrates of gold evaporated on mica (template stripped gold), prepared in house (see Note 3).
3. 3′ or 5′ Dithiol modified Oligonucleotide (see Note 4).
4. Solution for the immobilization of thiol-oligonucleotides: Potassium dihydrogen phosphate buffer (KH_2PO_4) 1 M.
5. 6-Mercaptohexanol solution: 1 mM in UHP water.
6. Hybridization solution: ~1nM solution of 10 nm gold nanoparticles conjugated to complementary oligonucleotide (18, 19) in 0.01 M phosphate buffer pH 7, 0.3 M NaCl.
7. Washing solution to remove salt from surface: ammonium acetate 0.03 M in UHP water.

2.3. Functionalization of Silicon Oxide Surfaces with Oligonucleotides

1. Ultrahigh purity (UHP) water was purified using a UHQ filtration system and used with resistivity >18 Ω/cm.
2. Silicon wafers (Si/SiO_2) were purchased from Virginia Semiconductors Inc. and were of the type: (a) ⟨111⟩ orientation, (b) resistance of 1–10 Ω/cm, (c) 100 nm (+5%) thick oxide layer on both sides, and (d) polished on one side.
3. Piranha solution: Concentrated H_2SO_4 and H_2O_2, (70:30). Caution: Solution dangerously attacks organic matter!
4. RCA (Radio Corporation of America) solution (UHP water; 30%H_2O_2; NH_4OH in a ratio 5:1:1).
5. APTMS solution: 0.5 mM solution of (3-aminopropyl)-trimethoxysilane (APTMS) in anhydrous ethanol (5 ml).
6. PDITC solution: 0.2% solution of 1,4-diphenylenediisothiocyanate (PDITC) in 10% pyridine/DMF.
7. 3′- or 5′-Amino modified oligonucleotide.
8. Solution used for the immobilization of amino-oligonucleotides: 10 μM oligonucleotide-NH_2 in 1 M Tris–HCl (pH 7) with 1% *N*,*N*-diisopropylethylamine.
9. Amine passivation solution: 50 mM 6-amino-1-hexanol, 150 mM *N*,*N*-diisopropylethylamine in UHP water.
10. Hybridization solution: 20 μM solution of the rhodamine-labeled oligonucleotide in 5× SSC containing 1% SDS and 20% formamide.

2.4. Functionalization of Glass Slides with Oligonucleotides

1. Glass slides functionalized with poly-(Lys) (MENZEL-GLASER, P4981-602677-CE).
2. PDITC solution: 0.2% solution of 1,4-phenylene diisothiocyanate (PDITC) in 10% pyridine/dimethylformamide.
3. Solution used for the immobilization of amino-oligonucleotides: A solution of oligonucleotide (10 μM) in sodium borate buffer (0.1 M, pH 8.0).
4. Passivation solution: a solution of succinic anhydride (1 g) in *N*-methylpyrrolidinone (50 ml) and sodium borate buffer (5.6 ml, 0.2 M, pH 8).
5. Prehybridization solution: a solution of 1% bovine serum albumin (BSA) in 6× SSC (saline sodium citrate).
6. Hybridization solution: a 20 μM solution of the rhodamine-labeled oligonucleotide in 5× SSC containing 1% SDS and 20% formamide.

3. Methods

As discussed, there is a large interest in the use of biomolecules covalently bound to surfaces for a variety of biomedical applications. Covalent attachment of oligonucleotides to surfaces requires modification of the oligonucleotides and often modification of the surface utilized. There are a large number of protocols to immobilize oligonucleotides to surfaces (3–5, 13). In this work we describe robust methods for the immobilization of synthetic oligonucleotides onto gold, glass and silicon oxide surfaces. As recent examples of the potential applications developed from our group using the methodologies described here we underline the formation of patterns on silicon oxide substrates using hairpin oligonucleotides carrying photolabile groups (18) and the use of thiolated oligonucleotides to form 2D-DNA arrays on gold surfaces (15) (Fig. 2).

3.1. Synthesis of Modified Oligonucleotides

1. Oligonucleotide sequences were prepared using solid-phase methodology and 2-cyanoethyl phosphoramidites as monomers. The syntheses were performed on a DNA synthesizer (Applied Biosystems Model 3400) using either 0.2 or 1 μmol scales.
2. The solid supports used more frequently for the introduction of amino or thiol groups at the 3′-ends are shown in Fig. 1. For the introduction of an amino group, the 3′-amino-modifier C7 CPG is used. This support is a controlled pore glass (CPG) support functionalized with an amino-diol, having the amino group protected with the 9-fluorenylmethyloxycarbonyl (Fmoc) group (see Note 5, (20)). For the introduction of the thiol group, the 3′-thiol-modifier C3S-S CPG is used. This support is a CPG support functionalized with the hydroxypropyldisulfide protected with the dimethoxytrityl (DMT) group.

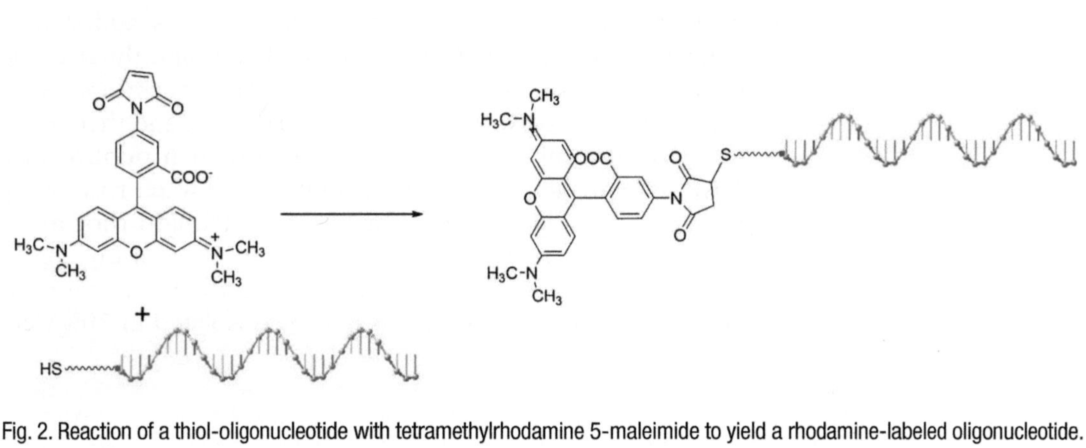

Fig. 2. Reaction of a thiol-oligonucleotide with tetramethylrhodamine 5-maleimide to yield a rhodamine-labeled oligonucleotide.

3. The introduction of amino or thiol groups at the 5′-end needs the use of a special phosphoramidite (Fig. 1). For the introduction of an amino group at the 5′-end, the MMT-amino modifier C6 phosphoramidite was used (21). This reagent is a derivative of 6-aminohexanol with the monomethoxy (MMT) group for the protection of the amino group (see Note 6). For the introduction of a thiol group at the 5′-end the 5′-thiol modifier C6S-S phosphoramidite was used.
4. After the assembly of sequences, ammonia deprotection was performed with 30% ammonia aqueous solution overnight at 55°C. If oligonucleotides carrying a free thiol group are required, they were deprotected with 1 ml of 0.1 M dithiothreitol (DTT) in 30% ammonia aqueous solution (see Note 4).
5. Oligonucleotides carrying amino groups were purified by reverse-phase HPLC. HPLC solutions are described above (Subheading 2.1). Columns: Nucleosil 120C18 (10 μm), 200 × 10 mm. Flow rate: 3 ml/min. Condition A: 20 min linear gradient from 15 to 80% (DMT on) B. Condition B: 20 min linear gradient from 0-50% B (DMT off). The DMT or MMT-oligonucleotides were purified using condition A (22). The major peak containing the DMT- or MMT-oligonucleotide eluted at 10–12 min.
6. The eluates were collected and concentrated to dryness and the residue treated with 80% acetic acid in water. After 30 min, 1 ml of water is added and the acetic acid was extracted with ethyl ether (three times). The resulting aqueous solution was repurified by HPLC using conditions B.
7. Oligonucleotides carrying free thiol groups and DTT were purified in a NAP-5 or NAP-10 (Sephadex G25) column to the DTT and used immediately without further purification.
8. The synthesis of rhodamine-labeled complementary oligonucleotide was performed by reaction of the thiol-oligonucleotide with tetramethylrhodamine 5-maleimide (see Note 7). The deprotected thiol-oligonucleotide (0.5 ml) was added to a large excess of maleimido-rhodamine (1 mg) and the reaction mixture was maintained for 4 h at room temperature. After which the solution was concentrated to dryness and then redissolved in 0.5 ml water. The solution was then purified in a NAP-5 (Sephadex G-25) column eluted with water to separate the desired conjugate from the excess maleimido rhodamine and the conjugate was further purified by HPLC using conditions A (DMT on). The product carrying the fluorescent oligonucleotide (retention time 8 min) was isolated in 70% yield (Fig. 3).

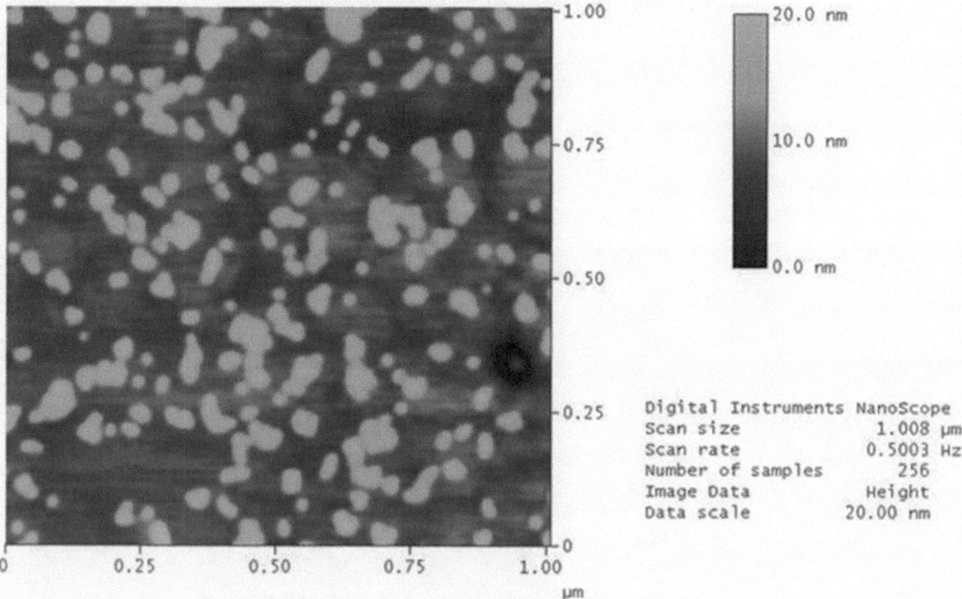

Fig. 3. AFM image of a gold substrate functionalized with an oligonucleotide and hybridized with complementary oligonucleotide labeled with 10 nm gold nanoparticles.

3.2. Immobilization of Oligonucleotides on Gold Substrates

1. Gold substrates are freshly cleaved from support or used freshly cleaned (see Note 3).
2. A 3 µM solution of the dithiol oligonucleotide in 1.0 M KH_2PO_4 is prepared and a 10-µl drop is cast onto the gold substrate.
3. The gold substrates are incubated in a humid environment for 10 h. Then, they are washed with millipure H_2O to remove any noncovalently bound oligonucleotides.
4. A 15 µl drop of 1 mM 6-mercaptohexanol is then cast onto the gold surface. After 1 h, the solution is washed with millipure H_2O.
5. Substrates are hybridized with 10 nm gold particles carrying a complementary oligonucleotide for 2 h at room temperature. Then, they are washed three times with 0.3 M NaCl, 0.01 M phosphate buffer pH 7 and once with 0.03 M ammonium acetate (Fig. 4).

3.3. Functionalization of Self-Assembled Monolayers (SAM) on Silicon Oxide Surfaces

1. Silicon wafers were cut to a square of size approximately 1 cm × 1 cm using a diamond tipped scriber. The substrates were then rinsed with ethanol to remove of any dust that was produced during the cutting process.
2. The cut silicon was then immersed in piranha solution at 90–100°C for 60 min. Once cooled, the piranha solution was rinsed off the substrate with UHP water.

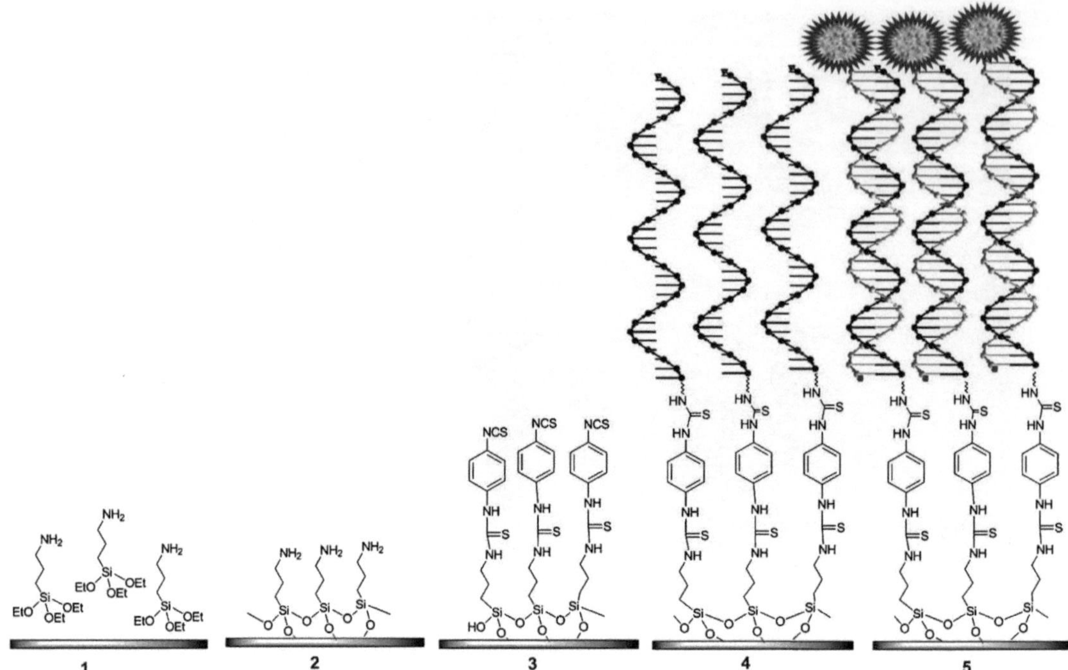

Fig. 4. Outline of the reactions involved in the functionalization of silicon oxide surfaces with oligonucleotides.

3. Then silicon wafers are then sonicated in RCA solution for 60 min. Sonication in RCA at this stage functionalizes the surface with hydroxyl groups to allow monolayer formation. Repeated rinsing of the substrate in UHP water finishes the cleaning procedure. The substrates are then stored in UHP water and used within 2 days to minimize loss of surface hydroxyl groups.

4. The cleaned silica substrates were transferred from UHP H_2O into anhydrous ethanol by stepwise exchange (H_2O–EtOH) 3:1, 2:2, 1:3, 0:4 and finally anhydrous ethanol. The wafers were then immersed, under a nitrogen atmosphere, into a 0.5-mM solution of (3-aminopropyl)-trimethoxysilane (APTMS) in anhydrous ethanol (5 ml) and sonicated at room temperature for 1 h. The substrates were then rinsed for 20 s each with ethanol and chloroform. This was followed by sonication of the wafers twice in fresh ethanol followed by a final sequential rinsing with ethanol and chloroform. Each sample was then dried under a steady stream of nitrogen and cured at 120°C for 30 min under vacuum to promote cross-linking of the SAMs.

5. APTMS Si SAMs were incubated for 2 h in PDITC solution. After incubation, samples were washed with methanol and acetone and stored in vacuum desiccators.

6. Oligonucleotide-NH_2 solution was then deposited onto substrates and incubated at 37°C for 2 h. Samples were then washed with 1% ammonia solution and copious amounts of

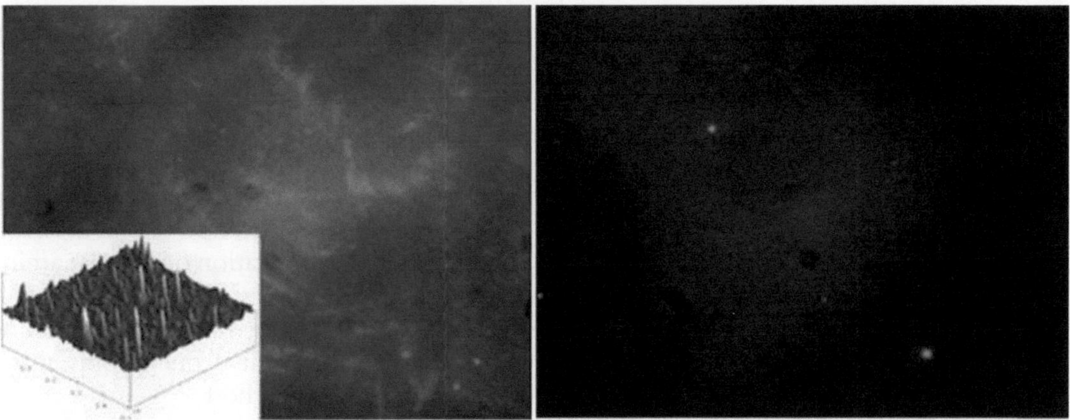

Fig. 5. Fluorescent image of a silicon oxide substrate functionalized with an oligonucleotide (*right*) and hybridized with complementary oligonucleotide labeled with rhodamine (*left*). *Inset* in the *right side* shows the AFM image of the silicon surface functionalized with double stranded oligonucleotides (after hybridization).

UHP H_2O. Samples were then treated with the passivation solution for 2 h. Samples were then dried under a thin stream of N_2 and stored in vacuum desiccators.

7. Samples of APTMS Si SAMs functionalized with the oligonucleotide were washed with 1× SSC. Hybridization with the complementary oligonucleotide labeled with rhodamine was performed for 2 h at room temperature. The resulting surface was washed with 1× SSC and water (Fig. 5).

3.4. Immobilization of Oligonucleotides on Glass Substrates

1. Glass slides functionalized with poly-(Lys) were cleaned ten times with acetone and dried for 45 min at 90°C.
2. The resulting glass slides were treated with PDITC solution for 2 h at room temperature. The resulting glass slides were washed with methanol (3×) and acetone (2×). Glass slides were used immediately.
3. A solution of the amino-oligonucleotide (~3 μM) was added to the surface of the glass slides. The reaction was incubated for 2 h at 37°C and the glass slides washed with 1% NH_4OH and water (30 ml × 5).
4. The resulting slides were post-processed to passivate the charged surface with a solution of succinic anhydride. The slides were rinsed by agitating in UHP H_2O and ethanol 95%, centrifuged for 6 min at 1,500 rpm and dried with a stream of argon.
5. The slides were prehybridized with the bovine serum albumin (BSA) solution for 10 min at room temperature. Hybridization with the complementary oligonucleotide labeled with rhodamine was performed with the hybridization solution for 2 h at room temperature. Fluorescence was measured on a fluorescence scanner.

4. Notes

1. Outside the scope of this short procedure is the electrostatic binding of DNA to surfaces, also the use of the many other useful substrates that may be utilized. The type of surface chosen is dependent on the required use and the type of characterization utilized. Described here is the fabrication of very flat amine monolayers on high-grade silicon wafers; this was utilized to allow characterization by AFM. Using prefabricated amine terminated glass slides may be sufficient if one is only using fluorescent microscopy as detection method.
2. There are several companies that provide regular phosphoramidite, ancillary reagents and solid supports for oligonucleotide synthesis such as Applied Biosystems, Sigma. In addition to standard reagents, more elaborated phosphoramidites and solid supports can be found at companies specialized on oligonucleotide reagents such as Link technologies (Scotland), Glen Research (Sterling, VI, USA), Berry & Associates (Dexter, MI, USA), TriLink Biotechnologies (San Diego, CA, USA), Nedken (Foster City, CA, USA), ChemGenes (Wilmington, MA, USA), Prime Synthesis (Aston, PA, USA), Metkinen (Finland).
3. There are many suitable gold substrates commercially available. Suitability of one substrate over another depends on application. Some suppliers include Arandee (Wether, Germany), Asylum Research (Santa Barbara, CA, USA) and Sigma Aldrich. Some require cleaning steps to remove organic and inorganic build-up, such cleaning procedures are usually supplied by the manufacturer. During our research we have used gold coated mica substrates prepared in house using a method first described by Hegner et al. (23), which have the advantage of not requiring a cleaning step and can be used 'as is' when cut from a support.
4. It is worth discussing the use of disufide and thiol modified oligonucleotides. Disulfide oligonucleotides such as those in Fig. 2 have the advantage that they are stable to long-term storage and spontaneously react with gold surfaces. Alkylthiol-protected thiol strands normally bear disulfide protecting groups that must be removed before conjugation. This is done by adding a cleavage agent such as DTT or DTE which cleaves the disulfide bond; however, DTT or DTE needs to be removed from the free thiol for it to react with a gold surface or any other suitable reactive group. This is done by size exclusion gravity chromatography (NAP-5 and NAP-10 columns). Often, if a thiol oligonucleotide is purchased from a commercial supplier, it will contain DTT or DTE.
5. During the addition of the MMT-amino-modifier C6 is important to avoid the capping reaction because the MMT-amino group

can be acetylated giving an unreactive acetyl derivative of the oligonucleotide (24, 25). Moreover, some authors described low efficiency of the removal of the MMT group. To solve this problem new amino linkers have been described (26) but they are not commercially available. Alternatively the trifluoroacetyl (TFA) derivative of 6-aminohexanol (5′-TFA-amino modifier C6 phosphoramidite) can be used. The TFA group is removed during the ammonia so HPLC purification based on the presence of the MMT group is not possible.

6. Some authors have described that the Fmoc group protecting the amino function is partially lost during the synthetic process and becomes acetylated during capping conditions (27). In shorter sequences we have also found a side product (5–10%) with a molecular weight 100 units higher than expected. We assign this side product to a succinylated amino oligonucleotide. This side compound can be produced by internal aminolysis of the succinyl ester. The use of a commercially available phthaloyl linker (3′-PT-amino modifier C6 CPG) is a good alternative to increase the yield of final product (27).
7. In addition to the conjugation with fluorescent derivatives carrying maleimido groups there are several phosphoramidites and solid supports that can be found at companies specialized on oligonucleotide reagents that allow the rapid preparation of fluorescently labeled oligonucleotides.

Acknowledgments

This work is supported by Spanish Ministry of Education (NAN2004-09415-C05-03, BFU2007-63287/BMC), Generalitat de Catalunya (2005/SGR/00693), Instituto de Salud Carlos III (CIBER-BNN), and European Communities (NANO-3D NMP4-CT2005-014006, DYNAMO contract 028669 (NEST), FUNMOL NMP4-SL-2009-213382). B.M. thanks the SFI-CRANN/UCD for the predoctoral fellowship. We are thankful to Elena Martínez and Xavier Sisquella of Scientific Park of Barcelona (PCB) for providing us template stripped gold.

References

1. Engel, E., Michiardi, A, Navarro, M., Lacroix, D., and Planell J.A. (2007) Nanotechnology in regenerative medicine: the materials side. *Trends Biotech* **26**, 39–47.
2. Cheng M.M-C., Cuda, G. Bunimovich, Y.L., Gaspari, M., Heath, J.R., Hill H.D., Mirkin, C.A., Nijdam, A.J., Terracciano, R., Thunday, T., and Ferrari M. (2006) Nanotechnologies for biomolecular detection and medical diagnostics. *Curr Op Chem Biol* **10**, 11–19.
3. Sassolas, A., Leca-Bouvier, D., and Blum, L. J. (2008) DNA biosensors and microarrays. *Chem Rev* **108**, 109–139.

4. Liu, J., Cao, Z., and Lu, Y. (2009) Functional nucleic acids sensors. *Chem Rev* **109**, 1948–1998.
5. Dittrich P.S. and Manz, A. (2006) Lab-on-a-chip microfluidics in drug discovery. *Nat Rev Drug Discov* **5**, 210–218.
6. Ito, Y. (2000) Surface micropatterning to regulate cell functions. *Biomaterials* **20**, 2333–2342.
7. Stevens, M.M., and George, J.H. (2005) Exploring and engineering the cell surface interface. *Science* **310**, 1135–1138.
8. Wilkins, C.D.W., Riehle, M., Wood, M., Gallagher, J., and Curtis, A.S.G. (2002) The use of nanomaterials patterned on a nano- and micro-metric scale in cellular engineering. *Mater Sci Eng C* **19**, 263–269.
9. Teixeira, A.I., McKie, G.A., Foley, J.D., Bertics, P.J., Nealey, P.F., and Murphy, C.J. (2006) The effect of environmental factors on the response of human corneal epithelial cells to nanoscale substrate topography. *Biomaterials* **27**, 3945–3954.
10. Chertok, B., Moffat, B.A., David, AE, Yu, F., Bergemann, C, Ross, B.D. and Yang, V.C. (2008) Iron oxide nanoparticles as a drug delivery vehicle for MRI monitored magnetic targeting of brain tumours. *Biomaterials* **29**, 487–496.
11. Murthy, N., Campbell, J., Fausto, N., Hoffman, A.S. and Stayton, P.S. (2003) Design and synthesis of pH-responsive polymeric carriers that target uptake and enhance the intracellular delivery of oligonucleotides. *J Control Release* **89**, 365–374.
12. Chertok, B., Moffat, B.A., David, A.E., Yu, F., Bergemann, C., Ross, B.D., and Yang, V.C. (2008) Iron oxide nanoparticles as a drug delivery vehicle for MRI monitored magnetic targeting of brain tumors. *Biomaterials* **29**, 487–496.
13. Pirrung, M.C. (2002) How to make a DNA chip. *Angew Chem Int Ed* **41**, 1276–1289.
14. Huang, E., Satjapipat, M., Han, S., and Zhou, F. (2001) Surface structure and coverage of an oligonucleotide probe tethered onto a gold substrate and its hybridization efficiency for a polynucleotide target. *Langmuir* **17**, 1215–1224.
15. Garibotti, A.V., Sisquella, X., Martínez, E., and Eritja R (2009) Assembly of two-dimensional DNA crystals carrying N^4-[2-(*tert*-butyldisulfanyl)ethyl]-cytosine residues. *Helv Chim Acta* **92**, 1466–1472.
16. Manning, M., and Redmond, G. (2005) Formation and characterization of DNA microarrays at silicon nitride substrates. *Langmuir* **21**, 395–402.
17. Manning, M., Galvin, P., and Redmond, G. (2002) A robust procedure for DNA microarray fabrication and screening in the molecular biology laboratory. *Am Biotech Lab* **20**, 16–17.
18. Manning, B., Leigh, S.J., Ramos, R., Preece, J., and Eritja, R. (2009) Fabrication of patterned surfaces by photolithographic exposure of DNA-hairpins carrying a novel photolabile group. *J Exp Nanoscience* **5**, 26–39.
19. Mirkin, C.A., Letsinger, R.L., Mucic, R.C., and Storhoff, J.J. (1996) A DNA-based method for rationally assembling nanoparticles into macroscopic materials. *Nature* **382**, 607–609.
20. Nelson, P.S., Kent, M., and Muthini, S. (1992) Oligonucleotide labelling methods. 3. Direct labeling of oligonucleotides employing a novel, non-nucleosidic, 2-aminobutyl-1, 3-propanediol backbone. *Nucleic Acids Res* **20**, 6253–6259.
21. Sinha, N.D., and Cook, R.M. (1988) The preparation and application of functionalised synthetic oligonucleotides. III Use of H-phosphonate derivatives of protected aminohexanol and mercaptopropanol or mercaptohexanol. *Nucleic Acids Res* **16**, 2659–2669.
22. Güimil García, R., and Eritja, R. (1997) Preparation, purification and analysis of synthetic oligonucleotides suitable for microinjection experiments. *Microinjection and transgenesis. Strategies and protocols* (A. Cid-Arregui, A. García-Carrancá, Eds) Springer Laboratory Manuals, Springer, Berlin pp 95–112.
23. Hegner, M., Wagner, P., and Semenza, G. (1993) Ultralarge atomically flat template-stripped Au surfaces for scanning probe microscopy. *Surf Sci* **291**, 39–46.
24. Zaramella, S., Yeheskiely, E., and Strömberg, R. (2004) A method for solid-phase synthesis of oligonucleotide 5′-peptide-conjugates using acid-labile alpha-amino protection. *J Am Chem Soc* **126**, 14029–14035.
25. Ocampo, S. M., Albericio, F., Fernández, I., Vilaseca, M., and Eritja, R. (2005) A straightforward synthesis of 5′-peptide oligonucleotide conjugates using N$^{\alpha}$-Fmoc-protected amino acids. *Org Lett* **7**, 4349–4352.
26. Kojima, N., Takebayashi, T., Mikami, A., Ohtsuka, E., and Komatsu, Y. (2009) Efficient synthesis of oligonucleotide conjugates on solid-support using an (aminoethoxycarbonyl) aminohexyl group for 5′-terminal modification. *Bioorg Med Chem Lett* **19**, 2144–2147.
27. Petrie, C. R., Reed, M. W., Adams, A. D., and Meyer, R. B. Jr (1992) An improved CPG support for the synthesis of 3′-amine tailed oligonucleotides. *Bioconjug Chem* **3**, 85–87.

Chapter 7

Fabrication of PEG Hydrogel Microwell Arrays for High-Throughput Single Stem Cell Culture and Analysis

Stefan A. Kobel and Matthias P. Lutolf

Abstract

Microwell arrays are cell culture and imaging platforms to assess cells at a single cell level and in high-throughput. They allow the spatial confinement of single cells in microfabricated cavities on a substrate and thus the continuous long-term observation of single cells and their progeny. The recent development of microwell arrays from soft, biomimetic hydrogels further increases the physiological relevance of these platforms, as it substantially enhances stem cell survival and the efficiency of self-renewal or differentiation. This protocol describes the microfabrication of such hydrogel microwell arrays, as well as the cell handling and imaging.

Key words: Stem cells, Single cell analysis, Poly(ethylene glycol), PEG, Hydrogel, Time-lapse microscopy, Soft lithography, Microcontact printing

1. Introduction

Stem cells are inherently heterogeneous cell populations that call for high-throughput single cell analyses. Although single cell transplantation experiments (1), fluorescence-activated cell sorting (FACS) (2) or reporter gene expression dynamics (3) were successfully utilized to document behavioral variations within stem cell populations, most current in vitro methodologies fail to shed light on important long-term behaviors such as cell proliferation or differentiation because of difficulties in tracking single cells in vitro.

Microwell arrays are emerging cell culture and imaging platforms with a microtopography to capture cells. They are produced by microfabrication processes such as photolithography or soft lithography and consist of miniature cavities with diameters and depths of tens to hundreds of microns. Depending on the diameter

Melba Navarro and Josep A. Planell (eds.), *Nanotechnology in Regenerative Medicine: Methods and Protocols*, Methods in Molecular Biology, vol. 811, DOI 10.1007/978-1-61779-388-2_7, © Springer Science+Business Media, LLC 2012

of the microwells, microwell arrays can comprise several hundreds to thousands of microwells per square centimeter arranged on a regular grid. Cells can then be seeded onto these arrays and will sediment randomly onto the bottom of the microwells (4). Because microwell arrays are typically made of a cell-nonadhesive substrate such as agarose (5), poly(ethylene glycol) (PEG) hydrogels (6–8), or poly(dimethylsiloxane) (PDMS) (9, 10), cells cannot overcome the topographical barrier and therefore they remain trapped inside the microwells. Moreover, the high density of single cells in microwell arrays significantly improves the throughput of imaging-based single-cell analyses compared to standard multi-well plates (11).

Not surprisingly, microwell arrays have, for example, been successfully used to study in a high-throughput manner the proliferation rates of single adult rat hippocampal progenitor cells (12), neural stem cells (NSCs) (13), or hematopoietic stem cell (HSCs) (7, 9). The long-term microscopy combined with image analysis and retrospective cell fate analysis allowed for example to identify differences in proliferation rates of long-term HSCs and multipotent progenitors or to screen for effects of various growth factors on stem cell function (7). Of note, the applicability of microwells is not restricted to proliferation analyses of stem cells but can also be used for high-throughput image cytometry (14) or the screening of single hybridoma cells (15). Another important application of microwell arrays is the generation of multicellular spheroids such as embryonic bodies (EBs) (6, 16), neurospheres (NS) (13) or cancer spheroids (5). Compared to classical methods to produce multicelluar spheroids, i.e., hanging drops or scraping of monolayers (in the case of EBs), microwell arrays not only enhance the throughput, but also yield more homogeneous, size-controlled cell aggregates or prevent merging of individual, clonally derived spheroids.

A wide range of methods and materials can be used for the fabrication of microwell arrays. The first generation of microwell arrays were made by photolithography, exposing photoresists on glass with UV through photomasks, and thus offered relatively limited possibilities to biofunctionalize the microwell arrays (12). Subsequently, soft lithography and micromolding approaches were developed whereby a negative stamp (i.e., a stamp with micropillars) is first molded against PDMS or PEG hydrogels. Upon complete crosslinking, these substrates irreversibly replicate the features of the microstamp that can be removed. These techniques not only allowed to micropattern soft hydrogels, that were shown to enhance stem cell culture (7, 13), but also to locally functionalize microwells with biomolecules by integrating microcontact printing (μCP) approaches into the microwell array fabrication (7, 10). Notably, the local tethering of microwells such as cell-adhesive proteins or growth factors (7), extends the advantages of microcontact printed substrates to clinically important, nonadherent stem cells, such as

NSCs (13) or HSCs which would not be possible on flat, purely microcontact-printed cell culture substrates (17).

This chapter describes (a) the fabrication by micromolding of PEG microwell array using a PDMS stamp, (b) the functionalization of the microwells with collagen, and (c) how to image by microscopy single cells on the PEG microwell array. The protocol is based on the photopolymerization of acrylated PEG, a robust and simple method to make PEG hydrogels from commercially available precursor materials circumventing time-consuming chemical synthesis (7, 18, 19). The protocol also describes how to reliably integrate these PEG microwell arrays into multi-chambered microscopy slides which enables live-cell imaging using time-lapse microscopy and allows to perform multiple experiments in parallel.

2. Materials

2.1. Microfabrication of PDMS Master

1. 4" ⟨100⟩ silicon test wafers single-side polished (Siltronix, France).
2. GM1070 SU8 Photoresist (Gersteltec, Switherland) (light sensitive).
3. 1-Methoxy-2-propyl acetate (PMA, ACS grade).
4. Chrome Blank 5 in. soda lime mask (SLM) (Nanofilm, Westlake Village, USA) (light sensitive).
5. Clean room facility equipped for writing masks and for photolithography with the following equipment: a plasma oven (Tepla 300, PVA TePla, Germany), mask writer (Heidelberg DWL200, Heidelberg Instruments GmbH, Germany), mask developer (Suess DV10), mask aligner (Suess MA6, both Suess MicroTec AG, Germany), wet benches for chrome etching and resist stripping, spin coater (Sawatec LSM200), and programmable hot plates (Sawatec HP401Z, both Sawatec AG, Principality of Liechtenstein).
6. Mask design program, e.g., CleWin 4.0 (PhoeniX Software, The Netherlands).

2.2. Fabrication of the PDMS Stamp

1. Poly(dimethyl siloxane) elastomer kit (Sylgardl 184 Silicone Elastomer Kit, Dow Corning).
2. Flat-bottom glass beaker with an inner diameter of 11–12 cm (to host a 4-in. wafer).
3. Scalpel, ideally with a crescent-shaped scalpel blade (#12).

2.3. Functionalization of Glass Slides

1. Isopropyl alcohol (ACS grade).
2. Acetone (ACS grade).

3. Heptane (ACS grade).
4. Standard microscopy glass slides.
5. Plasma cleaner femto (Diener electronic GmbH+Co, Germany).
6. 3-(Trichlorosilyl)propyl methacrylate (TPM, Sigma-Aldrich, Switzerland).
7. Carbon tetrachloride (ACS grade).

2.4. Inking of the Stamp

1. Acrylamide/Bis-acrylamide, 30% solution, 37.5:1 ratio, electrophoresis grade.
2. *N,N,N,N′*-Tetramethylethylenediamine (TEMED).
3. 1 mg/ml Ammonium persulfate solution (APS) in ddH_2O (stored at −20°C).
4. 1.50 M Tris–HCl, pH 8.8.
5. 10 mM Hydrochloric acid (HCl).
6. Bovine collagen solution 3 mg/ml (Sigma-Aldrich, Switzerland).
7. 1× Phosphate-buffered saline (PBS, to store the remaining acrylamide gel).

2.5. Fabrication of the PEG Microwell Arrays

1. 20% (w/v) Poly(ethylene glycol) diacrylate average Mn 2000 (Sigma-Aldrich, Switzerland) in PBS (see Note 1) with 2% (v/v) 2-hydroxy-2-methyl propiophenone (Sigma-Aldrich, Switzerland) (store solution at 4°C and protected from light).
2. Ultraviolet (UV) lamp HBO 100 (Zeiss, Germany).
3. Elastosil E 41 (one-component silicon rubber, Wacker Chemie, Switzerland).
4. LabTek II 8 chamber slides (Nunc, Switzerland).
5. Oriented Polyester, Plastic Shim Stock, 0.004 in. Thick 5 in. × 20 in., Color Tan (SmallParts Inc., USA).
6. 1× PBS (to store microwell arrays).

2.6. Cell Seeding and Time-Lapse Imaging

1. Clear polyolefin, advanced optical sealing tape for microscopy (Nunc, Switzerland).
2. Zeiss Observer Z1 (Zeiss, Germany), equipped with an motorized stage (Ludl, Austria) and an environmental chamber (Life Imaging Services, Switzerland).
3. Temperature controller "Brick" (Life Imaging Services, Switzerland).
4. CO_2 and humidifier "Cube" (Life Imaging Services, Switzerland).
5. MetaMorph 7.5 image acquisition software (Visitron, Germany).

3. Methods

The procedure to culture single stem cells on micromolded PEG microwell arrays includes three main steps. First, a PDMS microstamp needs to be replicated from a microfabricated master via soft lithography (Fig. 1a). This master is produced by photolithography

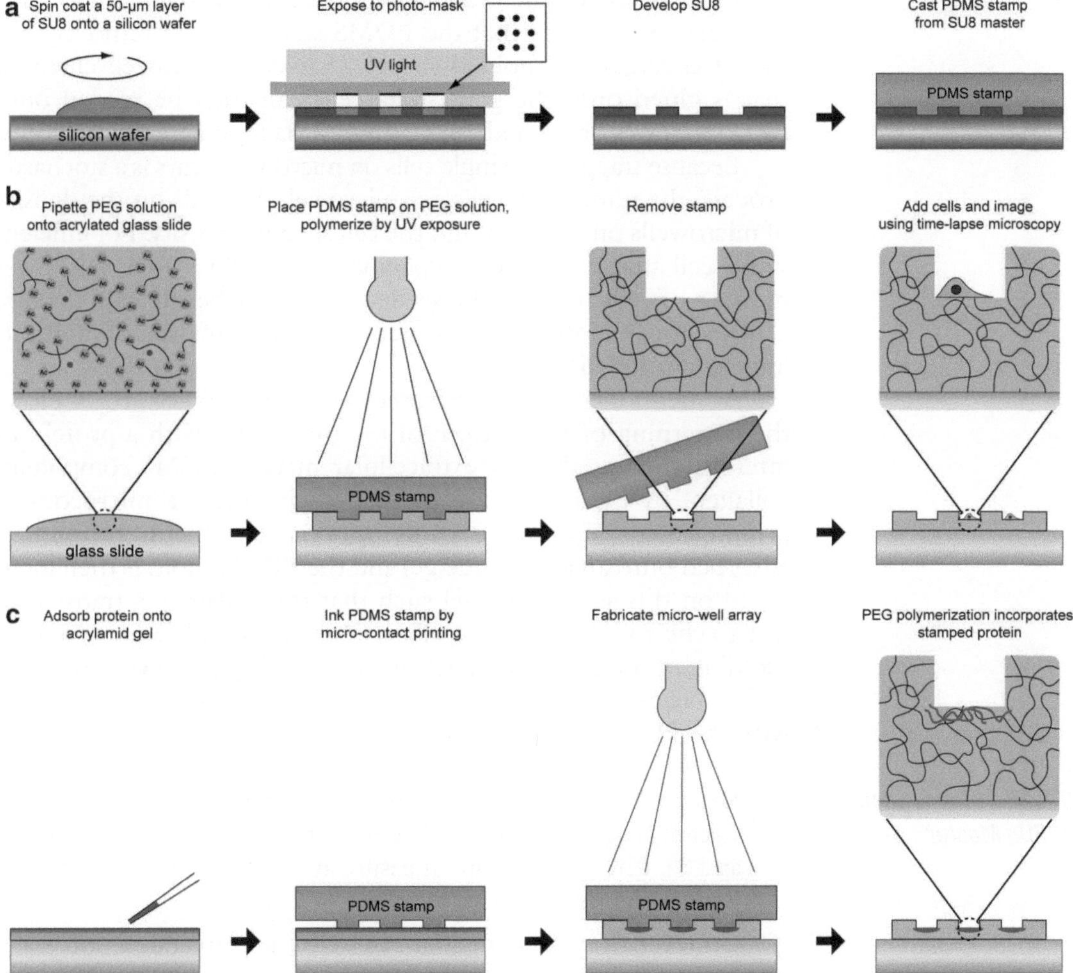

Fig. 1. The fabrication of microwell arrays and their functionalization with proteins. (**a**) Fabrication of the PDMS stamp. First, a SU8 master is produced by spin coating a 100-µm layer of SU8 onto a silicon wafer that is then exposed to a photomask. After development of the SU8 structures, a PDMS stamp can be molded from the SU8 and then be used for the fabrication of PEG microwell arrays (**b**). Therefore, a drop of PEG-diacrylate with a photoinitiator is applied onto an acrylated glass slide, covered with the PDMS stamp and exposed to UV light. The UV exposure triggers the polymerization of the PEG, its covalent attachment to the glass surface. The irreversible incorporation of the shape of the PDMS stamp finally allows the removal of the PDMS stamp and the seeding of single cells onto the microwell arrays. (**c**) This process also allows the functionalization of the bottom of microwells with adhesive proteins such as collagen by integrating a microcontact printing step. First, collagen is adsorbed onto an acrylamide gel from where it is transferred onto the PDMS stamp. During the micromolding of PEG microwell arrays with this inked PDMS stamp, the collagen is transferred into the PEG gel and locally polymerizes due to the change of pH.

in a clean room facility and results in a PDMS stamp consisting of an array of micropillars. In a second step, this PDMS stamp, which optionally can be coated with a protein of interest, is used to micromold PEG-diacrylate (Fig. 1b). To do so, the PEG precursor is mixed with a photoinitiator and applied onto an acryl-functionalized glass slide. After placing the PDMS stamp onto the PEG precursor, the sandwich is exposed to ultraviolet (UV) light at 350 nm. This leads to the free-radical polymerization of the acrylate groups of PEG and to the covalent attachment of the hydrogel to the acrylated glass slide, such that the PDMS stamp can be removed and the PEG retains its molded shape. Thanks to multiwell chamber that is glued onto the glass slide, cells can then be seeded onto these microwell arrays and can be imaged via time-lapse microscopy.

Because trapping of single cells on microwell arrays is a stochastic process, the number of cells per microwell depends on the density of microwells on the array and the cell seeding density. For efficient single cell analyses, these two parameters should match, but lower number of cells may also be seeded (as it may be the case for rare cells) or much higher numbers allowing the formation of multicellular spheroids (6).

As mentioned above, micromolding of microwell arrays allows the patterning of the bottom of the microwells with a protein of interest, such as the key extracellular matrix (ECM) component collagen, by expanding the method to integrate a microcontact printing step (Fig. 1c). In this optional treatment, collagen is adsorbed onto an acrylamide gel and the PDMS mold is then incubated on this acrylamide gel such that the collagen is transferred just to the top of the micropillars. During the subsequent micromolding of the PEG microwell array, the collagen polymerizes due to the change of pH and will form an interpenetrating network with the PEG gel.

3.1. Microfabrication of SU8 Master

1. Design a photomask using the mask writing program CleWin. Create arrays of circles with a diameter of 150 μm (see Notes 2 and 3). The arrays should measure about 4 × 5 mm (see Note 4). Repeat the designs with an offset of 2–3 mm.
2. Convert the file using the Heidelberg conversion software. Invert the design in the conversion settings (see Notes 5 and 6).
3. In the clean room facility, write the photomask, develop it with a mask developer, etch it in a chrome etch bath and strip the remaining photoresist from the mask.
4. Immediately prior use, clean silicon wafers in an oxygen plasma for 7 min at 500 W.
5. Pour a drop with a diameter of 3–4 cm of SU8 GM10170 onto the wafer and spin-coat it at 1,700 rpm for 40 s to obtain a 50-μm layer of SU8 (see Note 7).

6. Prebake the wafer on programmable a hotplate 30 min at 130°C, ramping the temperature at 4°C/min (see Note 8).
7. Align the photomask and the wafer using a mask aligner, and expose the wafer with 400 mJ/cm^2 with UV light (290–390 nm). Exact exposure dose depends on the mask design and may need to be determined experimentally.
8. After exposure, bake the wafer on a programmable hot plate for 40 min at 100°C (ramp 4°C/min).
9. Develop the wafer for 2–3 min in PMA under slight agitation, wash in a second PMA bath for another 2 min, and then rinse twice in IPA. Let dry under a fume hood.

3.2. Fabrication of the PDMS Stamp

1. Prepare 20–30 g of PDMS by mixing prepolymer and curing agent at ratio 10:1 (w/w) in a plastic cup. Mix rigorously and degas until all air bubbles are removed.
2. Place wafer in a glass container lined with a double layer of thin foil, pour PDMS onto the wafer to a thickness of 3–4 mm. Work in a clean and dust-free environment.
3. Bake the PDMS replica at 80°C for >4 h, remove the wafer from the oven, and allow to cool down.
4. Cut the border of the PDMS with a scalpel blade, carefully peel off the PDMS from the wafer and then cut the individual arrays slightly larger than an array using a scalpel. Store unused stamps in a clean Petri dish, protected from dust (see Note 9).

3.3. Acrylation of Glass Slides

1. Clean glass slides first with detergent, then wash with ddH$_2$O, acetone and isopropyl alcohol. Dry with air gun or in fume hood.
2. Plasma clean glass slides in an oxygen plasma at ~200 mTorr, 100 W for 5 min.
3. In the meantime, prepare a 1-mM solution of TPM (3-(trichlorosilylpropyl) methacrylate) in a 4:1 mixture of heptane and carbon tetrachloride.
4. Immediately after plasma cleaning, bath the glass slides in the TPM solution for 5 min.
5. After the treatment, wash slides with heptane, acetone, and water, dry with air gun or in fume hood.
6. Acrylated slides can be stored at 4°C until further use.

3.4. Functionalization of PDMS Stamp (Optional)

1. Cast a 0.8-mm thick acrylamide gel made of 4.4 ml acrylamide/bis-acrylamide, 2.5 ml Tris–HCl pH 8.8, 3 ml ddH$_2$O, 100 μl APS, and 10 μl TEMED using a standard gel casting system.
2. After polymerization, disassemble the gel casting system in a sterile hood. Cut a 2 × 4 cm piece from acrylamide gel, transfer

it onto a clean Petri dish and add 2–3 ml of 10 mM HCl for 20 min (see Note 10). Remaining gels can be stored in PBS at 4°C for up to 2 weeks.

3. Remove buffer with a pipette without touching the gel surface. Then slightly tilt the Petri dish to carefully adsorb the remaining liquid droplets with a kimwipe from the border of the acrylamide gel.
4. Add 100 μl of the collagen solution (3 mg/ml) equally onto the acrylamide gel, distribute homogeneously on the hydrogel with a pipette tip.
5. Let the collagen adsorb onto the acrylamide by evaporation until there are no visible droplets on the acrylamide gel left.
6. Place eight of the cut PDMS stamps on the collagen soaked acrylamide gel with the pillars. Mind the orientation of the PDMS, the micropillars of the stamp have to be in contact with the gel. Incubate for 30 min in the closed Petri dish.

3.5. Fabrication of the PEG Microwell Arrays

1. Apply some Elastil E 41 glue onto a piece of Parafilm or tin foil. Fold Parafilm and distribute the silicone glue homogeneously until a thin film is obtained.
2. Remove the chamber from the LabTek Chamber slide with the provided tools, remove remaining adhesive from the chamber walls and briefly put the chamber walls on the silicone glue.
3. Place the LabTek chamber on the acrylated glass slide, press firmly to remove potential air bubbles in the glue and to ensure complete sealing of the chamber, but avoid lateral movement of the chamber to keep the glass at the bottom of the wells clean from glue. Let the rubber glue cure at room temperature for >2 h.
4. Cut small pieces (2 × 5 mm) from the 100-m plastic spacer and place them on the bottom of the chambered glass slide such that they can hold the PDMS stamp.
5. Place 20–30 μl of the PEG precursors solution into all wells without introducing air bubbles to the PEG solution.
6. Gently place the (inked) PDMS stamp onto the PEG drop. Do not press strongly to avoid squeezing out the PEG.
7. Expose the sample for 60 s to an UV source at 100 mW/cm^2 (at 350 nm). Exact exposure times may be determined experimentally.
8. Carefully remove the PDMS stamp with tweezers and add PBS onto each microwell.
9. Sterilize the hydrogel microwell array twice with UV in a cell culture hood.
10. Store the arrays at 4°C in a Petri dish sealed with Parafilm for >12 h to allow the hydrogel to swell. They can be stored under these conditions for up to a week.

3.6. Cell Seeding and Time-Lapse Imaging

1. Preheat the incubator of the microscope and the sample holder to 37°C for 2–4 h.
2. Wash arrays 3× with medium for 30 min.
3. Prepare your cells according to your protocols.
4. To seed cells homogeneously on the microwell array, remove the medium (see Note 11) and immediately seed cells onto the array (see Note 12). Cover with a precut sealing tape.
5. Carefully place sample on the microscope. Cover sample with the environmental chamber and turn on CO_2 and humidity controller (see Note 13). Let equilibrate the sample for >1 h.
6. After thermal equilibration, acquire positions of single cells using the "multi-dimensional acquisition" application of MetaMorph.
7. Set illumination and time-lapse settings and launch the time-lapse experiment (see Note 14).

4. Notes

1. The density of PEG is approximately 1.2 kg/l and needs to be compensated when preparing the PEG precursor solution.
2. There are two main aspects to be considered when designing the mask. First, the critical dimension (i.e., the most difficult feature to microfabricate) is not the size of the microwells, but the wall separating them. Walls with a large aspect ratio (i.e., very high, but thin PEG structures) are fragile and difficult to mold because the hydrophobicity of the PDMS hinders the filling of narrow gaps of the mold with the PEG precursor solution. In our experience, an aspect ratio of 1–2 ensures complete filling; thus, the microwells should not be more than twice as high as the offset between each microwell (8). Hydrogel pillars or closed structures where air cannot escape from the PDMS stamp during the molding are similarly difficult to produce.
3. Second, PEG hydrogels intrinsically absorb water after polymerization and swell. This swelling leads to an enlargement of PEG structures and accordingly to a diminishment of the diameter and the depth of the microwells. For this reason, the diameters of microwells should be designed approximately 50% larger than the intended size (depending on the wall thickness between the microwells and the PEG concentration).
4. Note that the size of a microwell array (4 × 5 mm) is designed to fit into a well in an 8-well LabTek chamber, but can be easily adjusted to other cell culture devices.

5. We routinely use a mask writer to fabricate glass photomasks with a maximal resolution of 1 μm. However, the fabrication of microwell arrays normally does not need a high spatial resolution and may also be done using transparencies with a high-quality laser printer as described previously (17). Skip steps 2 and 3 of this section in this case.
6. As the fabrication of microwells arrays comprises two molding steps (PDMS replica, PEG micromolding) and because each molding step inverts the topography, the wafer must have the same topography as the PEG structures, i.e., it must consist of SU8 microwell arrays (the PDMS stamp will then host arrays of micropillars and the PEG hydrogel contain the microwell arrays again). To obtain the right topography in SU8 (a negative photoresist that polymerizes upon UV-exposure), the area of the microwells (the dots on the mask design) must remain the chrome on the photomask. Our setup requires inverting the design during the conversion (black becomes white and vice versa). However, if transparencies are used, this may not be the case. In this case, ensure that the mask consists of arrays of dots to obtain PEG microwell arrays.
7. Other thicknesses can be achieved by changing the photoresist or the spinning rate. This may also require adjusting the baking and exposure times of the entire process. Check the manufacturer's homepage (www.gersteltec.ch) for additional details.
8. We prebake the wafers at higher temperatures than indicated by the manufacturer to decrease baking times.
9. PDMS is very hydrophobic and thus easily attracts dust. The best way to remove dust or PDMS cutting shred is by gently applying a tape (e.g., Magic tape from 3 M) on the PDMS. Dust will now stick to the tape and can easily be removed together with the tape. Also note that, although the cleaning of PDMS stamps was described previously (17), we do not recommend reusing PDMS stamps to avoid cross-contaminations.
10. This step is required to exchange buffer and prevent the gelation of the collagen on the acrylamide gel. It may also be possible to print other proteins, but for that the buffer need to be adjusted and protein initially functionalized with a heterofunctional acryl-PEG-NHS linker to covalently bind the protein to the hydrogel (20).
11. Work quickly to avoid the complete drying of the hydrogel, which can induce air bubbles on the surface of the gel.
12. The number of cells to be seeded depends on the density of microwells in the array their size and on your application. Because cell trapping in the microwells is random, the number of cells follows a Poisson distribution for large microwells. For single cell proliferation analyses one cell per 2–3 microwells

typically is sufficient to avoid too many microwells with no or more than one cell, but may also be lower for rare primary cells like hematopoietic stem cells.

13. Typically, air flow rates of 15–20 l/h and a humidity of 90–95% are sufficient to avoid evaporation of the sample. Also avoid wetting of the sealing tape, which will increase evaporation and the risk that the sample dries out. Also ensure that the temperature of the sample and the microscope is well equilibrated to avoid focus drifts during the time-lapse.
14. For slowly moving and proliferating cells, such as HSCs, time intervals of 4 h are sufficient to assess the proliferation kinetics of single cells, but may need to be higher for other cell types or other readouts. Do not frequently image cells in fluorescent channels to avoid phototoxicity. If a fluorescent readout is available, cell counting may be automated (14, 21). However, for brightfield images, automated cell counting is difficult and not reliable and thus is mostly done manually.

Acknowledgments

We thank Dr. Samy Gobaa, Katarzyna Mosiewicz and Andrea Negro for valuable discussions.

References

1. Dykstra, B., Kent, D., Bowie, M., McCaffrey, L., Hamilton, M., Lyons, K., Lee, S. J., Brinkman, R., and Eaves, C. (2007) Long-term propagation of distinct hematopoietic differentiation programs in vivo. *Cell stem cell* **1**, 218–29.
2. Chang, H. H., Hemberg, M., Barahona, M., Ingber, D. E., and Huang, S. (2008) Transcriptome-wide noise controls lineage choice in mammalian progenitor cells. *Nature* **453**, 544–7.
3. Singh, A. M., Hamazaki, T., Hankowski, K. E., and Terada, N. (2007) A heterogeneous expression pattern for Nanog in embryonic stem cells. *Stem cells (Dayton, Ohio)* **25**, 2534–42.
4. Charnley, M., Textor, M., Khademhosseini, A., and Lutolf, M. P. (2009) Integration column: microwell arrays for mammalian cell culture. *Integrative Biology* **1**, 625–34.
5. Napolitano, A. P., Dean, D. M., Man, A. J., Youssef, J., Ho, D. N., Rago, A. P., Lech, M. P., and Morgan, J. R. (2007) Scaffold-free three-dimensional cell culture utilizing micromolded nonadhesive hydrogels. *BioTechniques* **43**, 494, 96–500.
6. Hwang, Y. S., Chung, B. G., Ortmann, D., Hattori, N., Moeller, H. C., and Khademhosseini, A. (2009) Microwell-mediated control of embryoid body size regulates embryonic stem cell fate via differential expression of WNT5a and WNT11. *Proceedings of the National Academy of Sciences of the United States of America* **106**, 16978–83.
7. Lutolf, M. P., Doyonnas, R., Havenstrite, K., Koleckar, K., and Blau, H. M. (2009) Perturbation of single hematopoietic stem cell fates in artificial niches. *Integrative Biology* **1**, 59–69.
8. Kobel, S., Limacher, M., Gobaa, S., Laroche, T., and Lutolf, M. P. (2009) Micropatterning of hydrogels by soft embossing. *Langmuir : the ACS journal of surfaces and colloids* **25**, 8774–9.
9. Kurth, I., Franke, K., Pompe, T., Bornhäuser, M., and Werner, C. (2009) Hematopoietic stem and progenitor cells in adhesive microcavities. *Integrative Biology* **1**, 427–34.
10. Ochsner, M., Dusseiller, M. R., Grandin, H. M., Luna-Morris, S., Textor, M., Vogel, V., and Smith, M. L. (2007) Micro-well arrays for 3D

shape control and high resolution analysis of single cells. *Lab on a chip* 7, 1074–7.

11. Rieger, M. A., Hoppe, P. S., Smejkal, B. M., Eitelhuber, A. C., and Schroeder, T. (2009) Hematopoietic cytokines can instruct lineage choice. *Science (New York, NY)* **325**, 217–8.
12. Chin, V. I., Taupin, P., Sanga, S., Scheel, J., Gage, F. H., and Bhatia, S. N. (2004) Microfabricated platform for studying stem cell fates. *Biotechnology and bioengineering* **88**, 399–415.
13. Cordey, M., Limacher, M., Kobel, S., Taylor, V., and Lutolf, M. P. (2008) Enhancing the reliability and throughput of neurosphere culture on hydrogel microwell arrays. *Stem cells (Dayton, Ohio)* **26**, 2586–94.
14. Roach, K., King, K. R., Uygun, B., Kohane, I., Yarmush, M. L., and Toner, M. (2009) High throughput single cell bioinformatics. *Biotechnology progress* **25**, 1772–9.
15. Ogunniyi, A., Story, C., Papa, E., Guillen, E., and Love, J. (2009) Screening individual hybridomas by microengraving to discover monoclonal antibodies. *Nature protocols* **4**, 767–82.
16. Ungrin, M. D., Joshi, C., Nica, A., Bauwens, C., and Zandstra, P. W. (2008) Reproducible, ultra high-throughput formation of multicellular organization from single cell suspension-derived human embryonic stem cell aggregates. *PloS one* **3**, e1565.
17. Peerani, R., Bauwens, C., Kumacheva, E., and Zandstra, P. W. (2009) Patterning mouse and human embryonic stem cells using micro-contact printing. *Methods in molecular biology (Clifton, NJ)* **482**, 21–33.
18. Karp, J. M., Yeh, J., Eng, G., Fukuda, J., Blumling, J., Suh, K. Y., Cheng, J., Mahdavi, A., Borenstein, J., Langer, R., and Khademhosseini, A. (2007) Controlling size, shape and homogeneity of embryoid bodies using poly(ethylene glycol) microwells. *Lab on a chip* 7, 786–94.
19. Revzin, A., Russell, R. J., Yadavalli, V. K., Koh, W. G., Deister, C., Hile, D. D., Mellott, M. B., and Pishko, M. V. (2001) Fabrication of poly(ethylene glycol) hydrogel microstructures using photolithography. *Langmuir* **17**, 5440–47.
20. Hume, P. S., and Anseth, K. S. Inducing local T cell apoptosis with anti-Fas-functionalized polymeric coatings fabricated via surface-initiated photopolymerizations. *Biomaterials* **31**, 3166–74.
21. Kachouie, N., Kang, L., and Khademhosseini, A. (2009) Arraycount, an algorithm for automatic cell counting in microwell arrays. *BioTechniques* **47**, x–xvi.

Chapter 8

Preparation of Polyelectrolyte Nanocomplexes Containing Recombinant Human Hepatocyte Growth Factor as Potential Oral Carriers for Liver Regeneration

Catarina Ribeiro, Ana Patrícia Neto, José das Neves, Maria Fernanda Bahia, and Bruno Sarmento

Abstract

The large number of cytokines and growth factors implicated in the regulation of liver regeneration has led to the possibility of using these molecules in therapy, namely, in the case of recombinant human hepatocyte growth factor (rhHGF). The importance and potential clinical usefulness of rhHGF has been extensively studied and documented, with results suggesting that this molecule could be a powerful tool toward increased success in hepatic regenerative therapy. However, the peptidic nature of this drug presents several challenges toward its effective administration and targeting. The possibility of encapsulating rhHGF in dextran sulfate/chitosan nanoparticles to allow its oral administration and direct liver-targeting is discussed in this manuscript. Details of a rapid and simple method for the preparation of such rhHGF-loaded nanocomplexes are presented. Beyond the practical aspects of the method, characterization techniques and main experimental features of obtained nanocarriers are also briefly analyzed and discussed.

Key words: Hepatocyte growth factor, Dextran sulfate, Chitosan, Insulin, Oral administration, Nanotechnology, Regenerative medicine

1. Introduction

Hepatocyte growth factor (HGF), or scatter factor (SF), is a heterodimeric protein produced by nonparenchymal cells with 82–85 kDa, composed of a 69 kDa α-subunit (containing four kringle domains) and a 34 kDa β-subunit, linked by a disulfide bond (1, 2). It possesses mitogenic, motogenic, and morphogenic activity, being implicated in organogenesis and tissue regeneration. Among others, HGF has been long recognized as a cytokine with potential therapeutic application in enhancing liver regeneration

Melba Navarro and Josep A. Planell (eds.), *Nanotechnology in Regenerative Medicine: Methods and Protocols*, Methods in Molecular Biology, vol. 811, DOI 10.1007/978-1-61779-388-2_8,

following hepatic injury or transplantation of small size grafts (3). Since its original isolation in 1984 several in vitro and animal in vivo studies have confirmed the beneficial effects of HGF in liver regeneration after surgical treatment (liver transplant and partial resection) or in treating liver diseases such as acute and fulminant hepatitis, septic liver failure, ischemic injury, cirrhosis, liver fibrosis, fatty liver, or liver injury related with Wilson's disease (reviewed in (4–6)). Also, the combined meta-analysis of several human clinical trials seems to indicate that HGF may be safe and effective for subacute and early stage liver failure therapy (7).

Although mostly recognized as a potent mitogen for hepatocytes implicated in liver regeneration, HGF is able to elicit different responses in different cells and tissues. The observed effects can be benign, beneficial or even harmful, being of particular concern the role of HGF as a regulator of carcinogenesis and cancer invasion and metastasis (8–11). These events are related with the expression of HGF receptor, Met tyrosine kinase, in many different tissues and cell lines, which can be freely accessible when HGF is administered intravenously, subcutaneously, intraperitoneally, or intramuscularly as a conventional injectable solution. One possible strategy that has been tested to circumvent or at least minimize unwanted effects at sites other than the hepatic tissue is delivering HGF directly to the liver, usually by administering an injectable solution through the portal vein. Pharmacokinetic studies in rats showed that the direct administration of recombinant human HGF (rhHGF) in the portal vein elicits higher liver accumulation and reduced distribution of the drug to other body sites, when compared to the intravenous administration of the same dose (0.1 mg/kg) (12). Additionally, local delivery of HGF may also reduce the total required dose for pharmacological activity which can represent an interesting economical advantage. Studies in dogs evidenced that doses of HGF over 100-fold less than those used by the intravenous route were as effective in enhancing regeneration following major hepatic resection (13, 14). However, invasive procedures of HGF administration present substantial hazard or are not realistic for most situations where this growth factor could be beneficial. Alongside the lack of targeted hepatic delivery, other disadvantage of administering HGF in solution is the inability of providing sustained amounts of the drug; this difficulty can be achieved by suitable formulation of sustained release systems incorporating HGF (15).

Nanotechnology resources in regenerative medicine have included nanoparticles for drug delivery, specially targeting and controlled delivery of biomolecules fundamental to regeneration and differentiation of tissues (16). Nanotechnology approaches to delivery systems can enhance the effect of therapeutic agents, such as cytokines, e.g., HGF. In one recent study (17), Li et al. incorporated this cytokine in sterically stabilized liposomes bearing a cyclic Arg-Gly-Asp peptide, this modification allowing targeting the liver by specific recognition of collagen type VI receptors present in hepatic stellate cells. In vivo results from a rat model demonstrated that these nanoparticulate systems were able to improve the efficiency of HGF in reducing chemically induced liver cirrhosis. However, the need for intraperitoneal administration of liposomes may be a strong setback for future clinical application of these nanosystems. Nanoparticles are also known to enable the intracellular delivery of molecules and the possibility of reaching targets that are usually inaccessible due to pharmacokinetic drawbacks. Examples of nanoparticles for delivery systems include polymeric nanoparticles, solid lipid nanoparticles, liposomes, micelles, and dendrimers. Biodegradable polymers are the most commonly used materials in drug delivery. Both biocompatible natural (18, 19) and synthetic (16) polymers can be used to formulate nanocarriers that exhibit different release properties. One of the most advantageous features of these systems is the depot capacity of bioactive drugs they provide. Once in the local of action, the release of the incorporated drugs can be tailored at very low rate, which suits potent pharmacologic molecules such as HGF, thus requiring few quantities of active molecules to exert physiologic effect.

Oral absorption of nanoparticles is today a well-known and accepted phenomenon, resulting in its passage across the intestinal barrier to blood, lymph, and other tissues. The nanoparticles must first adhere to the surface of the intestine and, only then, cross the intestinal wall due to intracellular uptake by the absorptive cells of the intestine or paracellular uptake or uptake by the M cells of the Peyer's patches (20). However, this last lymphatic absorption mechanism avoids presystemic metabolism by the liver (hepatic first-pass effect) since drugs are not drained to liver but rather to the post-liver blood circulation. But when drugs intended for hepatic action are considered, portal route that takes blood to the liver is preferable. Thus, hydrophilic, mucoadhesive nanoparticles made of natural polymers such as chitosan are preferable to avoid specific M cells targeting.

One suitable approach to passively target the hepatic tissue is administering HGF by the oral route, especially formulated into nanoparticulate systems. Indeed, the normal physiology of intestinal tract implies that absorbed molecules are carried by the hepatic portal system directly to the liver, providing a passive target delivery strategy. However, as for other proteins and peptides, HGF is destroyed by the acidic environment and trypsin content of gastrointestinal fluids (1, 2). In the last decade nanotechnology-based systems have been intensively studied for their ability to deliver proteins and peptides of therapeutic value by the oral route (21). Our group has been engaged for some time now in this last approach being particular emphasis given to the development and characterization of protein or peptide-loaded polysaccharides nanoparticles and microparticles for oral drug administration (22–27).

Various systems have been shown promising due to their ability to modulate drug release at different pH values, protect labile molecules from external harsh conditions, provide mucoadhesive properties and enhance intestinal penetration of peptide/protein-based drugs. In vivo results obtained for different polyelectrolyte nanocomplexes containing a model peptidic drug, i.e., insulin, appear to support the rationale behind this strategy, making it an innovative and promising subject of study (28, 29).

In this chapter we present a procedure for the preparation of dextran sulfate/chitosan polyelectrolyte nanocomplexes containing recombinant human hepatocyte growth factor (rhHGF) to be used in liver regeneration therapy by the oral route. Additionally, a brief discussion of characterization methodologies is included. When appropriate, parallelism between the described protocol (including characterization methodologies) and previously published work by our group is established.

2. Materials

2.1. Reagents

1. Low-molecular-weight chitosan (Chit) (≈50 kDa), 85% deacetylated (Sigma-Aldrich, St. Louis, MO, USA) (see Note 1).
2. Oligochitosan (MW ≈ 5 kDa) (Sigma-Aldrich, St. Louis, MO, USA) (see Note 1).
3. High-molecular-weight dextran sulfate (DS) (≈500 kDa) (pK Chemicals A/S, Køge, Denmark) (see Note 1).
4. Low-molecular-weight dextran sulfate (≈8 kDa) (Sigma-Aldrich, St. Louis MO, USA) (see Note 1).
5. Recombinant human hepatocyte growth factor (>98% purity) (Invitrogen, Camarillo, CA, USA) (see Notes 2 and 3).
6. Glacial acetic acid (Merck KGaA, Darmstadt, Germany).
7. HGF Human ELISA Kit (Invitrogen, Camarillo, CA, USA) (see Note 4).
8. Millipore #2 paper filters (Millipore, Billerica, MA, USA).
9. DTS1060 disposable zeta cell (Malvern Instruments Ltd, Worcestershire, UK).

2.2. Equipment

1. 50-ml and 100-ml glass beakers.
2. Cylindrical magnetic stirrer bars (2.9 cm length, 0.6 cm section diameter).
3. 25-ml screw capped cylindrical bottles (7.0 cm height, 3.3 cm diameter).
4. 30-ml centrifuge tubes.
5. 5-ml semistoppered glass vials with slotted rubber closures.

6. SM26 magnetic stirrer hotplate (Stuart Scientific, Staffordshire, UK).
7. Mini-S 840 peristaltic pump (Ismatec SA, Labortechnik-Analytik, Glattbrugg, Switzerland).
8. Accu-Rated™ PVC peristaltic pump tubing, 455 mm length, 1.52 mm internal diameter, 3.2 mm outside diameter, flow rate 1.4 ml/min (Ormantine USA Ltd., Palm Bay, FL, USA).
9. PowerWaveX microplate spectrophotometer (BioTek, Winooski, VT, USA).
10. Lyoquest freeze-dryer (Telstar, Madrid, Spain).
11. Zetasizer Nano ZS (Malvern Instruments Ltd, Worcestershire, UK).
12. FEI Quanta 400 FEG Scanning Electron Microscope (SEM) (FEI Company, Hillsboro, OR, USA).
13. Bomem IR-spectrometer (ABB Bomem, Quebec City, Canada).

3. Methods

3.1. Preparation of Solutions

3.1.1. rhHGF Solution

1. Centrifuge the vial of rhHGF (10 μg) at 2,000 × *g* for 10 min at 4°C to bring the contents to the bottom.
2. Reconstitute the vial with 100 μl of water (see Note 5) and gently shake to yield a 0.1% (w/v) solution.
3. Store at 2–8°C for up to 1 week or freeze at −20°C for greater amounts of time before usage (thaw at room temperature when needed).

3.1.2. 0.3% (w/v) DS Solution, pH 3.2

1. Sprinkle 150 mg of DS slowly over 50 ml of water under gentle magnetic stirring.
2. Adjust the pH of DS solution to 3.2 by using 0.1 M HCl or 0.1 M NaOH.
3. Cover the beaker with parafilm to avoid water evaporation and stir overnight.
4. Filter by Millipore #2 paper filter and store at 2–8°C until use.

3.1.3. 0.2% (w/v) Chit Solution, pH 5.0

1. Sprinkle 100 mg of Chit slowly over 50 ml of 1% (w/v) acetic acid solution under gentle magnetic stirring.
2. Cover the beaker with parafilm to avoid water evaporation and stir overnight.
3. Adjust the pH of Chit solution to 5.0 by using 0.1 M NaOH.
4. Filter by Millipore #2 paper filter and store at 2–8°C until use.

3.2. Preparation of Nanoparticles

DS/Chit nanoparticles are prepared by polyelectrolyte complexation by using a custom-made apparatus as depicted in Fig. 1. The production protocol comprises the following steps (see Notes 6, 10, 11 and 12):

1. Add 10–100 μl of rhHGF solution (corresponding to 1–10 μg of rhHGF) to 10 ml of 0.3% (w/v) DS solution, pH 3.2 and mix under magnetic stirring (600 rpm for 5 min).
2. Add dropwise (1 ml/min) 10 ml of 0.2% (w/v) Chit solution, pH 5.0 to the previous solution under magnetic stirring (600 rpm).
3. Stir for an additional 15 min at 600 rpm (curing time).

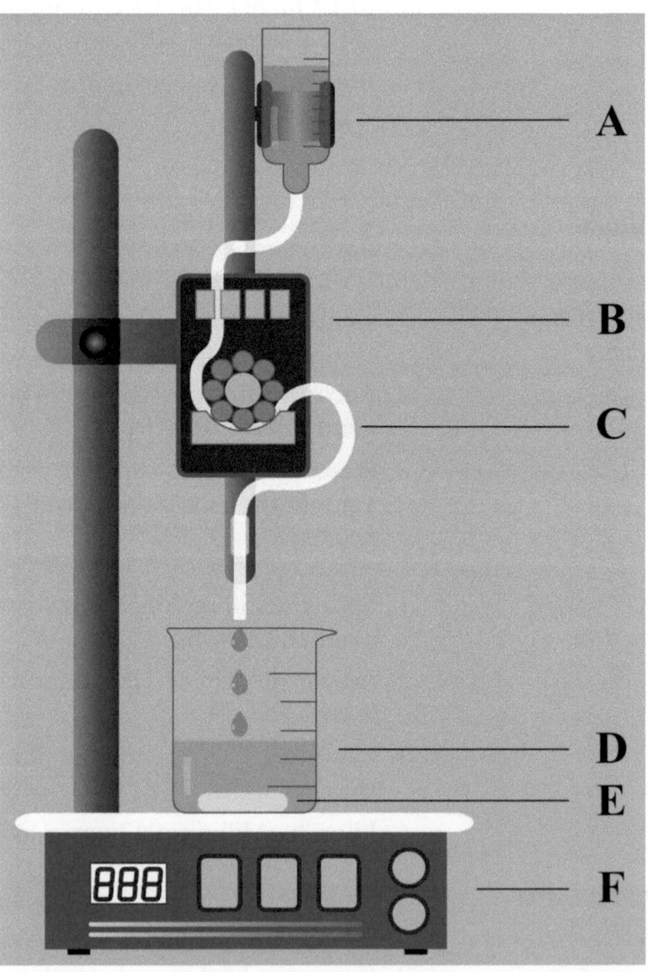

Fig. 1. Schematic representation of the custom-made apparatus used in the preparation of rhHGF-loaded DS/Chit nanoparticles. The apparatus comprises the following components: (A) plastic barrel containing 10 ml 0.2% Chit solution pH 5.0, (B) peristaltic pump, (C) tubing (455 mm × 1.52 mm ∅), (D) beaker containing 10 ml 0.3% DS solution pH 3.2 and rhHGF, (E) magnetic stirring bar, and (F) magnetic stirrer.

4. Place the nanoparticle suspension in a 30-ml centrifuge tube and collect nanoparticles by centrifugation at 17,000 rpm (50,000 × *g*) for 45 min at 4°C.
5. Collect and reserve supernatant for rhHGF determination (see below for details).
6. Store recovered nanoparticles at 2–8°C or freeze-dry (see Note 7) until further use.

3.3. Nanoparticle Characterization

Several techniques have been found useful for assessing drug-loaded nanocarrier properties, in particular those for protein and peptide delivery (30, 31). In this section, a brief list and discussion of critical characterization tests is provided.

3.3.1. Determination of rhHGF-Loaded Nanoparticles Association Efficiency and Loading Capacity

1. A commercially available ELISA kit intended for the in vitro quantitative determination of human HGF in human serum, EDTA plasma, buffered solution, or cell culture medium is used to indirectly determine the amount of associated rhHGF (minimum detectable dose = 20 pg/ml).
2. Use the ELISA kit according to manufacturer's instructions (32).
3. Assay whole or diluted supernantants obtained after nanoparticle centrifugation (see above for details) for rhHGF.
4. Calculate the total amount of associated protein by the following equation:

$$\mathrm{AE\%} = \frac{\text{Initial amount of rhHGF}\,|\,\text{Free rhHGF in supernatant}}{\text{Initial amount of rhHGF}} \times 100$$

   where AE% is the association efficiency in percentage of rhHGF.

5. Calculate loading capacity percentage (LC%) as follows:

$$\mathrm{LC\%} = \frac{\text{Initial amount of rhHGF}\,|\,\text{Free rhHGF in supernatant}}{\text{Total weight of nanoparticles}} \times 100$$

   Total weight of nanoparticles is assessed by freeze-drying an aliquot (approximately 50 mg of the hydrated pellet) of hydrated nanoparticles obtained after isolation.

3.3.2. Determination of Nanoparticle Size and Zeta Potential

1. Determine nanoparticle hydrodynamic radius and zeta potential by dynamic light scattering (DLS) using a Zetasizer Nano ZS.
2. Diluted samples (1:10) with aqueous 0.9% KCl solution.
3. Measure samples in triplicate at a scattering angle of 173° and 25°C, using previously activated and water-flushed DTS1060 disposable zeta cell. Produced nanoparticles must preferentially present mean particle size in the range of 500 nm and negative zeta potential of around –30 mV.

3.3.3. Nanoparticle Morphology by Scanning Electronic Microscopy

1. Mount samples of nanoparticle dispersions on aluminum stubs.
2. Air-dried nanoparticles before being observed.
3. Observe nanoparticles using an accelerating voltage of 10 kV, without any coating.
4. Figure 2 presents SEM microphotograph of DS/Chit nanoparticles obtained by the described protocol, evidencing clustered (due to Chit bridging), spherical-like particles with smooth surfaces, typical of this type of systems (24, 25). Nanoparticles must present unimodal size in agreement with DLS analysis (around 500 nm).

3.3.4. In Vitro Release Profile of rhHGF

The release profile of proteins from nanocomplexes is assessed by mimicking the natural gastrointestinal pH pathway, from the highly acidic pH of the stomach to the nearly neutral pH of the proximal intestine (see Notes 11). Figure 3 provides an example of the release profile for DS/Chit nanoparticles containing a model peptidic drug (insulin), as previously reported (28). The protocol allows obtaining biorelevant data concerning the oral administration of developed nanosystems:

1. Place 200 mg of nanoparticles collected after centrifugation into 25 ml screw capped cylindrical bottles containing 20 ml hydrochloride acid buffer at pH 1.2 (USP30-NF25) at 37°C (±1°C) (see Notes 8 and 9).

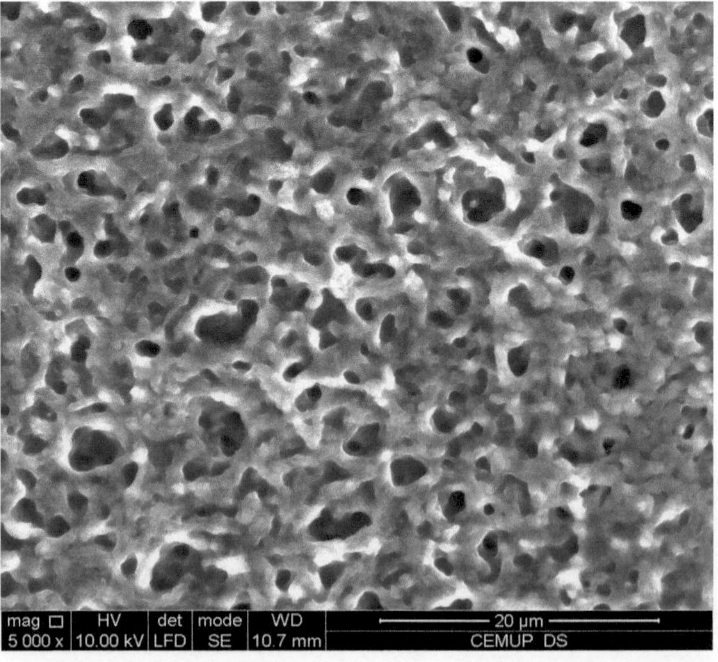

Fig. 2. Typical morphology of DS/Chit nanoparticles as obtained by SEM.

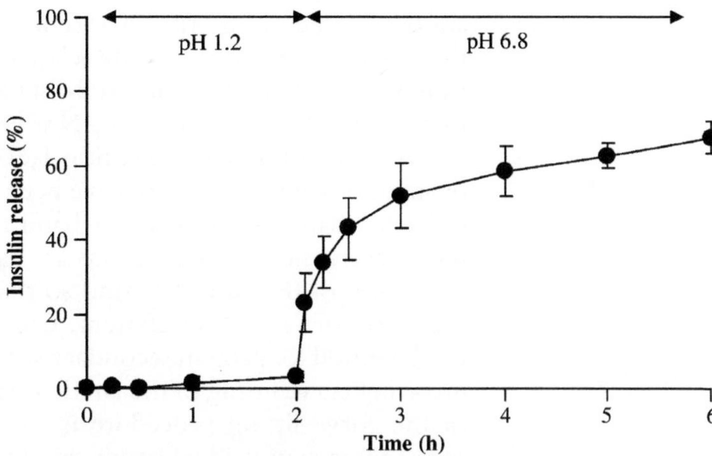

Fig. 3. Insulin release from insulin-loaded DS/Chit nanoparticles produced with a DS:Chit mass ratio of 1.5:1 in simulated gastric media (HCl buffer, pH 1.2) for 2 h followed by additional 4 h in simulated intestinal media (phosphate buffer, pH 6.8) at 37°C ± 1°C (mean ± SD, $n = 3$). Reprinted with permission from ref. 27. Copyright 2007 American Chemical Society.

2. Incubate under magnetic stirring (100 rpm) for 2 h.
3. At predetermined times, collect 0.4 ml samples and separate supernatant from nanoparticles by centrifugation (20,000 × *g* for 15 min, at 4°C).
4. Restore initial volume with 0.4 ml of fresh medium after each sample collection.
5. At 2 h, collect the total amount of medium and separate supernatant from nanoparticles by centrifugation (20,000 × *g* for 15 min, at 4°C).
6. Place the nanoparticles collected after centrifugation into 25 ml screw capped cylindrical bottles containing 20 ml phosphate buffer at pH 6.8 (USP30-NF25) at 37°C (±1°C).
7. Incubate under magnetic stirring (100 rpm) for an additional 4 h.
8. Proceed as for steps 3 and 4.
9. Determine drug content in the supernatant samples using the ELISA kit.

3.3.5. Fourier Transform Infrared (FTIR) Spectroscopy to Assess Structural Disposition of rhHGF

FTIR spectroscopy can provide information about the secondary structure content of proteins (33). The lost of protein structure that may take place during formulation can lead to the lost of activity, thus, the evaluation of protein arrangement and its preservation are of fundamental importance to the biological activity. The types of secondary structure include the α-helices and β-sheets, which allow the amides to hydrogen bond very efficiently with one another. Amide I absorption arise from the amide bonds that link the amino acids and is directly related to the backbone conformation with a major contribution from C=O stretching vibration and minor contribution from the C–N stretching vibration. Absorption for this band occurs in the region 1,600–1,700 cm^{-1}. Studies with proteins of known structure have been used to correlate systematically the shape of the Amide I band to secondary structure content. Several numerical methods are used to increase the apparent resolution of the Amide I band so that estimates can be made of the secondary structure content. Curve fitting is the most widely used method for protein secondary structure quantification, mainly involving curve fitting of the amide I band (34). The basic principle of the curve fitting procedure is to resolve the original protein spectrum into individual bands that fit the spectrum (33).

IR-spectra are measured using a Bomem IR-spectrometer (Bomem, Quebec City, Canada). rhHGF spectra of free rhHGF and encapsulated rhHGF are obtained according to a double subtraction procedure (35) and unloaded nanoparticles and water vapor spectra are collected under identical conditions for blank subtraction. The second derivative spectra are obtained with a seven-point Savitsky–Golay derivative function and the baseline is corrected using a three to four point adjustment. In addition, the spectra are area-normalized in the amide I region from 1,710 to 1,590 cm^{-1} using the Bomem-GRAMS software (Galactic Industries, Salem, NH) and area-overlap compared to an rhHGF aqueous standard solution by using an appropriate software (Origin® software). All samples are run in triplicate and the data presented are the average of three measurements.

4. Notes

1. Other suppliers of Chit and DS may offer different molecular weights and/or degrees of deacetylation. Such variables may interfere on the final properties of nanoparticles and thus carefully evaluated.
2. Used rhHGF (>98%) is produced in baculovirus-infected High-5 cells and purified sequential chromatography, possessing molecular weight of 80 kDa (an α-chain with 463 amino acid residues and a β-chain with 234 amino acid residues) and endotoxin content of less than 0.1 ng/μg.
3. Although the described method is considered optimized for rhHGF from the supplier stated in the materials section, this protein is also available from other suppliers such as Sigma-Aldrich, Inc. (St. Louis, MO, USA), R&D Systems (Minneapolis, MN, USA), PeproTech Inc. (Rocky Hill, NJ),

and ProSpec-Tany TechnoGene Ltd. (Rehovot, Israel). Therefore, differences among products from different suppliers, namely, rhHGF physical–chemical properties (e.g., isoelectric point, molecular weight, or origin), other added excipients, and storage and handling conditions for optimal stability, should be taken into account and eventually justify minor adjustments to the described preparation procedure.

4. ELISA kits for rhHGF assay are also available from other suppliers such as B-Bridge International, Inc. (Mountain View, CA, USA), R&D Systems (Minneapolis, MN, USA) and Otsuka Pharmaceutical Co. (Tokushima, Japan). Precaution should be taken when comparing different kits since significantly different results may be observed. See ref. 36 for more guidance on rhHGF ELISA kit selection.
5. Type I ultrapure water with a resistivity of 18.2 MΩ cm and total organic content of less than five parts per billion should be used. This standard is referred to as "water" throughout this manuscript and is produced in-house by means of a Simplicity® UV Ultrapure Water System (Millipore, Billerica, USA).
6. Allow all solutions attaining room temperature (20–25°C) if previously stored under refrigeration (2–8°C).
7. Freeze-drying of obtained nanoparticles by dispensing the remaining pellet, with or without cryoprotectives (trehalose or sucrose corresponding to 10–30% of the total weight of nanoparticles), in 5-ml semistoppered glass vials with slotted rubber closures and add up to 2 ml of water, followed by freezing during 24 h on the shelves of the lyophilization chamber at its minimum temperature (–85°C). Sublimation lasts 48 h at a vacuum pressure of 4×10^{-5} atm and without heating, being maintained at the condenser surface temperature of –60°C. Finally, glass vials are sealed under anhydrous conditions and stored until being rehydrated by using the same initial volume (1 ml) of water.
8. In vitro release profile studies must be performed under sink conditions maintaining a volume of dissolution media that is five to ten times greater than the volume at the saturation point of the drug contained in the drug delivery system being tested. The maximum 10 μg of rhHGF per 20 ml of release medium guarantees those conditions.
9. Temperature maintenance (37°C ± 1°C) during the in vitro release profile studies is obtained using a thermostatic water bath.
10. This protocol can be adapted for other therapeutic molecules, either peptidic in nature or not, and for other delivery routes (25, 28).

11. Because of the high cost of rhHGF, it is advisable to proceed with previous experiments using other peptide/protein models to become familiar with the protocol and optimize several parameters (concentration of DS and Chit solutions, DS:Chit mass ratio, polymer:protein mass ratio, pH of polymer solutions, and curing time) (25). Our experience indicates that insulin provides good hints for preliminary development of rhHGF-loaded DS/Chit nanoparticles. However, molecules with higher similarity with rhHGF, namely, for isoelectric point (p*I* of HGF ≈ 9.5 (37)) or molecular weight, may provide further insights.
12. Unless specified, all procedures are performed at room temperature (20–25°C) (25, 28).

Acknowledgments

The authors acknowledge the financial support from Fundação para a Ciência e a Tecnologia, Portugal (PTDC/SAU-FCF/104492/2008 and SFRH/BPD/35996/2007).

References

1. Nakamura, T., Nawa, K., Ichihara, A., Kaise, N., and Nishino, T. (1987) Purification and subunit structure of hepatocyte growth factor from rat platelets, *FEBS Lett* **224**, 311–316.
2. Nakamura, T. (1991) Structure and function of hepatocyte growth factor, *Prog. Growth Factor Res* **3**, 67–85.
3. Nakamura, T., Nishizawa, T., Hagiya, M., Seki, T., Shimonishi, M., Sugimura, A., Tashiro, K., and Shimizu, S. (1989) Molecular cloning and expression of human hepatocyte growth factor, *Nature* **342**, 440–443.
4. Funakoshi, H., and Nakamura, T. (2003) Hepatocyte growth factor: from diagnosis to clinical applications, *Clin Chim Acta* **327**, 1–23.
5. Mizuno, S., and Nakamura, T. (2007) Hepatocyte growth factor: a regenerative drug for acute hepatitis and liver cirrhosis, *Regen Med* **2**, 161–170.
6. Taub, R. (2004) Liver regeneration: from myth to mechanism, *Nat Rev Mol Cell Biol* **5**, 836–847.
7. Cui, Y. L., Meng, M. B., Tang, H., Zheng, M. H., Wang, Y. B., Han, H. X., and Lei, X. Z. (2008) Recombinant human hepatocyte growth factor for liver failure, *Contemp Clin Trials* **29**, 696–704.
8. Jiang, W., Hiscox, S., Matsumoto, K., and Nakamura, T. (1999) Hepatocyte growth factor/scatter factor, its molecular, cellular and clinical implications in cancer, *Crit Rev Oncol Hematol* **29**, 209–248.
9. Jiang, W. G., Martin, T. A., Parr, C., Davies, G., Matsumoto, K., and Nakamura, T. (2005) Hepatocyte growth factor, its receptor, and their potential value in cancer therapies, *Crit Rev Oncol Hematol* **53**, 35–69.
10. Pan, W., Yu, Y., Yemane, R., Cain, C., Yu, C., and Kastin, A. J. (2006) Permeation of hepatocyte growth factor across the blood-brain barrier, *Exp Neurol* **201**, 99–104.
11. Gong, R. (2008) Multi-target anti-inflammatory action of hepatocyte growth factor, *Curr Opin Investig Drugs* **9**, 1163–1170.
12. Ido, A., Moriuchi, A., Kim, I., Numata, M., Nagata-Tsubouchi, Y., Hasuike, S., Uto, H., and Tsubouchi, H. (2004) Pharmacokinetic study of recombinant human hepatocyte growth factor administered in a bolus intravenously or via portal vein, *Hepatol Res* **30**, 175–181.
13. Ueno, S., Aikou, T., Tanabe, G., Kobayashi, Y., Hamanoue, M., Mitsue, S., Kawaida, K., and Nakamura, T. (1996) Exogenous hepatocyte growth factor markedly stimulates liver regeneration following portal branch ligation in dogs, *Cancer Chemother Pharmacol* **38**, 233–237.

14. Kobayashi, Y., Hamanoue, M., Ueno, S., Aikou, T., Tanabe, G., Mitsue, S., Matsumoto, K., and Nakamura, T. (1996) Induction of hepatocyte growth by intraportal infusion of HGF into beagle dogs, *Biochem Biophys Res Commun* **220**, 7–12.
15. Oe, S., Fukunaka, Y., Hirose, T., Yamaoka, Y., and Tabata, Y. (2003) A trial on regeneration therapy of rat liver cirrhosis by controlled release of hepatocyte growth factor, *J Control Release* **88**, 193–200.
16. Engel, E., Michiardi, A., Navarro, M., Lacroix, D., and Planell, J. A. (2008) Nanotechnology in regenerative medicine: the materials side, *Trends Biotechnol* **26**, 39–47.
17. Li, F., Sun, J.-Y., Wang, J.-Y., Du, S.-L., Lu, W.-Y., Liu, M., Xie, C., Shi, J.-Y. (2008) Effect of hepatocyte growth factor encapsulated in targeted liposomes on liver cirrhosis, *J Control Release* **131**, 77–82.
18. Dang, J. M., and Leong, K. W. (2006) Natural polymers for gene delivery and tissue engineering, *Adv Drug Deliv Rev* **58**, 487–499.
19. Malafaya, P. B., Silva, G. A., and Reis, R. L. (2007) Natural-origin polymers as carriers and scaffolds for biomolecules and cell delivery in tissue engineering applications, *Adv Drug Deliv Rev* **59**, 207–233.
20. Ponchel, G., and Irache, J. (1998) Specific and non-specific bioadhesive particulate systems for oral delivery to the gastrointestinal tract, *Adv Drug Deliv Rev* **34**, 191–219.
21. Soares, A. F., Carvalho Rde, A., and Veiga, F. (2007) Oral administration of peptides and proteins: nanoparticles and cyclodextrins as biocompatible delivery systems, *Nanomed* **2**, 183–202.
22. Sarmento, B., Ferreira, D., Veiga, F., and Ribeiro, A. (2006) Characterization of insulin-loaded alginate nanoparticles produced by ionotropic pre-gelation through DSC and FTIR studies, *Carbohyd Polym* **66**, 1–7.
23. Sarmento, B., Ferreira, D. C., Jorgensen, L., and van de Weert, M. (2007) Probing insulin's secondary structure after entrapment into alginate/chitosan nanoparticles, *Eur J Pharm Biopharm* **65**, 10–17.
24. Sarmento, B., Martins, S., Ribeiro, A., Veiga, F., Neufeld, R., and Ferreira, D. (2006) Development and comparison of different nanoparticulate polyelectrolyte complexes as insulin carriers, *Int J Pept Res Ther* **12**, 131–138.
25. Sarmento, B., Ribeiro, A., Veiga, F., and Ferreira, D. (2006) Development and characterization of new insulin containing polysaccharide nanoparticles, *Colloids Surf B Biointerfaces* **53**, 193–202.
26. Sarmento, B., Ribeiro, A. J., Veiga, F., Ferreira, D. C., and Neufeld, R. J. (2007) Insulin-loaded nanoparticles are prepared by alginate lonotropic pre-gelation followed by chitosan polyelectrolyte complexation, *J Nanosci Nanotech* **7**, 2833–2841.
27. Martins, S., Sarmento, B., Souto, E. B., and Ferreira, D. C. (2007) Insulin-loaded alginate microspheres for oral delivery - Effect of polysaccharide reinforcement on physicochemical properties and release profile, *Carbohyd Polym* **69**, 725–731.
28. Sarmento, B., Ribeiro, A., Veiga, F., Ferreira, D., and Neufeld, R. (2007) Oral bioavailability of insulin contained in polysaccharide nanoparticles, *Biomacromolecules* **8**, 3054–3060.
29. Sarmento, B., Ribeiro, A., Veiga, F., Sampaio, P., Neufeld, R., and Ferreira, D. (2007) Alginate/chitosan nanoparticles are effective for oral insulin delivery, *Pharm Res* **24**, 2198–2206.
30. Hall, J. B., Dobrovolskaia, M. A., Patri, A. K., and McNeil, S. E. (2007) Characterization of nanoparticles for therapeutics, *Nanomed* **2**, 789–803.
31. Peltonen, L., and Hirvonen, J. (2008) Physicochemical characterization of nano- and microparticles, *Curr Nanosci* **4**, 101–107.
32. Human HGF ELISA Kit Manual. Invitrogen Corporation (Camarillo, CA, USA). Available from URL: http://tools.invitrogen.com/content/sfs/manuals/KAC2211%20(pr072)%20Hu%20HGF%20(BSE)Rev%201.0.pdf [Accessed 2009 Aug 15].
33. Van de Weert, M., Hering, J. A., and Haris, P. I. (2005) Fourier Transform Infrared Spectroscopy, In *Methods for Structural Analysis of Protein Pharmaceuticals* (Jiskoot, W., and Crommelin, D., Eds.), pp 131–166, AAPS Press, Arlington, VA.
34. Jorgensen, L., Vermehren, C., Bjerregaard, S., and Froekjaer, S. (2003) Secondary structure alterations in insulin and growth hormone water-in-oil emulsions, *Int J Pharm* **254**, 7–10.
35. Dong, A., Huang, P., and Caughey, W. S. (1990) Protein secondary structures in water from second-derivative amide I infrared spectra, *Biochemistry (Mosc)* **29**, 3303–3308.
36. Fukuta, K., Adachi, E., Matsumoto, K., and Nakamura, T. (2009) Different reactivities of enzyme-linked immunosorbent assays for hepatocyte growth factor, *Clin Chim Acta* **402**, 42–46.
37. Gherardi, E., Gray, J., Stoker, M., Perryman, M., and Furlong, R. (1989) Purification of scatter factor, a fibroblast-derived basic protein that modulates epithelial interactions and movement, *Proc Natl Acad Sci U S A* **86**, 5844–5848.

Chapter 9

Electrospinning Technology in Tissue Regeneration

Oscar Castaño, Mohamed Eltohamy, and Hae-Won Kim

Abstract

Electrospinning is one of the most versatile and effective tools to produce nanostructured fibers in the biomedical science fields. The nanofibrous structure with diameters from tens to hundreds of nanometers largely mimics the native extracellular matrix (ECM) of many tissues. Thus far, a range of compositions including polymers and ceramics and their composites/hybrids have been successfully applied for generating electrospun nanofibers. Different processing tools in electrospinning set-ups and assemblies are currently developed to tune the morphology and properties of nanofibers. Herein, we demonstrate the electrospinning process and the electrospun biomaterials for specific use in tissue regeneration with some examples, involving different material combinations and fiber morphologies.

Key words: Electrospinning, Nanostructured fibers, Nanofibers, Polymer, Composites, Ceramic, Tissue regeneration

1. Introduction

In this chapter, electrospinning methodology for the production and nanostructuration of micro and nanofibers is described. Historically, fiber is a term that comes from the Latin term *fibra* or *fillum*, which means thread. Here, fiber will be taken to mean a continuous filament or discrete elongated piece of a particular material. Micro and nano, taken from ancient Greek, with the prefix micro (μικρός) mean small and the prefix nano (νᾶνος) means dwarf, and denote 10^{-6} and 10^{-9}, respectively.

Although the electrospinning process was patented by Cooley and Morton in 1902 (1, 2) and several modification by Formhals (3), charge-fluids attraction phenomena were first described in the sixteenth century by the English physician William Gilbert, who reported the observation of a cone formation when a rubbed piece of amber was held above a spherical drop of water on a dry surface

Melba Navarro and Josep A. Planell (eds.), *Nanotechnology in Regenerative Medicine: Methods and Protocols*, Methods in Molecular Biology, vol. 811, DOI 10.1007/978-1-61779-388-2_9, © Springer Science+Business Media, LLC 2012

(4). In 1732, Gray and Mortimer (5) published a detailed description of the effects of electrostatic charges on water. In 1882, Lord Rayleigh described a theoretical evaluation of the limit of charge that a liquid droplet could be exposed without the formation of a jet. This value was known as "Rayleigh limit" (6). Following steps increasing the knowledge of the technique were performed by Larmor (7), Wilson and Taylor (8), Zeleny (9, 10), Nolan (11) and Macky (12). Additionally, in 1964, Taylor described the basis of the electrofluidic process by demonstrating through a theoretical treatment that a conductive fluid can remain in equilibrium when it is exposed to an electrical field (13).

In the last two decades, the electrospinning process, a versatile mean of nanotechnology, has opened a door to new approaches due to the possibility of the production of nanofibers of medical-available materials, ranging bioactive materials, polymers, biological entities or composites, for different applications, especially in tissue engineering (14, 15).

Chemical and biological properties of meshes prepared by electrospinning depend not only on the inherent properties of the deposited material, but also on the texture and size of the fibers. The degradation of fibers in fluid is increased as the fiber size is reduced. Mechanical properties also depend on the morphology of the fibers, especially nano or micropores, as well as the interspacing between each fiber. The pore size of the electrospun mesh ranges from hundreds of nanometers to a few micrometers, which is largely dependent on the fiber size. When considering the use of electrospun fiber for cell culture substrate, those pore sizes are not expected to be enough for cell migration and penetration. Therefore, several methodologies have been exploited to produce macrosized pores within the nanofibrous network. One example is to use sodium chloride particles in the course of electrospinning which can be further eliminated in water, remaining macropores. However, some disintegrations of the fibrous structure are encountered. Nonetheless for the small-sized pores, cells have often been to migrate relatively well through the porous network, which is mainly due to the flexibility of the polymer nanofibers in response to the cell movement. As the electrospun nanofibers represent mainly thin sheet form their potential use as three-dimensional scaffolds for large-sized defect regions still remain a challenge. As the fiber size and the pores associated with it are the main extrinsic factors affecting the properties of electrospun materials, controlling the fiber size is of special importance. In this sense, the properties of slurry should be adjusted appropriately, and in most cases, the selection of proper solvent is considered important for the electrospinning of biopolymers. Table 1 presents some organic solvents and their physical properties which are generally used for the electrospinning of biopolymers.

Table 1
Typical organic solvents used and their properties for the electrospinning of biopolymers

	ε	Bp (°C)	Density (g/cm³)	Vapor pressure (kPa 20°C)
Tetrahydrofuran	7.5(27°C)	65	0.883 (25°C)	17.3
Chloroform	4.8 (25°C)	61	1.478 (25°C)	21.1
Dichloromethane	9.9 (25°C)	40	1.266 (20°C)	46.7
2,2,2-Trifluoroethanol	27.7 (25°C)	74	1.384 (20°C)	6.9
1,1,1,3,3,3-Hexafluoro-2-propanol	16.7 (25°C)	59	1.460 (20°C)	13.6
1,4-Dioxane	2.2 (25°C)	101	1.034 (20°C)	3.6

Although many electrospinning applications are focused on electrical, magnetic, and catalyst properties, in biological environments, fibers are usually prepared as a biodegradable support for cell adhesion, migration, and differentiation. Therefore, degradation properties of the electrospun nanofibers should also be considered. Compared to bulk degradable polymers, the nanofibrous form is much highly degradable because of its large surface area, which being in direct contact with water and allowing its rapid attach by diffusion. Biodegradable polymers have chains that are cleaved by hydrolysis in physiological condition, and the degradation products are eliminated from the body through metabolic pathways. Examples of biodegradable polymers are either in synthetic base such as polylactides (PLA), poly(L-lactide)(PLLA), poly(D,L-lactide) (PLDLA), polycaprolactone, polyglycolides (PGA) and their copolymers, poly(D,L-lactide-co-glycolide) (PLGA), or natural base, including collagen, gelatin, chitosan, and alginate. These polymers basically degrade into their respective monomers or oligomers via hydrolysis in aqueous environments (16). In this sense, molecular weight and crystallinity are variables that must to be taken into account. Surface properties also play a key role in obtaining a good protein adsorption prior to cell attachment.

To gain better biological responses, many research groups biofunctionalized the electrospun nanofibers by adding molecules such as drugs and growth factors, directly within the spinning solution or after modification of the electrospun fiber surface (17). This enables the nanofibers to mimic the extra cellular matrices or to elicit therapeutic functions.

1.1. Electrospinning Apparatus and Fiber Design

A typical electrospinning set-up involves a syringe with a controlled gauge size, an infusion pump to control the slurry flow rate that exits from the tip, a high-voltage power supply ranging from 0 to 30 or 60 kV, and a grounded collector that consists of a flat metallic surface or a metallic rotary mandrel.

The electrospinning process basically consists of a jet produced by applying a high voltage (typically in the order of a few to tens of kV) onto a fluid through the charged metallic tip. When the charge is higher than the fluid surface tension, it produces the aforementioned jet, which usually goes to the positive or grounded conductive collector, which is held at around 10–20 cm distance. A stationary cone originates from the meniscus on the top of the fluid before producing the fiber, and is known as Taylor cone (Fig. 1).

This phenomenon occurs because when a small fraction of a liquid with conductive properties is submitted to an electric field, the form of the fluid starts to change its shape due to the increase in its surface charge. Then, when electric field intensity increases and it forces the droplet to exceed its surface tension, a cone begins to form. The jet of liquid produced exemplifies the beginning of the electrospinning process (18).

Multiple variations in the fabrication set-up can be applied for the production of different fiber structures. The main variations are found in modifying the way of injection, type of collector, and atmosphere.

1.1.1. Injector/Nozzle

Multinozzle fiber deposition is commonly used to obtain a homogeneous distribution of textures (oriented) fibers with wider area or to allow a better control over pore size of fibers for cell migration into the meshes (19). It is possible to use different nozzles with an individual pump and flow rate or to adopt multinozzle tips at constant flow rate and voltage, and to braid them to obtain suprafiber structures (Fig. 2). Core–shell depositions involve the

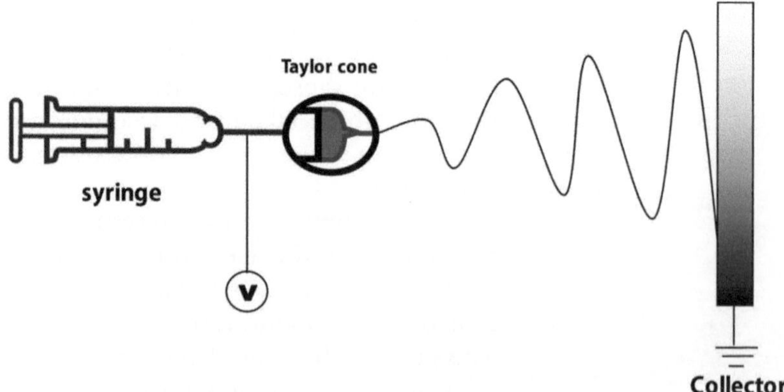

Fig. 1. Electrospinning apparatus and fiber jets from the nozzle under an electrostatic field.

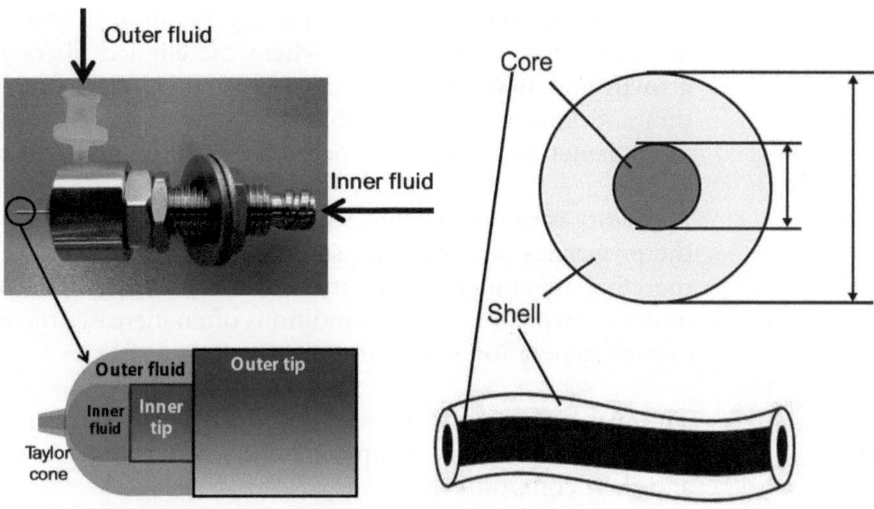

Fig. 2. Nozzle set-up for core–shell nanofiber configuration.

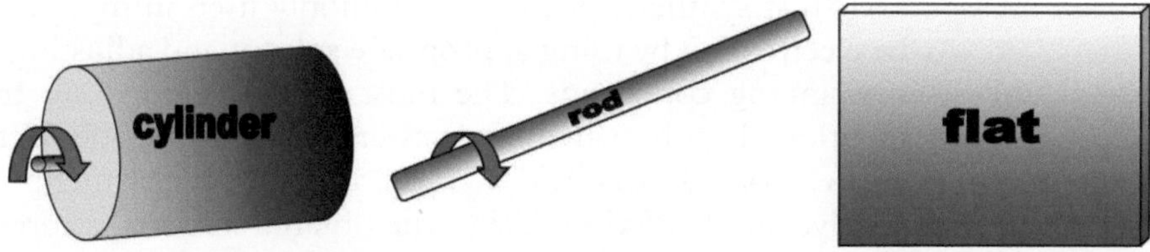

Fig. 3. Collectors used to obtain woven and nonwoven morphologies, as well as small tubes.

use of a concentric nozzle from two fluid sources to obtain a concentric joint of fibers of different natures. A combination of properties is usually involved in the use of this deposition mode. For example, to improve the mechanical flexibility of a fiber, a polymer can be used inside as a support while a bioactive compound can be on the outside part of the nozzle, which is the first portion to come into contact with the physiological fluid.

1.1.2. Collector

Other modifications can be performed, for example, the way that fibers are collected. One is the use of a grounded flat conduction surface in conjunction with a rotary drum for the production of textured meshes or a rotary rod for the production of thin tubes (Fig. 3). Rotary speed also influences the alignment of the fibers, with higher speeds leading to highly textured fibers. Moreover, to produce homogeneously distributed fibers on a collector, the collector position can be moved periodically (see Note 1).

Various collector designs, such as parallel set-up of metal pins or lines, are also possible to produce aligned fibers. The aligned fibers find some specific uses in the regeneration of tissues, such as nerve, tendon, and ligament, where the aligned fibers guide cell growth and tissue arrangement. Collecting fibers onto a rotary thin and long metal bar enables the generation of vascular grafts with diameters of a few micrometers or centimeters in size.

1.1.3. Atmosphere

Humidity is one important variable in the sense that it influences the properties of a solution, such as viscosity and surface tension, therefore affecting the resultant fiber morphology. Electrospinning under controlled moisture conditions often increases the possibility of pore generation within fibers (20).

1.2. Electrospun Biomaterials

The materials used to obtain scaffolds for tissue regeneration by the electrospinning technique are biopolymers, bioactive ceramics, and their composites/hybrids.

1.2.1. Biodegradable Polymers

Biodegradable polymers are the class of polymers that are degradable and can be applied in a biologic environment without involving toxic reactions, which can be classified into either synthetic or natural polymers.

Most synthetic polymers commonly used in medical uses can be electrospun by using appropriate solvent and adjusting the electrospinning conditions. The most well developed are the aforementioned poly-α-hydroxyl esters, including poly(lactic acid) (PLA), poly(glycolic acid) (PGA), and their copolymer poly(lactic-co-glycolide) (PLGA). Using the organic solvents, as presented in Table 1, those fibers with sizes of a few hundreds of nanometers to a few micrometers can be well produced. Polycaprolactone, which is more flexible but less degradable than PLA, has also been well studied for the production of nanofibers. Polyurethanes are also shaped into nanofibers that present very good mechanical properties and an appropriate biocompatibility (21).

Polymers that are produced in nature by organisms have been the subject of interest in the biomedical field. They are mostly degraded by specific enzymes that are present in aqueous physiological fluids (22, 23). Natural biopolymers are attractive due to their biocompatibility, sensitivity, and precise selectivity. However, they must be treated before use to eliminate any bacterial and viral contaminants present in biological solutions in order to avoid any infectious diseases. Furthermore, the production of these polymers is limited and their extraction and purification procedures are still very expensive.

Three subgroups of natural biopolymers can currently be electrospun (21):

- Proteins (polyamides) produced from amino acid monomers. Examples include collagen, elastin, casein, keratin, silk, albumin, wheat gluten, and wool.

- Polysaccharides produced from sugar monomers. Examples include cellulose, chitin (chitosan), chondroitin sulfate, hyaluronic acid, starch, glycogen, α-amylose, and amylopectin.
- Polyesters produced by micro-organisms such as polyhydroxy-alkanoate (PHA), polyhydroxybutyrate (PHB), and poly(hydroxybutyrate–hydroxyvalerate) (PHB/HV).

1.2.2. Bioactive Ceramics

The more commonly used bioactive ceramics are calcium phosphates and bioactive glasses, which are attractive for bone regeneration. Utilizing bioactive ceramics in the form of nanofibers reaps up the compositional benefit and morphological trait. Unlike polymers, these ceramics are not easy to spin into fiber form. The most common way is to use the sol–gel based solutions of the ceramic compositions. Some examples have recently been reported, which include hydroxyapatite and bioactive glasses. Kim et al. (24) successfully produced bioactive glass by electrospinning with a composition of ($70SiO_2$–$25CaO$–$5P_2O_5$) using the sol–gel process and polyvinylbutyral (PVB) as a binder polymer. The electrospun fibers need to be calcined above ~500°C to obtain a pure vitreous network and to eliminate organic residues. Previously, in vitro studies have shown that these glass nanofibers have a higher bioactivity than the same glass in bulk form due to their high-surface area.

While they show a great potential to guide tissue cells to differentiate into bone cells and direct osteoconduction, and ultimately help the regeneration of bony tissues, the most significant challenge in utilizing the bioactive ceramics is avoiding the weakness in mechanical integrity. For this, the composite approach with biopolymers has been more intensively developed.

1.2.3. Polymer–Polymer Blends

Tuning the properties such as mechanical strength and degradability is possible by blending polymers with different characteristics. Electrospun blends of PLA with PCL show better properties in terms of flexibility and degradation rate, with respect to their individual polymer.

1.2.4. Polymer–Inorganic Composites

Recently, the composite approach has become a significant issue, particularly finding good candidate matrices for hard tissue regeneration. Ideally, inorganic nanocomponents can be incorporated within the polymer matrix and generated into a nanofiber. However, the most significant problem in this process is the dispersion of the hydrophilic nanoparticles within the relatively hydrophobic polymer. To overcome this, some elegant methodologies have recently been developed. One method is to use surfactants in order to disperse hydroxyapatite nanoparticles embedded in a PLA matrix (24). The produced composite fibers showed good spinnability and maintained fiber morphology without beads formation. Moreover, bone cell response including cell proliferation and differentiation was also significantly improved on the composite fibers when compared to those on bare polymer fiber. Another strategy to obtain a homogeneous dispersion of the inorganic nanoparticles into the polymer matrix is based on the in situ precipitation of hydroxyapatite in a gelatin matrix, mimicking the bone composition. Consequently, ultrafine apatite nanocrystals were evenly distributed within the gelatin amino acids sequence, which facilitated the creation of electrospinning jets and fiber production (25). A different way to obtain homogeneous hybrid fibers without the use of inorganic particles is the xerogel approach, through sol–gel method in order to attain an organically modified glass. Kim and coworkers reported and study where gelatin and a siloxane were previously mixed and then produced into nanofiber by electrospinning. Because of the sol–gel approach, the inorganic phase was hybridized within the polymer network enabling the ease-of-spinnability without bead formation. Furthermore, the added siloxane enhanced in vitro bone bioactivity and bone cell differentiation, providing a potential and promising hybrid nanofiber for bone regeneration (26).

In this chapter, some specific examples to produce nanofibers by electrospinning, from the inorganics and polymers to their composites in random or aligned form, are described.

2. Materials

2.1. Materials for Producing Randomly Oriented PLA Nanofibers

1. PLDLA (95DL/5 L) (Purasorb, Purac).
2. Chloroform (Aldrich, >99,5%).
3. Conventional aluminum foil.
4. Glycerol (Aldrich for molecular biology, ≥99%).
5. Infusion pump KDS100 (KD Scientific).
6. Flat type SUS collector 400 × 400 mm (NanoNC).
7. High Voltage Power source, maximum output voltage:~30 kV, maximum output current:~2 mA (NanoNC).
8. Stainless steel tips of gauge of 25 and 5 in. long (EFD®).

2.2. Materials for Producing Biopolymer–Bioinorganic Composite-Oriented Nanofiber

1. Commercially obtained ceramic nanoparticles or lab-made ceramic nanoparticles by sol–gel method.
2. PLDLA (95DL/5 L) (Purasorb, Purac).
3. Chloroform (Aldrich >99.5%).
4. Conventional aluminum foil.
5. Glycerol (Aldrich for molecular biology, ≥99%).
6. Infusion pump KDS100 (KD Scientific).

7. Drum-type collector, Ø90 mm × W200 mm, 312.41 × g maximum speed (NanoNC).
8. High Voltage Power source, maximum output voltage:~30 kV, maximum output current:~2 mA (NanoNC).
9. Stainless steel tips of gauge of 25 and 5 in. long (EFD®).

2.3. Materials for Producing Core–Shell Structured Biopolymer/Hybridized Nanofiber

1. Commercially obtained siloxane or a lab-made xerogel obtained by sol–gel method.
2. PLDLA (95DL/5 L) (Purasorb, Purac).
3. Polycaprolactone average Mw ~65,000, average Mn ~42,500, pellets (Aldrich).
4. Chloroform (Aldrich >99.5%).
5. Tetrahydrofuran anhydrous, ≥99.9%, inhibitor-free (Sigma-Aldrich).
6. Conventional aluminum foil.
7. Glycerol (Aldrich for molecular biology, ≥99%).
8. Two infusion pumps KDS100 (KD Scientific).
9. Dual Concentric Nozzle, 22–30 G, stainless steel (NanoNC).
10. Stick-type collector, Ø5 mm × W200 mm, 312.41 × g maximum speed (NanoNC).
11. High Voltage Power source, maximum output voltage: ~30 kV, maximum output current:~2 mA (NanoNC).
12. Stainless steel tips of gauge of 25 and 5 in. long (EFD®).

3. Methods

3.1. Method for Producing Randomly Oriented PLA Nanofibers

In this case, a PLA must be dissolved in organic solvents of low dielectric constant, such as chloroform ($CHCl_3$) at 3 wt%. A continuous and stationary deposition is achieved following the protocol hereafter:

1. Dissolve poly(D,L-lactide) (PLDLA) in a relative nonpolar and volatile solvent such as chloroform. Concentration should be around 3 wt%. Once PLDLA is added on $CHCl_3$, 2 h of vigorous stirring at around 40°C are needed (see Note 2).
2. An aluminum foil will be used in this case to collect a nonwoven sample. Usually, nanofiber meshes are very sticky and for that reason a small amount of glycerol as lubricant is recommended in order to separate the meshes without problems (see Note 3).
3. Once the polymer is completely dissolved, fill a syringe with the polymer solution. For a 3% PLDLA in $CHCl_3$ solution, viscosity reaches around 200–300 mPa s at 25°C (see Note 4).

Use precision stainless steel tips of gauge of 25 and 5 in. long (EFD®) (see Note 5).

4. Attach the syringe to the infusion pump. The correct diameter must be adjusted to have a good control of the flow rate (see Note 6).
5. Connect the power source to the tip in a safe way otherwise it is easy to produce short-circuits (see Note 7). Increase voltage gradually 7.5–8 kV. If we observe the tip at this point, perfect Taylor cone must appear from the meniscus of the polymer solution that exits the tip. If there is more than one means that the voltage is too high. In contrast, if a drop of solution is permanently formed, the flow rate is too high. Finally, if a polymer drop is evaporated in the tip and dried, among other variables, it can mean that the voltage is so high that the tip is becoming warmer and warmer. A compromise is needed (see Note 8).
6. A "full moon" must appear in the aluminum foil collector. The moon gradually increases its diameter and thickness from the center to the periphery like a cone.

3.2. Method for Producing Biopolymer–Bioinorganic Composite-Oriented Nanofiber

1. The preparation of bioinorganic nanoparticles is required. In this example, glass nanoparticles are obtained by alkoxide-based sol–gel reaction in a relative nonpolar solvent such as 1,4-dioxane. However, 1,4-dioxane is not the best solvent for electrospinning of PLA-based composite solutions. Therefore, a further drying in a rotary evaporator or filtering step using submicron pore filters should be performed.
2. Disperse nanoparticles in $CHCl_3$ using a powerful ultrasonic probe (20 kHz and 100 W). Polymer (PLDLA 3 wt%) is added to the solution and dissolved under a vigorous stirring (see Note 9).
3. Attach the syringe in the infusion pump and adjust the correct diameter in the infusion pump to have a good control of the flow rate, as described in the previous example.
4. Connect the tip to the power source exactly as in the previous example (see Note 7). In this case, a slightly higher flow rate is required because here the pulling force of the drum is added to the electrostatic one (see Note 10).
5. Attach the aluminum foil to the metallic rod to be spun. Rotate drum at high speed > $49.99 \times g$ to allow fibers to be well aligned, perpendicular to the axis of cylindrical rotating drum. However, excessive rotary speeds lead to fiber rupture (27) (see Notes 1 and 10).
6. Finally, a homogeneous semitranslucent sheet is obtained due to the difference in the light diffraction with the nonwoven mesh example.

3.3. Method for Producing Core–Shell Structured Biopolymer/Hybridized Nanofiber

1. As in the second example, a hybrid organometallic network can be obtained by sol–gel or commercially purchased. In this example, a hybrid organometallic network is synthesized by alkoxides via a sol–gel reaction and then partially hydrolyzed and condensed to obtain a xerogel. This xerogel is mixed with a poly(caprolactone) solution of 16 wt% in tetrahydrofuran by magnetic stirring until a slightly opaque dispersion is obtained. Hereafter, the obtained slurry is referred to as slurry 1.
2. As in the first example, dissolve poly (D,L-lactide) (PLDLA) in a relatively nonpolar and volatile solvent such as $CHCl_3$ to a final concentration of about 3 wt%. Once PLDLA is added to the $CHCl_3$, the sampled must be stirred vigorously for 2 h at around 60°C. This solution is hereafter referred to as slurry 2.
3. Fill Syringe 1 and Syringe 2 with slurries 1 and 2, respectively, and connect both syringes in their different infusion pumps to obtain different rates according to their fluid features.
4. Aluminum foil must be attached to the metallic rod to be spun. It is recommended that a lubricant that is also biocompatible, such as glycerol, be spread on the surface. It is also important to split the aluminum foil into two to remove the formed tube from the metallic rod and the aluminum foil. This enables the tube to be obtained at the end of the process by just pulling each side (see Note 11).
5. Adjust both pumps between 0.5 and 1 ml/h. It is highly recommended to use the same type of syringe for both slurries (see Note 12).
6. Connect the inner tip to the power source exactly as in the first example (see Note 7).
7. Finally, a homogeneous slightly white colored tube is obtained with a partially nonwoven fiber structure. If a textured morphology is needed, an increase in the spinning speed to approximately 49.99 × *g* is required, as in the previous example. Other possibility is to alternate speeds such as 100 and 49.99 × *g* to obtain a mix of woven and nonwoven morphologies (see Note 1).

4. Notes

1. Speed of the rotary mandrel strongly influences the alignment of the fibers. Low speeds (<12.5 × *g*) usually leads to nonwoven morphologies (randomly oriented) while higher speeds (~49.99 × *g*) should force their orientation. However, depending on the material, mechanical properties, thickness, and distance between tip and collector, there is a critical speed limit. Surpassing this speed may provoke the rupture of fibers.

2. High molecular weight polymers are not easily dissolved in certain solvents, unless temperature is applied. However, these solvents usually have a high volatility. In order to work safe and avoid solvent evaporation, try to use a temperature between 40 and 50°C and a hermetic flask.

3. To pull out fibers from the aluminum foil, a lubricant is highly recommended. A wide variety of lubricants can be used. However, this wide range is significantly reduced if we consider biocompatibility as a main requirement. That is the main reason why glycerol can be chosen as a lubricant. Try to spread out a small amount of lubricant on the aluminum surface prior to deposition.

4. Relative viscosity is one of the most useful parameters in order to control spinnability of the solution and it is highly recommended to measure it in order to have a good control of the electrospun fibers. Our experience showed that a slurry with an approximate viscosity between 100 and 1,000 Cp (mPa s) is perfectly suitable for electrospinning. Other values can also be spinnable, but quality and size distribution of fibers will be affected.

5. There is a wide spectrum of tips or needles for electrospinning. They may have different materials, lengths, morphologies, and gauge (diameter of the needle following the French gauge system). Usually, the choice of the right diameter depends on the viscosity and the nature of the polymer and solvent to be used. The diameter of the tip should be as small as possible taking into account that it cannot be blocked during deposition.

6. Several models have an internal data base of the more common syringe fabricants. Glass syringes are better to have an exhaustive control of the flow rate and of the total deposited volume. However, they are more expensive, brittle, and it is easier to block them due to an undesired solidification of the polymer. Nonetheless, there are also propylene syringes with a good control of the volume and enough precision.

7. We recommend to use a flat washer or a crocodile clip. In addition, isolation between the power supply and the earth must be tight otherwise the fiber will be blown, and deposition will occur outside the scope of the collector.

8. We strongly recommend the use of distances between tip and collector of around 10 and 15 cm. It must be enough to evaporate all the solvent of the fiber. If shorter distances are used, remaining solvent may partially dissolve previously formed fibers. Longer distances should be balanced with higher voltages. However, it implies an increase of the temperature of the

tip, which may result in undesired early evaporation of solvent and clogging of the needle tip.

Tip is usually made of stainless steel, whose thermal conductivity (~16 W/K/m) is relatively low when compared with other metals such as aluminum (~250 W/K/m) or silver (~430 W/K/m). Consequently, excessive voltage values dissipation may raise the temperature of the tip to a point where solvent evaporates instantaneously.

9. Ultrasonic probes are a useful tool to homogenize and disperse particle agglomeration. A powerful sonicator is recommended (~20 kHz and 100–300 W). However, few seconds should be applied to avoid an excessive evaporation of solvent due to the fast heating provoked by ultrasounds.
10. Due to the pulling force sometimes fibers are broken if their resistances to rupture are not high enough. In this case, an increase of viscosity obtained by increasing polymer concentration or by increasing the flow rate in order to electrospin a higher amount of material is recommended.
11. To obtain a clean electrospun tube take the following steps: firstly, lubricant is needed; Secondly, aluminum coil that covers the metallic rod can be cut in two parts in order to liberate the tube when pull each other in opposite senses as shown in Fig. 4.
12. Core–shell nanofibers are easily obtained if the same pump and same flow is used. However, different kinds of fluids require different conditions. In order to compare speeds, the same diameter of the syringes is recommended.

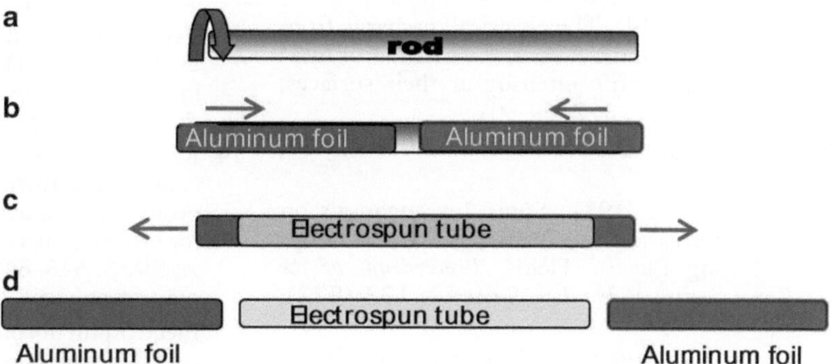

Fig. 4. Scheme for electrospun tubes fabrication: (**a**) fine metallic spinning rod with the desired diameter should be used as support; (**b**) the rod is covered by two pieces of aluminum foil greased with glycerol; (**c**) after rod removing, the two pieces of aluminum should be pulled in opposite senses; (**d**) the tube is obtained without any damage.

Acknowledgments

This work was supported by the WCU program through the NRF funded by the MEST, Korea (R31-10069). Dr. O. Castaño thanks the Ramón y Cajal programme of the MICINN.

References

1. Cooley, J.F. (1902) Apparatus for electrically dispersing fluids *US Patent Specification,* **692631**.
2. Morton, W.J. (1902), Method of dispersing fluids *US Patent Specification,* **705691**.
3. Formhals, A. (1934), Process and apparatus for preparing artificial threads *US Patent Specification,* **1975504**.
4. Gilbert, W. (1628), De Magnete, Magneticisque Corporibus, et de Magno Magnete Tellure, London, Peter Short.
5. Gray, S., and Mortimer C. (1732), A letter concerning the electricity of water, *Phil. Trans,* **37**, 227–260.
6. Rayleigh, L. (1882), On the Equilibrium of Liquid Conducting Masses charged with Electricity, *Philosophical Magazine,* **14**, 184–186
7. Larmor J. (1898) *Note on the complete scheme of electrodynamic equations of a moving material medium, and on electrostriction Proc. R. Soc,* **63**, 365–372.
8. Wilson C.T., and Taylor G.I. (1925), The bursting of soap bubbles in a uniform electric field, *Proc. Cambridge Philos. Soc,* **22**, 728–730.
9. Zeleny, J. (1917), Instability of electrified liquid surfaces". *Physical Review,* **10, 1**, 1–6.
10. Zeleny J. (1914), The electrical discharge from liquid points and a hydrostatic method of measuring the electric intensity at their surfaces, *Physical Review,* **3, 2**, 69–91.
11. Nolan, J.J. (1926) *Proc. R. Ir. Acad. Sect,* **A 37**, 28–39.
12. Macky W. A. (1931) Some Investigations on the Deformation and Breaking of Water Drops in Strong Electric Fields, *Proceedings of the Royal Society of London,* **Series A 133 (822)**, 565–587.
13. Taylor, G. (1964) Disintegration of Water Droplets in an Electric Field, *Proc. Roy. Soc. London*, **Ser. A 280, 1382**, 383–397.
14. Doshi, J. and Reneker, D. H. (1995), Electrospinning process and applications of electrospun fibers, *J. Electrost,* **35**, 151–160.
15. Teo W.E., and Ramakrishna S. (2006), A review on electrospinning design and nanofibre assemblies, *Nanotechnology,* **17**, R89–R106.
16. Gunja N. J. (2006), Biodegradable Materials in Arthroscopy, *Sports Medicine & Arthroscopy Review,* **14**, 3, 112–119.
17. Liao, S., Li, B., Ma, Z., Wei, H., Chan, C., and Ramakrishna, S. (2006) Biomimetic electrospun nanofibers for tissue regeneration, *Biomed. Mater,* **1** R45.
18. Gómez, A., and Tang, K. (1994), Charge and fission of droplets in electrostatic sprays. *Physics of Fluids,* **6 (1)**, 404–414.
19. Yoshimoto, H., Shin, Y.M., Terai, H., and Vacanti, J. P. (2003), A biodegradable nanofiber scaffold by electrospinning and its potential for bone tissue engineering, *Biomaterials,* **24, 12**, 2077–2082.
20. Casper, C. L., Stephens, J.S., Tassi, N. G., Chase, D. B. and Rabolt, J. F. (2004) Controlling Surface Morphology of Electrospun Polystyrene Fibers: Effect of Humidity and Molecular Weight in the Electrospinning Process, *Macromolecules,* **37, 2**, 573–578.
21. McGill, D.B. and Motto, J.D. (1974) An industrial outbreak of toxic hepatitis due to methylenedianiline, *New England J Med,* **291**, 278–282.
22. Gunatillake, P.A., and Adhikari, R. (2003) Biodegradable synthetic polymers for tissue engineering, *Eur Cell Mater,* **20, 5**, 1–16.
23. Taylor, M.S., Daniels, A.U., Andriano, K.P., and Heller, J. (2004) Six bioabsorbable polymers: in vitro acute toxicity of accumulated degradation products. *J Appl Biomater,* **5**, 151–157.
24. Kim, H. W., Lee, H. H., and Knowles J. C. (2006), Electrospinning biomedical nanocomposite fibers of hydroxyapatite/poly(lactic acid) for bone regeneration, *J Biomed Mater Res Part A*, **79, 3**, 643–649.
25. Song, J.H., Kim, H.E., and Kim, H.W. (2008) Electrospun fibrous web of collagen–apatite precipitated nanocomposite for bone regeneration, *J Mater Sci: Mater Med,* **19, 8**, 2925–2932.
26. Song, J.H., Yoon B.H., Kim H.E., and Kim H.W. (2008), Bioactive and degradable hybridized nanofibers of gelatin-siloxane for bone regeneration, *J Biomed Mater Res B 84A*, **4**, 875–884.
27. Matthews, J.A., Wnek, G.E., Simpson D.G., and Bowlin, G.L., (2002), Electrospinning of collagen nanofibers, *Biomacromolecules,* **3**, 232–238.

Chapter 10

Protein Adsorption Characterization

M. Cristina L. Martins, Susana R. Sousa, Joana C. Antunes, and Mário A. Barbosa

Abstract

Protein adsorption from (aqueous) solutions onto a (solid) surface is a common process that takes place at biological interfaces. This phenomenon, that spontaneously occurs, changes the properties of the surface and can induce structural modifications on proteins.

Proteins in solution can be easily identified/quantified using classical biochemical methods. However, adsorbed proteins are more difficult to assess since they are always associated with a substrate. The selection of the analytical method depends on the type of substrate used, the amount of adsorbed protein, the type of solution (single protein solution vs. complex biological media), and the type of information that is demanded (quantification of the adsorbed protein, adsorption kinetics, conformation, and orientation of the adsorbed protein). Until now, none of the techniques available are capable by its own to characterize all the protein adsorption process. Therefore, a multitechnique analysis is required. During this chapter, the methodologies to measure human serum albumin to poly(ethylene terephthalate) using the three different techniques, radiolabeling, ellipsometry, and quartz crystal microbalance with dissipation – QCM-D, are described in detail. The specific preparation of polymeric surfaces to be used with each technique is also presented.

Key words: ^{125}I radiolabeling, Iodogen method, Protein adsorption, Ellipsometry, Quartz crystal microbalance, Biomaterial, Surface

1. Introduction

The understanding of the protein adsorption process is essential for the development of recent emerging areas such as biotechnology and biomedical science.

Protein adsorption can be monitored in solution or directly on the surface. In solution, several methods can be used to quantify (e.g., colorimetric assay, ultraviolet–visible spectroscopy), identify (sodium dodecyl sulfate-polyacrylamide gel electrophoresis (SDS-PAGE), Western blotting), and quantify–identify (enzyme linked immunosorbent assay – ELISA) proteins in solution (1). For that,

Melba Navarro and Josep A. Planell (eds.), *Nanotechnology in Regenerative Medicine: Methods and Protocols*, Methods in Molecular Biology, vol. 811, DOI 10.1007/978-1-61779-388-2_10,

the adsorbed protein must be detached from the surface or indirectly calculated by the difference between the protein amount in the solution before and after contact with the substrate. However, with this approach, conformation of adsorbed proteins and kinetics of protein adsorption onto surfaces is not possible. Moreover, the quantification of adsorbed proteins in solution has two main problems:

1. The detachment of proteins from a surface is extremely dependent on the degree of interaction of protein–surface.
2. The amount of adsorbed protein is usually below the detection limit of most of the techniques used.

The first problem (1) can only be overcome if all the protein can be desorbed from the surface. The second problem (2) can be solved by increasing the surface area of the materials (as in the case of powders, nanoparticles, microspheres, etc.), which is not always possible.

There are several techniques for detection and quantification of protein adsorption directly on the surface. Table 1 describes the most commonly used techniques for the study of protein adsorption at biomaterials surface (2–6).

Due to the high number and variety of techniques that can be used to study protein adsorption to surfaces, only three different techniques were selected to be described in detail in this chapter: radiolabeling (1), ellipsometry (2), and QCM-D (3). The principles of these techniques, as well as the methodologies to prepare polymeric surfaces for each specific technique are described in detail. As an example, human serum albumin (HSA) adsorption to poly(ethylene terephthalate) (PET) is considered.

1.1. Radiolabeling

Radiolabeling is one of the most useful and powerful experimental methods for the quantification of protein or peptide adsorption at solid–liquid interfaces. It consists of incorporating a radioactive nuclide into the molecular structure of the protein/peptide to be studied, and then the addition as a tracer, to the solution to be studied. This solution should contain the same protein (unlabeled) in buffer, or may include mixtures of proteins (including the protein of interest). Then, the contact between the material to be tested and the protein solution is promoted at appropriate experimental conditions (protein concentration, temperature, time of adsorption, etc.). At the end of the adsorption step the two phases are separated and the surface-bound radioactivity is counted (7).

In the case of proteins or peptides with tyrosine residues, the radiolabeling is most readily accomplished by labeling with ^{125}I. The preparation of labeled protein to which radioiodine is covalently bound to the phenolic ring of tyrosine aminoacids, constitutes the central part of the technique (Fig. 1).

Table 1
Techniques most commonly used for study protein adsorption at biomaterials surfaces (2–6)

Techniques	Principle	Information depth	Analytical sensitivity	Spatial resolution	Information
Radiolabeling techniques	Use of radioisotopes on certain residues of the protein	Uppermost molecular level	High	Poor; droplet size ca. 2 mm	Adsorbed amount; adsorption kinetics and isotherms; exchange of molecules on surfaces; specificity of adsorption
Immuno-fluorescence/ ELISA	Use of an unlabeled primary antibody of a particular protein and a fluorochrome/ enzimatic labeled secondary antibody	Uppermost molecular level	Very high	Poor	Detection of the location and relative abundance of a particular protein for which an antibody exists; conformation information of an adsorbed protein
Ellipsometry	Detection of polarization change of light reflected on a sample	100–300 nm (evanescent field depth)	0.1–1 nm	Lateral resolution ca. 2 μm	Adsorbed amount; adsorption kinetics and isotherms; conformation information of the adsorbed layer, if antibodies are used
SPR	Influence of adsorbed molecules on electron charge density waves in the surface of a thin metal film (Au or Ag)	50–300 nm	0.1–1 ng/cm^2	10–50 μm for in situ measurements in the imaging mode	Adsorbed amount; adsorption kinetics and isotherms; conformation information of the adsorbed layer, if antibodies are used
ATR-FTIR	IR spectroscopy using evanescent field above the surface of a specially cut crystal in which the light propagates between the crystal surfaces	1–5 μm	1–10 mol%	ca. 5 μm in the microscope mode	Identification of the typical chemical functional groups, of proteins (amide I, II and III). Conformation information of the adsorbed protein layer

(continued)

Table 1 (continued)

Techniques	Principle	Information depth	Analytical sensitivity	Spatial resolution	Information
ToF-SIMS	Ion bombardment sputters secondary ions from the surface. The ions are characterized according to their mass-to-charge ratio using a time-of-flight mass spectrometer	<1 nm in "static mode"; 1 μm in the "dynamic mode"	Very high; detection limits 10^{11}–10^{17} atoms/cm^2	Lateral resolution >100 nm	Map of protein distribution; identification of different proteins; protein conformation; degree of denaturation of an adsorbed protein film; surface coverage
AFM	Atomic interaction between surface and tip are used to generate an atom-scale, electron density image of a surface.	Uppermost atomic or molecular level	Single atoms/ molecules	Depends on the shape and diameter of the tip 0.1 nm (z), 2 nm (x)	Protein visualization; dimensions of protein, adhesion, friction forces, molecular conformation, surface dynamics, topography
QCM-D	Piezoelectric technique, with capacity to detect surface binding mass and their viscoelastic properties	–	0.5–1 ng/cm^2	–	Adsorbed amount; adsorption kinetics and isotherms; conformation information of the adsorbed layer, if antibodies are used

SPR surface plasmon resonance, *ATR-FTIR* attenuated total reflection Fourier transform infrared spectroscopy, *ToF-SIMS* time-of-flight secondary ion mass spectrometry, *AFM* atomic force microscopy, *QCM-D* quartz crystal microbalance with dissipation

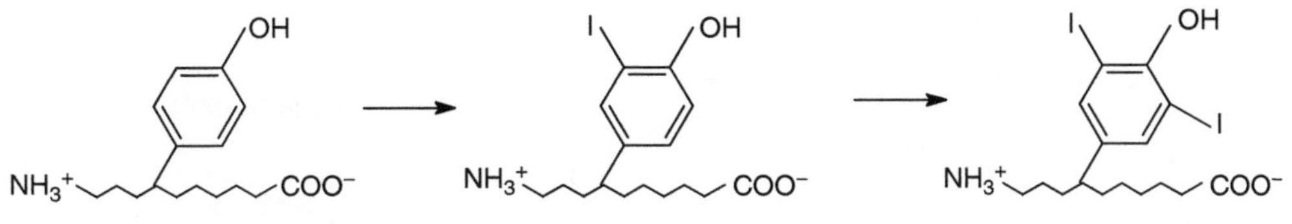

Fig. 1. Reaction of Iodination on tyrosine (adapted from ref. 5).

There are several radioiodination methods such as iodine chloride (ICl), Cloramine-T, and Iodogen method. However, Iodogen method is very simple, uses mild reaction conditions resulting in low oxidative protein damages, and gives high yields (5, 8).

Some examples of the use of radiolabeled proteins as tracers where ^{125}I is the nuclide of choice are protein adsorption to biomaterials, immunoassays, drug delivery, investigation of protein metabolism, and protein interactions such as hormone, hormone-receptor binding kinetics, and radiolabeling of proteins in situ in cells followed by cell disruption and solubilization.

Advantages of using the radiolabeling technique in protein adsorption can be listed as sensitive, simple, and enabling of the quantification of the amount of adsorbed protein (kinetics, isotherms), the desorption and exchange of proteins, and the study of competitive adsorption from mixtures of proteins. The great advantage of γ emitting isotopes, such as ^{125}I, relies on the fact that there is no special sample preparation involved and the radiation can be measured directly on the samples. The disadvantages of this technique are essentially due to the use of radiolabelling solutions. However, it is important to point out that there are two distinct levels of hazard: one during the protein radiolabelling and another, during the quantification of protein adsorbed at a much lower level of hazard.

1.2. Ellipsometry

Ellipsometry is an optical reflection-based technique that can be used to measure protein adsorption/desorption onto a surface in real-time and in a label-free environment (9–11). The substrates for ellipsometry must be reflective and flat being metals the most used. In the case of nonreflecting materials, as polymers, these must be deposited onto a suitable reflecting substrate (such as optically polished silicon slides (12) or flat slides sputter-covered with an optically reflecting metal film as gold (10, 13), titanium (14), etc.).

The fundamental principles behind ellipsometry are well described at Tompkins *et al.* (15). Ellipsometry is based on the change of the polarization state of light after reflection from a surface. Adsorption of a thin film (e.g., a protein layer) onto a reflective surface will shift the phase and amplitude of the reflected light. The changes are different for the two components of light polarized parallel (p) and perpendicular (s) to the plane of incidence. This change of polarization can be detected by the shift of the ellipsometric angles Δ and ψ (Fig. 2) measured with the ellipsometer (16).

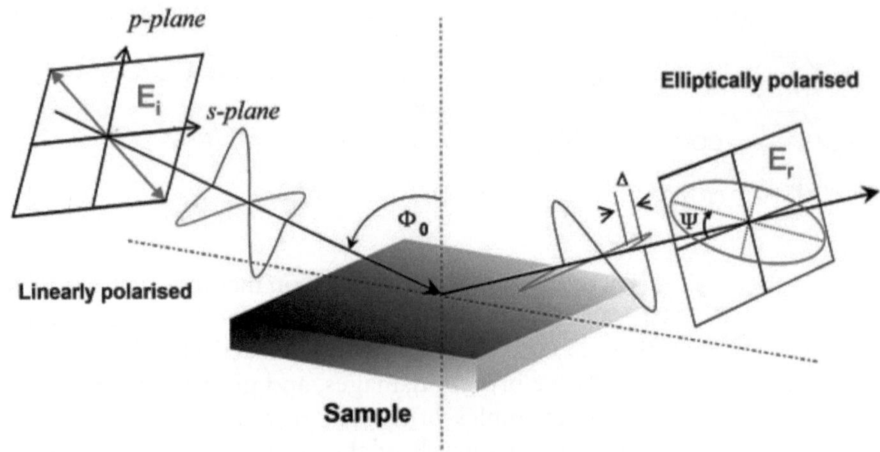

Fig. 2. Schematic illustration of the principle of ellipsometry, where Δ is the difference between the phase changes of the *p* and *s* components of the reflected light and Ψ is an angle which tangent gives the ratio of amplitude changes for the *p* and *s* components of the reflective light (16).

The total effect caused by reflection is described by the fundamental equation of ellipsometry:

$$\rho = R_p / R_s = \tan \psi \, e^{i\Delta} \tag{1}$$

where ρ is a function of the overall reflection coefficients of the parallel (R_p) and perpendicular (R_s) components. These coefficients depend on the wavelength of light, the angle of incidence, and the optical properties of the reflecting system (15).

From the change of Δ and ψ due to protein adsorption, the refractive index and the thickness of the protein film can be calculated. Since the refractive index of the adsorbed protein films is always close to n_{protein} ~1.5 (14), the thickness of the protein film can be calculated with an accuracy of ~0.01 nm (11). The amount of adsorbed protein per unit of area (Γ) can be determined using the equation described by de Feijter *et al.* (17):

$$\Gamma = d_{\text{protein}} \frac{n_{\text{protein}} - n_{\text{ambient}}}{\mathrm{d}n / \mathrm{d}c} \tag{2}$$

where n_{ambient} is the refractive index of the buffer and d*n*/d*c* is the refractive index increment. This method uses the assumption that the refractive index of a protein in solution is a linear function of the protein concentration (~0.180 mL/g for globular proteins) (17).

The combination with antibodies allows the quantification of the amount of a specific protein adsorbed from serum or plasma (11) and also to detect conformation changes of the adsorbed proteins (10).

The advantage of ellipsometry is its capacity to detect with high sensitivity (0.5 ng/cm^2) protein adsorption in situ and in real time without using any label molecule. This is also a common advantage to other techniques such as SPR and QCM-D. However, for ellipsometry, any flat and reflective substrates can be used. The disadvantages rely on the requirement of specific equipment, the use of high volume of protein solution (>3 mL), and the use of flat and reflective materials.

1.3. Quartz Crystal Microbalance with Dissipation (QCM-D)

The quartz crystal microbalance (QCM) is commonly employed as a sensing device for measuring protein adsorption, and, in general, film deposition from dry and liquid environments. An extension of the technique enables the measurement of energy dissipation during surface-related phenomena (18). The instrument, known as QCM with Dissipation (QCM-D) and commercialized by Q-Sense (Gothenburg, Sweden) has been widely used in protein adsorption studies. Currently, there are two different versions of QCM-D instruments available: a single channel version and a four-channel version; and, recently, an automatic system with four channels, which can run QCM-D experiments without the need for supervision. A combined QCM-D/ellipsometry setup enables simultaneous QCM-D and ellipsometric measurements on the same substrate.

The basis of QCM operation relates to quartz inherent property of piezoelectricity. Piezoelectric crystals are excited to ultrasonic vibrations by applying alternating voltages on opposite crystal faces. The heart of QCM-D is a sensor, a thin disk of crystalline quartz sandwiched between two electrodes, the top one being the sensor surface (19), as it can be seen in Fig. 3 (19).

As the QCM is piezoelectric, the oscillating electric field produces a mechanical oscillation, a standing wave, in the bulk of the quartz wafer. In the QCM, the mechanical shear oscillation is predominant and the displacements are parallel to the wafer surface. This shear wave oscillation is induced efficiently in AT-cut quartz wafers. A resonant frequency is obtained by inserting the crystal in an oscillating circuit where the electrical field oscillates at a frequency near the fundamental frequency of the crystal. The latter depends on the thickness, chemical structure, shape, and mass of the wafer (20). The fundamental frequency of the crystals used in this protocol was 4.95 MHz. Protein adsorption to the crystal results in changes in the fundamental frequency and overtones, which are odd-integral multiples of the fundamental frequency. The response at the fundamental frequency is generally neglected, since it is highly sensitive to the mounting conditions of the crystals, particularly the adjustment of the sealing O-rings.

As shown by Sauerbrey (1959), changes in the resonant frequency are simply related to the mass accumulated on the crystal, when the crystal is loaded with a rigid film that is evenly distributed over the active area of the crystal, does not slip or deforms with the oscillatory motion, and has a mass that is much smaller than that of

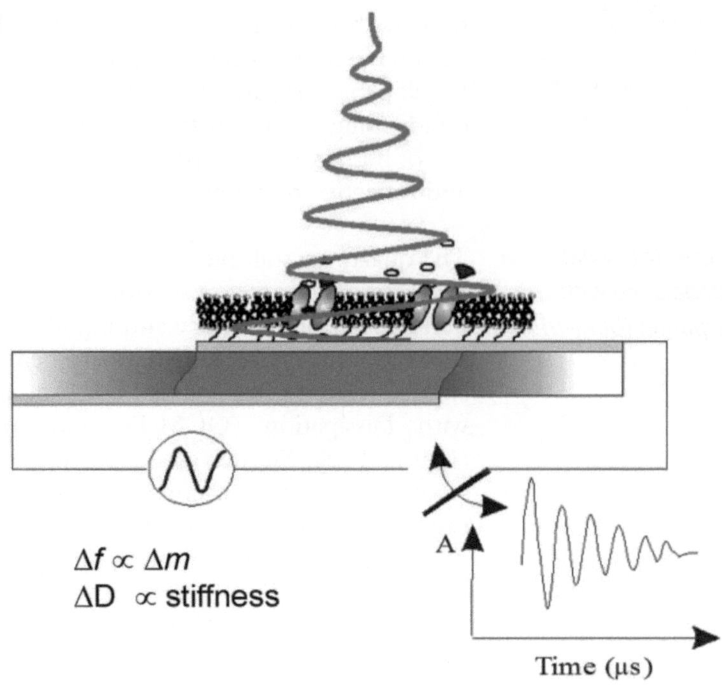

Fig. 3. Schematic representation of a QCM-D. An alternating RF voltage is applied to a quartz crystal coated with metal (e.g., gold). Adsorption of biomolecules occurs on the top surface. The change in frequency is proportional to the mass of adsorbed species, whereas the change in dissipation is a function of the stiffness of the adsorbed layer (19).

the quartz disk (14). The change in resonant frequency, Δf, is described by the Sauerbrey's equation:

$$\Delta m = -\frac{C\Delta f}{n} \tag{3}$$

where $\Delta m =$ mass change ; C = a constant that describes the sensitivity of the device to changes in mass; n = overtone number (1,3,…).

For the crystals used in this protocol C = 17.7 ng/cm^2/Hz at f = 4.95 MHz.

Deviations from Sauerbrey's relationship are typically observed when studying soft interfaces in liquids, which are interpreted in terms of layer viscoelastic properties. Therefore, for nonrigid layers, in addition to the measurement of frequency shift, it is necessary to measure the damping of the crystal oscillation. Simplistically, frictional losses occur in the crystal and the adsorbed material that lead to a damping of the oscillation, with decay in the amplitude of the piezolelectric resonator when the driving voltage is turned off, revealing dissipative properties of viscoelastic overlayers (18).

The dissipation parameter, D, is defined as

$$D = \frac{E_{\text{dissipated}}}{2\pi E_{\text{stored}}}, \tag{4}$$

where $E_{dissipated}$ = the energy dissipated (lost) during one oscillatory cycle; E_{stored} = total energy stored in the oscillating system (21).

The change in the dissipation factor, $\Delta D = (D - D_0)$, is measured when the material is adsorbed, while D_0 is the dissipation factor of the pure quartz crystal immersed in the solvent. The software Q-Tools, provided by Q-Sense with the QCM-D, contains the Sauerbrey relation, and the viscoelastic Maxwell and Voigt models, thus allowing for characterization of both rigid and soft films. The Maxwell model is usually applied to layers which are more viscous than elastic. The Voigt model is applicable for solid viscoelastic layers. Viscoelastic materials with more complex rheology (e.g., cells, membranes, liquid crystal polymers, etc.) can be described by a combination of these two basic viscoelastic schemes (22).

The QCM-D possesses a number of features that investigators have long recognized as advantageous, as the real-time nanogram sensitive detection capability for surface mass binding and a surface viscoelastic characterization capability for the bound mass. In addition, specialized surfaces may be prepared and monitored stepwise during preparation and they can be used as sensor platforms for studying biomolecular interactions. Also, the absence of any need for a labeling step to measure the mass of bound analyte being sensed is a considerable advantage of the method. The main drawbacks are the inherent nonspecificity, unless an adequate surface chemistry is used, and lack of a high-throughput capability.

2. Materials

2.1. Radiolabeling

CAUTION: In all cases where radioisotopes are used, depending on the quantity and nature of the isotope, precautions must be taken to ensure the safety of the researcher. All radioactive solutions handling must be performed by trained researchers. When working with radiochemicals such as ^{125}I, always observe specific safety precautions, including the use of rooms adapted to γ emitting radiation radioactives, external vented hood for γ radiation radioactives, shielding (Plexiglas impregnated with lead, lead barriers), and personal shielding such as gloves and overall with lead, dosimetry monitoring, and proper disposal of radioactive waste.

Oxidation converts iodide to a volatile and hazardous form, but protein/peptide-bound iodide is not volatile, although emits γ radiation and remains a potential hazard. Be sure to perform all steps that give rise to volatile iodide in an external vented hood (5, 7, 8).

2.1.1. Radioiodination of Protein/Peptide (see Note 1)

1. Iodogen (1,3,4,6-tetrachloro-3α,6α-diphenylglycouril), 1 mg/mL in 99.5% dichloromethane (see Note 2).
2. 0.01 M phosphate buffer solution (PBS, pH = 7.4).
3. Radioactive iodine (Na-^{125}I, 1 mCi, Perkin Elmer).
4. HSA solution (1 mg/mL) in PBS.
5. 0.25 M phosphate buffer: add 2.224 g of $Na_2HPO_4 \cdot 2H_2O$ and 1.725 g of $NaH_2PO_4 \cdot H_2O$ to a 50-mL falcon tube. Add Milli-Q water up to 40 mL, and adjust the pH to 7.5. Add deionized water up to 50 mL. Store at 4°C.
6. 0.05 M Phosphate buffer, by 5× dilution of 0.25 M phosphate buffer.
7. PD-10 desalting column (Sephadex G-25 M) for the separation of labeled protein from free ^{125}I.
8. γ-Counter (Wallac, model 1470 Wizard).
9. Geiger counter (Saint Gobain Crystals and Detectors, Series 900 mini-monitor).

2.1.2. Yield of Labeled Protein Determination

1. 5% (w/v) Bovine serum albumin (BSA) in PBS solution.
2. 20% (w/v) Trichloroacetic acid (TCA) in Milli-Q water.

2.1.3. Protein Adsorption Test

1. Iodinated PBS (PBSI), degassed PBS with NaI (0.01 M) (see Note 3).
2. Labeled HSA solution (2×) at a concentration of 0.2 mg/mL. Unlabeled HSA solution (0.2 mg/mL) with ca. 2% (v/v) ^{125}I-HSA to obtain a final activity of ~10^7 cpm/mg in PBSI.

2.2. Ellipsometry

2.2.1. Equipment

1. Imaging Null Ellipsometer, model EP^3, from Nanofilm Surface Analysis, operating polarizer–compensator–sample–analyzer (PCSA) mode, containing appropriate software, a laser working at a wavelength of 532 nm and at an angle of incidence of 60°.
2. Flow cell (3 mL with glass windows with an angle of 60°).
3. Teflon tubes.
4. Peristaltic pump (Ismatec, SA).
5. Spin coater (Specialty Coating Systems, Inc., model G3P-8).
6. Ultrasonic bath for cleaning the crystals (Bandelin, Sonorex Digitec).
7. Clean cabinet for preparing the films.

2.2.2. Solutions

1. 0.01 M PBS (pH = 7.4).
2. Filter (0.2 µm filter) and degas PBS buffer in order to avoid small air bubbles and other particles that can pass through the flow system.

3. 0.1 mg/mL HSA solution in filtered and degassed PBS.
4. 2% v/v Hellmanex II solution in Milli-Q water for cleaning the flow cell and tubes.

2.3. QCM-D

2.3.1. Equipment

1. Q-Sense E4 microbalance, which enables four tests has to be carried out simultaneously. Peristaltic pump (e.g., Ismatec IPC-N 4) with four channels, each one for an individual module. Several types of modules are available. The flow modules are made of aluminum and titanium, but the liquid sample only comes in contact with titanium. For more information refer to the website http://www.q-sense.com (see Note 4).
2. Spin coater (Specialty Coating Systems, Inc., model G3P-8).
3. UV/ozone generator for removing the polymer film (BioForce Nanosciences, UV/Ozone ProCleaner™).
4. Ultrasonic bath for cleaning the crystals (Bandelin, Sonorex Digitec).
5. Clean cabinet for preparing the films.

2.3.2. Sensors and Other Consumables

QSX 301 gold-coated (thickness 100 nm; Q-Sense) quartz crystals as substrates for polymer deposition. A large variety of surfaces is available, but some of them can be prepared in the laboratory. Here, we exemplify the preparation of a PET film on a gold substrate. The number of tests that can be done with each sample is variable. In order to prolong the life span of sensors handle and clean them with great care. They are the most expensive consumables for the QCM-D. Other consumables, like tubings, O-rings, ferrules, and adaptors can be purchased from Q-Sense or other companies.

2.3.3. Solutions

1. 2% v/v Hellmanex II solution for cleaning the metallic modules where the crystals are inserted and tubes.
2. Ammonia solution and hydrogen peroxide solution (30%) for cleaning the crystals.
3. Milli-Q water for all applications
4. Dry nitrogen or argon to dry modules and crystals.
5. Buffer and protein solutions are the same as in Subheading 2.2.2.

3. Methods

3.1. Radiolabeling

3.1.1. Sample Preparation

1. Cut the PET film (Trafoma A/S, Denmark) (23) into disks with a diameter of 8 mm, using a punch.
2. Rinse the PET disks with ethanol, Milli-Q water, and ethanol, for periods of 3 min in a ultrasonic bath.
3. Dry the samples overnight in a vacuum oven at 30°C.
4. Store the dried films in a desiccator until use.

3.1.2. Radioiodination of Protein/Peptide

On the day before protein/peptide radiolabeling:

1. Prepare all the solutions/materials that will be used (such as buffers, eppendorfs, racks, and RIA tubes, etc.).
2. Equilibrate the PD-10 column with 25 mL PBS and store at 4°C (see Note 5). Elute the residual PBS from the column until about 1.5 cm from the resin surface.

On the day of protein radiolabeling:

1. Prepare the iodogen solution and add 40 μL to an eppendorf.
2. Dry the solvent in a nitrogen stream, rotating the tube in order to obtain an iodogen film well distributed at the bottom of the eppendorf walls. At the radioactivity external vented hood.
3. Add 100 μL of 0.25 M PBS to the eppendorf with the iodogen film.
4. Add 10 μL of radioactive iodine to the eppendorf.
5. Add 10 μL of the protein to be labeled (HSA) (1 mg/mL) to the eppendorf.
6. Allow the mixture to stand on ice for 20 min.
7. Add 250 μL of 0.05 M PBS to the eppendorf. With the same tip transfer all the volume to the PD-10 column. The tip and the eppendorf should be discharged to the radioactive waste.
8. Add 2,130 μL of PBS (to perform a volume of 2,500 μL, according to the instructions of column supplier).
9. Recuperate the eluted fractions to eppendorf tubes (~500 μL), adding PBS to avoid the dryness of the resin. Usually, a total of 20 fractions will be sufficient.
10. Cap the column (top and end) and put it inside a falcon tube (50 mL), and dispose as radioactive waste.
11. Withdraw a 10 μL aliquot from each fraction and transfer it (along with the tip) to a RIA tube. Cap the RIA tube.
12. Count (counts per minute, cpm) the aliquots for radioactivity in the γ-counter.
13. The fractions of labeled protein/peptide with potential interest are usually in the first 5–6 eppendorfs with detected radiation (cpm).
14. Calculate the yield of iodination of these protein factions.

3.1.3. Yield of Labeled Protein Determination

The yield of iodination is performed using the TCA protein precipitation method.

1. Outside the external vented hood, add to an eppendorf, 45 μL of 5% BSA in PBS.

2. Inside the external vented hood, add 2 μL of each protein (see Note 6).
3. Add 50 μL of TCA to each eppendorf.
4. Let the tubes stand on ice for 20 min.
5. Centrifuge the tubes at $16{,}000 \times g$ for 10 min.
6. Carefully withdraw the supernatant (see Note 7) and transfer it to an identified RIA tube (labeled with the number of fraction/S). Cap the RIA tube.
7. Cut the end of the tube containing the precipitated protein and put it in another identified RIA tube (labeled with the number of fraction/P). Cap the RIA tube.
8. Count the aliquots for radioactivity in the γ-counter.
9. Compute the yield of iodination by the equation:

$$\text{Yield of iodination } (\%) = \frac{\text{activity}_{\text{precipitate}}(\text{cpm})}{\text{activity}_{\text{solution}}(\text{cpm}) + \text{activity}_{\text{precipitate}}(\text{cpm})} \quad (5)$$

10. Select the most adequate protein fraction (higher yield vs. higher activity) and use it as soon as possible (see Note 8).
11. Discard the radioactivity waste, all the other protein fractions that are not going to be used.

3.1.4. Protein Adsorption Test

On the day before protein adsorption test:

1. Place each PET sample in an eppendorf tube. Use at least four replicates for each condition.
2. Add 150 μL of degassed PBSI overnight (see Note 9).

On the day of protein adsorption test:

Determination of the activity of HSA solution

1. Withdraw a 5-μL aliquot from labeled HSA solution (prepared in Subheading 2.1.3, item 2) and transfer it (along with the tip) to a radioimmunoassay (RIA) tube (six replicates), to obtain the activity of the protein solution, $\text{activity}_{\text{solution}}$ (cpm/mL).
2. Count the aliquots for radioactivity in the γ-counter.

3.1.5. HSA Adsorption Tests

1. Add 150 μL labeled HSA solution (0.2 mg/mL) to each tube with the PET samples and incubate at room temperature for 60 min. The final HSA concentration will be 0.1 mg/mL.
2. Aspirate the solution using a micropipette.
3. Wash three times the surfaces with 300 μL PBS, to remove nonadsorbed protein.
4. Transfer the samples to RIA tubes and count the radioactivity using a γ-counter.

5. The amount of HSA adsorbed can be calculated by the equation:

$$\text{HSA}\ (\mu g / m^2) = \frac{\text{activity}_{\text{sample}}(\text{cpm}) \cdot |\text{HSA}|_{\text{solution}}\ (\mu g / mL)}{\text{activity}_{\text{solution}}(\text{cpm} / mL) \cdot \text{surface area}(m^2)} \quad (6)$$

where the $\text{activity}_{\text{sample}}$ is the radioactivity of the samples, $|\text{HSA}|_{\text{solution}}$ is the HSA concentration in solution, $\text{activity}_{\text{solution}}$ is the specific activity of the HSA solution.

3.2. Ellipsometry

3.2.1. Sample Preparation: Deposition of PET Thin Films (24)

1. Flat and optically reflected gold substrates (1×1 cm^2) (see Note 10).
2. Dissolve PET in 0.072% (w/w) trifluoroacetic acid (TFA).
3. Clean the gold substrates immersing them in "piranha" solution (seven parts concentrated H_2SO_4 and three parts 30% H_2O_2) for 5 min (*Caution*: *this solution reacts violently with many organic materials and should be handled with care*), followed by the sequential rinse with ethanol, water (Milli-Q), and ethanol (2 min in each solution) in an ultrasonic bath.
4. Dry the substrates with a gently stream of nitrogen.
5. Place the gold substrates onto the spin coater support.
6. Place a drop (100 μL) of the polymer-containing solution on the gold substrates.
7. Start the rotation of the gold substrates at 2,000 rpm.
8. Stop the rotation after 1 min.
9. Remove the PET films substrates and dry it with a stream of nitrogen (see Note 11).

3.2.2. Protein Adsorption Assay

1. Mount the substrate in the flow cell (Fig. 4).
2. Fill the flow cell with the PBS buffer (>3 mL).
3. Eliminate air bubbles in the system. This is a very important step since even small air bubbles can induce errors in the process.
4. Start the continuous injection of the buffer solution in to the flow cell in a constant flow rate (0.5 mL/min). Temperature can be controlled in most of the equipments (25°C).
5. Start to measure the ellipsometer angles Δ_0 and Ψ_0 of the substrate (baseline) in continuous (kinetics). It is important to have a reliable baseline before the initiation of the protein adsorption process. Baseline is usually stabilized after 40–60 min.
6. After baseline stabilization, stop the injection of the buffer and start the injection of the protein solution in to the flow cell (0.5 mL/min; 25°C). This process can be performed in continuous (during all the adsorption process) or in static conditions, after injection of enough protein solution to guarantee

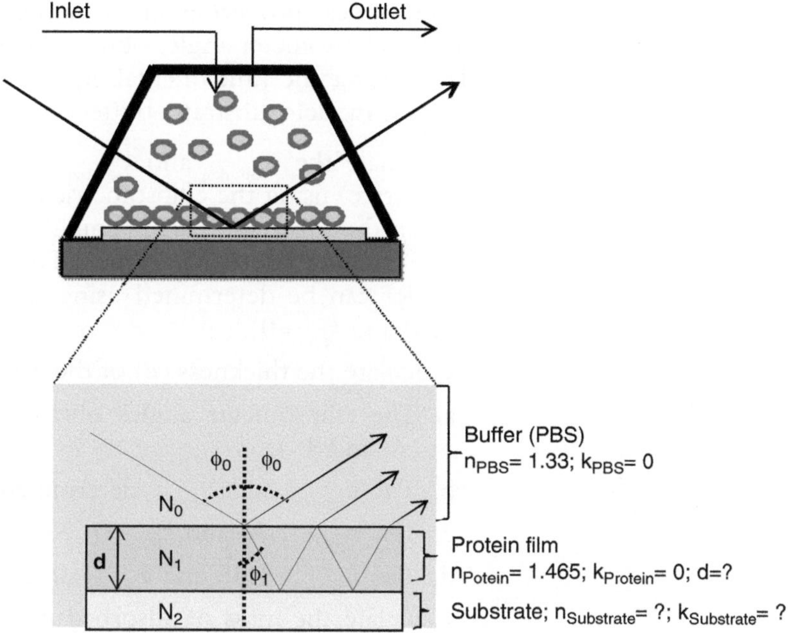

Fig. 4. A schematic drawing of the ellipsometer measurement cell and of the three-phase model, consisting of buffer (0), protein film (1) and substrate (2). Their respective complex refractive indexes are $N_0 = n_{pbs} + ik_{pbs}$, $N_1 = n_{protein} + ik_{protein}$, and $N_2 = n_{substrate} + ik_{substrate}$, where n corresponds to the refractive index, k to the extinction coefficient, and i the imaginary number (adapted from refs. 11, 14).

the saturation of the flow cell. During this process it is important to avoid air bubbles. During continuous flow, vials must be capped to prevent sample evaporation. At this step, the shift of ellipsometer angles (Δ and Ψ) can be observed in real time and the kinetic of protein adsorption can be determined in situ and in real time.

7. After 2 h, stop protein injection and inject the buffer solution into the flow cell (0.5 mL/min; 25°C). During this process, proteins that are loosely bound to the surface will be flushed off.
8. After 30 min (or when there is a stabilization of the adsorption process) stop the assay and remove the sample.

3.2.3. Calculations

Ellipsometer software allows the determination of the thickness (d) and refractive index of the protein layer ($n_{protein}$) using the ellipsometric angles of the substrate, before (Δ_0 and Ψ_0) and after protein adsorption (Δ_1 and Ψ_1). The fundamental equation of ellipsometry Eq. 10.1 for three-phase model (substrate; protein layer and buffer – Fig. 4) is applied (11, 14).

The calculation of the protein layer thickness (d) needs the data of the optical properties of the substrate ($n_{substrate}$ and $k_{substrate}$)

and the optical properties of the buffer. This can be calculated using the ellipsometer angles obtained before protein injection (Δ_0 and ψ_0) using the fundamental equation of ellipsometry for the two-phase model (substrate; buffer) (see Note 12).

1. Calculate the $n_{substrate}$ and $k_{substrate}$ of the substrate (PET/gold substrate) using the ellipsometric angles (Δ_0 and Ψ_0) obtained when the substrate reaches equilibrium with the buffer (baseline), before protein injection. The optical properties of the buffer can be determined using a refractometer (n_{PBS} ~1.33 (25) and $k_{PBS}=0$).
2. Calculate the thickness (d) of the protein layer (nm) using:
    a. The ellipsometer angles obtained after protein injection (Δ_1 and Ψ_1).
    b. The $n_{substrate}$ and $k_{substrate}$ determined at point 1.
    c. The $n_{PBS}=1.33$ and $k_{PBS}=0$ (25).
    d. The $n_{protein}=1.45$ and $k_{protein}=0$ (26, 27).
3. Calculate the mass of adsorbed protein per unit of area (Γ), using the Feijter Eq. 10.2 assuming $n_{protein}=1.465$ (26, 27), n_{Buffer} ~1.33 (25) and dn/dc ~ 0.18 mL/g (17).

3.2.4. Flow Cell Rinsing

1. Rinse the flow cell and the tubes with the detergent solution (2% Hellmanex® II, Hellma) in Milli-Q water with a flow rate of 1 mL/min during 30 min.
2. Rinse the detergent with Milli-Q water with a flow rate of 1 mL/min during 30 min.
3. Rinse the inlet tube with 150 mL of distilled deionized water.

3.3. QCM-D

3.3.1. Sample Preparation

1. Dissolve PET in 0.072% (w/w) TFA.
2. If using a clean sensor go directly to step 3. If not, clean the sensor first, according to Subheading 3.3.2.
3. Place the sensor on the spin coater support, making sure that the electrical contacts are facing downward and the crystal is adequately centered (off-centered sensors may spin off and break). In order to avoid contamination carry out this operation in a clean cabinet.
4. With a micropipette cover the sensor with the polymer solution, ensuring that the entire surface becomes covered (work from the center toward the periphery by depositing small drops; volume required: ca. 100 μL).
5. Start the rotation of the sensor at 2,000 rpm.
6. Stop the rotation after 1 min.

7. Remove the PET-coated sensor and dry it with a stream of dry nitrogen or argon (see Note 11).

3.3.2. Cleaning the Sensors

1. Place the face to be cleaned upwards in the UV/ozone cleaner for 10 min.
2. Prepare a solution with 10 mL of Milli-Q water, 2 mL of 25% ammonia and 2 mL of 30% hydrogen peroxide. Heat the solution in a glass beaker up to 75°C and immerse the sensor(s) for 5 min. While the solution is warming up cover the glass beaker with a curved glass, to avoid evaporation. Place the sensor(s) inside the cleaning solution using the Teflon holder provided by Q-Sense.
3. Rinse the sensor(s) with Milli-Q water and then twice in an ultrasonic bath, also with Milli-Q water, for 2 min.
4. Dry the sensor(s) (both sides) with a gentle stream of dry argon or nitrogen in a clean cabinet and store them (see Note 13).

3.3.3. Carrying Out the Test

1. Mount the sensor(s) in the flow modules in the clean cabinet, to avoid contamination. Manipulate the sensor(s) with tweezers, touching only the edges, in order to avoid damaging the PET film and scratching of the gold substrate. Make sure the sensors are directly touching the O-ring before closing the modules, otherwise you may break the sensor(s) when tightening the bolts.
2. Place the flow modules in the four-sensor chamber and connect the inlet and outlet tubings, making sure that they are firmly tight.
3. In the software provided by the manufacturer set the temperature and the other parameters. The instructions manual is very clear. Problems are not generally experienced if the operator follows the instructions.
4. Flush the cell with PBS for about one hour, which should enable the frequencies to reach a steady state. Use the flow rate around 30 mL/min (see Note 14).
5. After flushing with PBS, stop the peristaltic pump and change to the albumin solution (if the pump is not stopped, air bubbles will be introduced in the circuit). Restart the pump, using approximately the same pump speed as in step 4.

3.3.4. Cleaning the Flow Modules After the Experiment

1. Flush 10 mL of Milli-Q water from a 15 mL falcon tube. Use an intermediate-to-high pumping speed.
2. Remove sensors and O-rings in the clean cabinet, placing them in a Petri dish with Milli-Q water.
3. Dry the sensors (both sides) with a gentle flow of argon/nitrogen and store them in a desiccator until further use.

4. Rinse the titanium half of the flow module with Milli-Q water and dry with oil-free compressed air (or with argon/nitrogen). The other half, where the electrical contacts are located, should not be wetted.
5. Mount cleaning O-rings and cleaning crystals (crystals that may have been used in previous experiments but that may be scratched and unsuitable for further experiments) in the flow modules and tighten up the bolts (see Note 15).
6. Mount the flow modules again in the four-sensor chamber and flush them with a 10 mL solution of 2% Hellmanex. Use a medium pumping speed.
7. Place the O-rings used in the test in a plastic container with 2% Hellmanex for 15 min, then go to step 9.
8. Flush the flow modules with 20 mL Milli-Q water. Use a medium pumping speed. After the volume of rinsing water has finished make sure that no water remains trapped in the tubing (wait until no more drops come out), then go to step 10.
9. Empty the plastic container (step 7), rinse the cleaning O-rings with Milli-Q water, first by replacing the water three times and then in an ultrasonic bath for 2 min (repeat two times). Dry with argon/nitrogen and leave in the clean cabinet to dry further, before placing them again in the flow module (step 12).
10. Dismount the flow modules, remove and dry the cleaning crystals with argon or nitrogen. Remove the cleaning O-rings and dry them also. Store them in a box sealed with parafilm.
11. Dry the flow modules with oil-free compressed air (or argon or nitrogen), blowing it through the big holes and drying the surface gently (see Note 16).
12. Mount the test O-rings and assemble the flow modules until further usage.

3.3.5. Calculations

The data (frequency and dissipation for the various overtones and temperature) are plotted on the computer screen during the course of the experiment. The data are saved into a spreadsheet that can be used for the calculations. Q-Sense provides a very detailed step-by-step guide to calculate the amount of mass adsorbed using the Sauerbrey equation and for viscoelastic modeling using the Maxwell and Voigt approaches. Figure 5 illustrates the change in mass, Δm , and dissipation, ΔD , upon exposure of a TiO_2 surface to fibrinogen and to an antifibrinogen antibody (14).

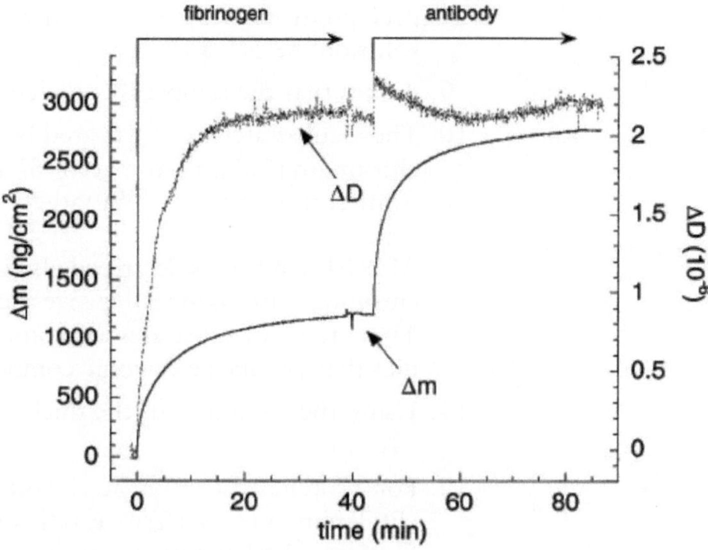

Fig. 5. Change in dissipation, ΔD , and in mass, Δm , upon exposure of a TiO_2-coated QCM-D sensor crystal to first fibrinogen (t= 0) and then, after rinsing in pure buffer, to a fibrinogen antibody (t= 45 min) solutions (14).

4. Notes

1. Alternative to this procedure, it is possible to obtain from the market the protein already labeled.
2. This solution must be prepared in the external vented hood, immediately before the iodination procedure and kept in ice until use.
3. Buffer with iodide, PBSI, is used to inhibit adsorption of free ^{125}I ions, present in trace amounts during protein adsorption experiments (28).
4. Care must be taken when cleaning the modules, particularly the aluminum part, to avoid its corrosion.
5. Do not let the column dry; PBS should be left at about 1.5 cm from the surface.
6. Since the labeled protein is in a small quantity, BSA is used to assure that all protein is precipitated.
7. If the supernatant is not carefully withdrawn, the precipitate may be aspirated and an erratic yield of iodination may be obtained.
8. These fractions can be stored inside a lead container at –20°C for a maximum of 1 week. However, since the iodine can be released from the protein, it is important to perform a TCA

precipitation test to verify the free iodine present in the solution, before use.

9. Assure that the sample is completely immersed in PBSI.
10. These substrates were prepared by deposition of a thin layer of chromium (2.3 nm) and gold (37 nm) by ion beam sputtering from chromium and gold targets (99.9% purity) onto silicon wafers (polished/etched, crystal orientation ⟨100⟩, from AUREL Gmbh) as described elsewhere (13). The thin layer of chromium was used to improve the adhesion of gold to silicon. These types of substrates are commercially available at companies that produce electronic components.
11. Using these conditions the thickness of the PET film will have ~18 nm.
12. For protein adsorption it is easier to consider the substrate (PET film + gold surface) as only one surface and not as a PET film deposited at a gold surface.
13. For cleaning other surface films the manufacturer provides a comprehensive list of methods.
14. If large oscillations in frequency occur this may be indicative of air bubbles trapped in the system or incorrect mounting of the sensor. This normally requires that you mount the sensor again and check the tube connections.
15. When mounting the crystals make sure that they are properly supported on the O-rings. Otherwise, they will break. When tightening the bolts turn them alternately, in order to avoid breaking the crystal.
16. Do not blow air with the O-rings mounted, because you may blow them away.

References

1. *Current Protocols in Protein Science* (1995), ed. J.E. Coligan: Wiley.
2. Castner, D.G. and B.D. Ratner (2002), *Biomedical surface science: Foundations to frontiers. Surface Science.* **500** (1–3):28–60.
3. Ratner, B.D., et al. (2004), *Biomaterials science - an introduction to materials in medicine.* 2nd edition ed. San Diego, CA, USA: Elsevier Academic Press.
4. Nakanishi, K., T. Sakiyama, and K. Imamura (2001), On the adsorption of proteins on solid surfaces, a common but very complicated phenomenon. *Journal of Bioscience and Bioengineering.* **91** (3):233–244.
5. Dawis, S. (1993), *Test Procedures for Blood Compatibility of Biomaterials*, ed. s. edition: Springer.
6. Rodahl, M., et al. (1995), Quartz-Crystal Microbalance Setup for Frequency and Q-Factor Measurements in Gaseous and Liquid Environments. *Review of Scientific Instruments.* **66** (7):3924–3930.
7. Schumacher, T.N.M. and T.J. Tsomides (1995), *In Vitro Radiolabeling of Peptides and Proteins* in *Current Protocols in Protein Science* C.J.E.e. al., Editor., John Wiley & Sons, Inc.
8. Iodine-125 (1993), *A guide to radioiodination techniques.* Little Chalfort: Amersham Life Science. 64.
9. Arwin, H. (2000), Ellipsometry on thin organic layers of biological interest: Characterization and applications. *Thin Solid Films.* **377–378**: 48–56.

10. Elwing, H. (1998), Protein adsorption and ellipsometry in biomaterial research. *Biomaterials.* **19** (4–5):397–406.
11. Tengvall, P., I. Lundstrom, and B. Liedberg (1998), Protein adsorption studies on model organic surfaces: an ellipsometric and infrared spectroscopic approach. *Biomaterials.* **19** (4–5):407–422.
12. Ying, P.Q., et al. (2003), Competitive protein adsorption studied with atomic force microscopy and imaging ellipsometry. *Colloids and Surfaces B-Biointerfaces.* **32** (1):1–10.
13. Martins, M.C.L., B.D. Ratner, and M.A. Barbosa (2003), Protein adsorption on mixtures of hydroxyl- and methylterminated alkanethiols self-assembled monolavers. *Journal of Biomedical Materials Research Part A.* **67A** (1):158–171.
14. Hook, F., et al. (2002), A comparative study of protein adsorption on titanium oxide surfaces using in situ ellipsometry, optical waveguide lightmode spectroscopy, and quartz crystal microbalance/dissipation. *Colloids and Surfaces B-Biointerfaces.* **24** (2):155–170.
15. Tompkins, H.G. and E.A. Irene (2005), *Handbook of Ellipsometry*: William Andrew Publishing and Springer-Verlab Gmbh & Co.KG.
16. Trinity College Dublin, U.o.D. http://www.tcd.ie/Physics/Surfaces/ellipsometry2.php. 2009.
17. Defeijter, J.A., J. Benjamins, and F.A. Veer (1978), Ellipsometry as a tool to study adsorption behavior of synthetic and biopolymers at air-water-interface. *Biopolymers.* **17** (7):1759–1772.
18. Rodahl, M. and B. Kasemo (1996), A simple setup to simultaneously measure the resonant frequency and the absolute dissipation factor of a quartz crystal microbalance. *Review of Scientific Instruments.* **67** (9):3238–3241.
19. Kasemo, B. (2002), Biological surface science. *Surface Science.* **500** (1–3):656–677.
20. O'Sullivan, C.K. and G.G. Guilbault (1999), Commercial quartz crystal microbalances - theory and applications. *Biosensors & Bioelectronics.* **14** (8–9):663–670.
21. Hook, F. and B. Kasemo (2007), *The QCM-D Technique for Probing Biomacromolecular Recognition Reactions*, in *Piezoelectric Sensors*, C. Steinam and A. Janshoff, Editors., Springer. pp.425–447.
22. Voinova, M.V., M. Jonson, and B. Kasemo (2002), 'Missing mass' effect in biosensor's QCM applications. *Biosensors & Bioelectronics.* **17** (10):835–841.
23. Holmberg, M. and X. Hou (2010), Competitive protein adsorption of albumin and immunoglobulin G from human serum onto polymer surfaces. *Langmuir.* **26** (2):938–942.
24. Ferreira, L., et al. (2005), Improving the adhesion of poly(ethylene terephthalate) fibers to poly(hydroxyethyl methacrylate) hydrogels by ozone treatment: Surface characterization and pull-out tests. *Polymer.* **46** (23):9840–9850.
25. Sousa, S.R., et al. (2007), Dynamics of fibronectin adsorption on TiO2 surfaces. *Langmuir.* **23** (13):7046–7054.
26. Benesch, J., A. Askendal, and P. Tengvall (2000), Quantification of adsorbed human serum albumin at solid interfaces: a comparison between radioimmunoassay (RIA) and simple null ellipsometry. *Colloids and Surfaces B-Biointerfaces.* **18** (2):71–81.
27. Kurrat, R., et al. (1998), Plasma protein adsorption on titanium: comparative in situ studies using optical waveguide lightmode spectroscopy and ellipsometry. *Colloids and Surfaces B-Biointerfaces.* **11** (4):187–201.
28. Du, Y.J., R.M. Cornelius, and J.L. Brash (2000), Measurement of protein adsorption to gold surface by radioiodination methods: suppression of free iodide sorption. *Colloids and Surfaces B-Biointerfaces.* **17** (1):59–67.

Chapter 11

Measuring Wettability of Biosurfaces at the Microscale

Conrado Aparicio, Yassine Maazouz, and Dehua Yang

Abstract

Determining the contact angle of a liquid on a solid surface is a simple method to assess the surface wettability. The most common method to measure the contact angle of a liquid consists of capturing the profile of a sessile drop of a few microliters on the surface using an optical system. Currently, this is a widely used technique to analyze wettability both in researched materials and in products of multiple technological fields. However, the drop dispensed by a traditional macroscopic contact angle meter is too big to assess the wettability properties of individual topographical features and/or chemical patterns at the micro/nanoscale. Recently, contact angle meters that can discharge drops that are microscopic, with volumes in the range of 1×10^{-3} to 10^{-5} μL have been developed. The novel microscopic contact angle meter uses a pneumatic injection system to discharge the drop of the liquid through a capillary of a few micrometers of internal diameter and a high-resolution ultrafast digital camera.

We have tested different biosurfaces – microimprinted polymers for biosensors, calcium-phosphate cements with different topographical microfeatures, orthodontic wires – and assessed the potential applicability in the field in comparison with the conventional macroscopic contact angle meters.

This protocol describes the basic tasks needed to test wettability on biosurfaces with a microscopic contact angle meter. The focus of the protocol is on the challenging methodological steps and those that differentiate the use of this equipment to the use of a traditional macroscopic contact angle meter.

Key words: Microscopic contact angle, Wettability, Biosurface, Biomaterial

1. Introduction

1.1. Wetting Phenomena and Contact Angle

Young's equation is the basis for a quantitative description of wetting phenomena (1). If a drop of a liquid is placed on a solid surface a finite contact angle is established (θ) (Fig. 1). A wetting contact line is formed. At this line three phases are in contact; i.e., the vapor, the liquid, and the solid. Young's equation relates the contact angle to the solid/liquid interfacial free energy (γ_{SL}), the liquid surface tension (γ_L), and the solid surface free energy $(\gamma_S) : \gamma_L \cos\theta = \gamma_S - \gamma_{SL}$.

Melba Navarro and Josep A. Planell (eds.), *Nanotechnology in Regenerative Medicine: Methods and Protocols*, Methods in Molecular Biology, vol. 811, DOI 10.1007/978-1-61779-388-2_11,

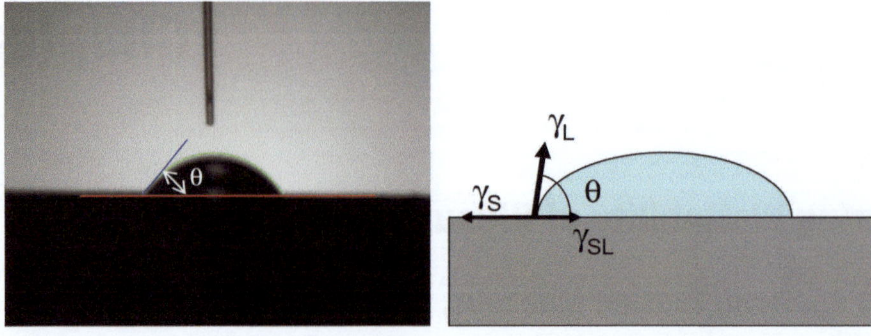

Fig. 1. *Left.* Contact angle (θ) of a droplet of 500 pL of ultrapure water after 50 ms of being in contact with an orthodontic wire obtained with a microscopic contact angle meter. The droplet was injected through a capillary with a tip with 5 μm of internal diameter. The orthodontic wire has 0.016 × 0.022 in. rectangular cross-section. *Right.* Representation of the equilibrium of interfacial tensions at the three-phase contact line between a drop of a liquid and a solid planar surface. γ_{SL} solid/liquid interfacial free energy, γ_L liquid surface tension, and γ_S solid surface free energy.

If the interfacial tension of the bare solid; i.e., the solid surface free energy, is higher than that of the solid/liquid interface, the contact angle is smaller than 90°, and thus the liquid partially wets the surface. The surface has good wettability properties and if the liquid is water, the surface is hydrophilic.

If the solid/liquid interface is energetically less favorable than the bare solid surface, the contact angle will be higher than 90°, the surface has poor wettability and if the liquid is water, the surface is hydrophobic.

Consequently, determining the contact angle of a liquid on a solid surface is a simple method to assess the wettability of a surface as well as its hydrophobic/hydrophilic characteristics. This has many practical meanings because minimal physical and/or chemical changes in the surface of the solid affect significantly its surface free energy and wetting properties, which is notably reflected in the observed contact angle. Among others, if a surface gets dirty, worn, or contaminated as well as if a certain surface manufacturing process is applied to it, the water contact angle will be modified in comparison to that obtained in the pristine surface.

By testing selected different liquids on the solid surface other relevant properties of the surface can be assessed, such as the surface free energy, the critical surface tension, the different components of the surface free energy – polar, apolar/dispersive, Lewis acid, Lewis base – or the work of adhesion (2).

1.2. Measuring Contact Angles at the Microscale

There are several methodologies available to measure the contact angle of a liquid on a solid surface (3). The most common method consists in capturing the profile of a sessile drop of a pure liquid on the surface using an optical system, as shown in Fig. 1. The contact angle is either determined with a goniometer or the image is captured by a camera and transferred to a computer that automatically fits a line in the contour of the profile. Traditionally, the drop of the liquid – in most cases water of a controlled quality – has a volume of a few microliters and thus, a size of several millimeters. Currently, this is a widely used technique to analyze wettability both in researched materials and in products of multiple technological fields.

However, the drop dispensed by a traditional macroscopic contact angle meter is too big to assess the wettability properties of individual topographical features and/or chemical patterns at the micro/nanoscale, such as those obtained with novel manufacturing processes based on nanotechnologies. Moreover, those droplets do not stay on the surface of final products of small size or with intricate morphologies and the contact angle on the actual piece is not measurable.

Recently, a few manufacturers have developed contact angle meters that can discharge drops that are microscopic, with volumes in the range of 1×10^{-3} to 10^{-5} μL, and that can be stabilized on the tested surfaces for a few seconds. This is a challenging task because the microscopic droplets rapidly evaporate from or absorb into the surface. The novel microscopic contact angle meter uses a pneumatic injection system to discharge the drop of the liquid through a capillary of a few micrometers of internal diameter and a high-resolution ultrafast digital camera that transfers the images to a software similar to the one used for traditional contact angle meters.

With this technology the contact angle meter of a drop of water on gaps between microscopic patterns, thinner filaments like dingle fibers and hairs, jet nozzles, or individual parts of microcircuits are possible.

1.3. Potential Application of a Microscopic Contact Angle Meter for Biotechnology and Biomaterials in Regenerative Medicine

The surface properties of synthetic and bioinspired materials are known to dictate their biological in vitro and in vivo performance. Current biomaterial applications and emerging areas, such as tissue engineering and regenerative medicine require a fuller understanding of biomolecular-surface and cell-surface interactions. Measurements of biomolecular interactions and cellular response to surfaces including adhesion, function, proliferation, migration, and differentiation have been shown to be modulated by the surface chemistry, topography, mechanical properties, and physicochemical properties of the materials (4). Among them, wettability and surface free energy of materials for biomedical applications, including those to play an important role in the development of therapies for regenerative medicine, have been largely investigated at the macroscale – 123 papers in the journal Biomaterials included the word “wettability” in its title or abstract during the last 10 years. In recent literature, reported values of the wettability of the materials under research are common place. Others, including research in our group (5, 6), have successfully attempted to relate specific wetting response of a biomaterial or values of its total surface energy or individual components to the adsorption of biomolecules and cell response.

Also important for regenerative medicine and biomedical devices products is the possibility of obtaining surfaces that have extreme wettability properties to elicit desired bioactivities, such as preventing bacterial colonization or accelerating bone regeneration (7).

One example of an advanced material with unique wettability properties that has been developed to attain multiple and very promising applications in tissue engineering and regenerative medicine is the use of poly(*N*-isopropylacrylamide) (PIPAAm) for a novel tissue reconstructive technique named cell sheet engineering (8). PIPAAm is a thermoresponsive polymer that changes from hydrophobic to hydrophilic when cooled down from 37°C to 20°C. Cells are cultured in vitro on Petri dishes at 37°C until they form a continuous cell layer. In this condition, cells attach to the hydrophobic culture surface via their extracellular matrix (ECM), and connect to each other via cell-to-cell junction proteins. The construct is then cooled to 20°C, which releases the interconnection between ECM and the now hydrophilic culture surface, but avoiding the disruption of the cell-to-cell connections. Consequently, the cells are detached as a contiguous cell sheet and the ECM retained underneath the cell sheets works as an adhesive agent. In this condition, the cell sheets can be used either to be implanted or manipulated to build more complex 3D constructs. Initial successful attempts to reconstruct and regenerate corneal surfaces, periodontal ligament, interior surfaces of organs of the urinary tract, and pulsing myocardial 3D tissue have been performed.

Controlling and measuring surface wettability and liquid spreading on surfaces at the microscopic scale is also of significant interest for a broad range of applications in the biomedical field. This includes surfaces with structures and features at the micro/nanolevel, such as DNA microarrays, digital lab-on-a-chip, top-down and bottom-up microstructured surfaces made of bioactive gels, and micropatterning of biomolecules and polymers for cell culture applications and biosensing (9–12) to name a few. In fact, recent advancements in surface engineering, with the fabrication of various micro/nanoscale topographic features, and selective chemical patterning on surfaces, have enhanced or prevent surface wettability with a high degree of control (13, 14). Moreover, wettability properties of different biomedical devices of small dimensions with microscopic features or intricate shapes are also of main relevance to determine their clinical performance. This is the case for orthodontic brackets, dental implants, contact lenses, and cardiovascular drug-eluting stents which are under development or already in the market.

This protocol describes the basic tasks needed to test wettability on biosurfaces with a microscopic contact angle meter. The focus of the protocol is on the challenging methodological steps and those that differentiate the use of this equipment to the use of a traditional macroscopic contact angle meter.

The technique and equipment described in the protocol are novel and there is almost no literature reporting results obtained using them (15, 16). We have tested different biosurfaces – microimprinted polymers for biosensors, calcium-phosphate cements with different topographical microfeatures, orthodontic wires, dental implants, cardiovascular stents, and metals coated with amphiphilic multiblock recombinant biopolymers in combined gradients of chain-length and multiblock configuration – and assessed the potential applicability in the field in comparison with the conventional macroscopic contact angle meters.

2. Materials

2.1. Liquids

1. Ultrapure water (Milli-Q Academic, Milli-Q, USA). This water is particle free, has a resistivity of 18.2 MΩ cm, a total organic content of less than 10 parts per billion, a content of microorganisms of less than 1 colony-forming unit per mL, and a surface tension of 72.8 mN/m at 25°C (17) (see Note 1).
2. Acetone (ACS Reagent, BDH®, VWR International, USA).
3. Water-repellent liquid for coating the tip of the water-dispenser capillary before dispensing the drops of ultrapure water (Glaco® Chemical Liquid Super Wiper, Japan) (see Note 2).

2.2. Devices

1. Water-dispenser capillaries (Kyowa Interface Science Co., Japan). Three different water-dispenser capillaries are available with tips of 50, 30, or 5 μm of internal diameter. This protocol presents the use of all of them.
2. Glass syringe (TLL, Hamilton Co., USA) equipped with a 28G 304 stainless steel needle of 51 mm in length with a 90°-bevel tip (N728 NDL, Hamilton Co., USA) (see Note 3).

2.3. Equipment

This protocol is based on the use of a microscopic contact angle meter (MCA-3, Kyowa Interface Science Co., Japan). The equipment needed to perform the experiments includes the following additional parts (Fig. 2):

1. A pneumatic compressor (D500SR, Paasche, USA).
2. A pneumatic automation dispenser controller (DMC-M100, Kyowa Interface Science Co., Japan).
3. A camera for zenithal view of the testing area linked to a TV screen.
4. A high-resolution camera for lateral view of the testing area, with 12× zooming capabilities, and a shooting speed of up to

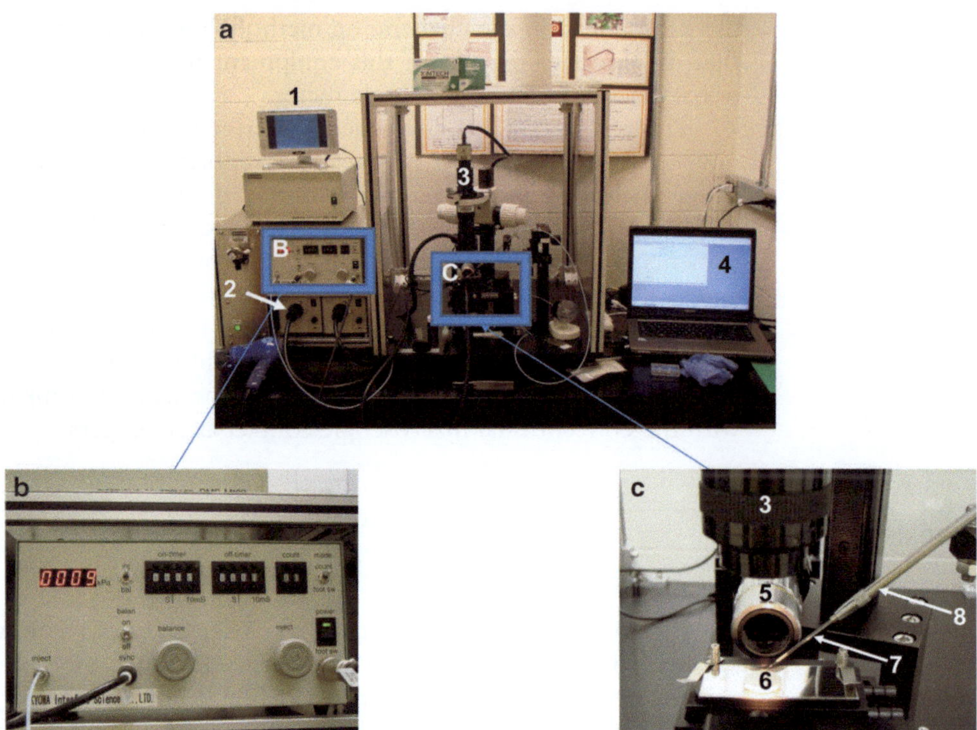

Fig. 2. General view (**a**) of the microscopic contact angle meter (MCA-3, Kyowa Interface Science Co., Japan) and close ups of the pneumatic automation dispenser controller (**b**) and the area of testing (**c**). TV screen (1), light source controllers (2), zenithal camera (3), FAMAS v3.1.3 software (4), high-resolution lateral camera (5), sample (6), capillary (7), and dispenser (8).

60 frames per second (M PLAN APO 10×/0.28, Mitutoyo Co., Japan).

5. FAMAS v3.1.3 software for analyzing water drops and determining water contact angles (Kyowa Interface Science Co., Japan).
6. Two light source controllers with optical fibers (KBEX-102, Kyowa Interface Science Co., Japan).
7. A transformer box linked to all parts.
8. A hair dryer.

3. Methods

This protocol provides with the essentials for testing the wettability of surfaces using the microscopic contact angle meter. The actual measurement of the contact angle meter and other related types of analysis of the wettability using the FAMAS v3.1.3 software is not covered and only some notes refer to the potential use of the software for alternative/further measurements and treatment of the data.

Thus, the protocol is divided into four tasks. Preparing the capillary, filling it with water, and appropriately positioning it before starting the session are three critical tasks to be performed prior to taking pictures of the water drops in contact with the biosurfaces and the actual measurement of the contact angle.

3.1. Capillary Preparation

The first task in order to get started with the equipment is to coat the external surface of the tip of the capillary with a water-repellent agent. The repellent agent reduces the surface energy of the exterior surface of the tip and thus it prevents the uncontrolled dripping of the water from the capillary tip as well as having the water held on the external surface of the tip by capillary effect (see Note 4).

1. On the pneumatic automation dispenser controller switch to "bal" the "inj/bal" button as well as switch to "on" the "Balance on/off" button (see Note 5).
2. Put medical examination gloves on of your hand size (see Note 6).
3. Take gently the capillary out of its box. Be very careful and do not touch or hit the tip of the capillary. The tip of the capillary is fragile and difficult to see, particularly in the case of the capillary with a tip with 5 μm of internal diameter.
4. Loosen the screw at the very end of the dispenser and introduce the end of the capillary opposite to the tip until slight resistance is felt. This resistance is felt because the capillary touches the silicone seal inside the dispenser (Fig. 3).
5. Force the capillary in until the capillary is inserted into the holder additional 5 mm (Fig. 3).
6. Check that the capillary cannot be moved by applying mild force and tight the screw at the end of the dispenser (Fig. 3).
7. Adjust the balance knob until an actual flow of air is going through the capillary. 20 kPa for capillaries with tips with an internal diameter of 50 and 30 μm and 40 kPa for capillaries

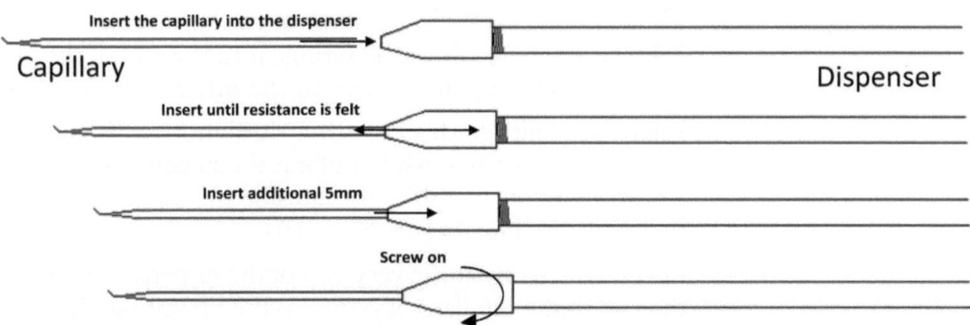

Fig. 3. Schematics showing the steps to follow for mounting the capillary into the dispenser. This corresponds to steps 4–6 of the "Subheading 3.1" and to steps 9–11 of the "Subheading 3.2".

with tips with 5 μm of internal diameter are appropriate starting settings for the air-flow pressure (see Note 7).

8. Immerge the capillary in the water-repellent solution for 10 s and check that small bubbles are coming out from the tip of the capillary. If bubbles are not coming out of the capillary, slightly and gradually increase the pressure until bubbles can be seen inside the water-repellent liquid (see Note 5).
9. Take the tip of the capillary out of the solution and dry the capillary with a hair dryer for at least 30 s.
10. Repeat steps 8 and 9 for at least three times if working with capillaries with tips with 50 or 30 μm of internal diameter and for at least six times if working with capillaries with tips with 5 μm of internal diameter.
11. Switch the "Balance on/off" button to "off."
12. After completing step 10, remove the capillary from the dispenser and proceed with the filling of the capillary.

3.2. Capillary Filling with Ultrapure Water

This task consists in filling the capillary with ultrapure water by using the auxiliary glass syringe with a long thin needle. It is a difficult task that requires precision and very good handling.

1. Dispense acetone in a cleaned beaker, fill the syringe through the needle with acetone from the beaker, and inject out all the acetone to a solvent-wasting container (see Notes 6 and 8).
2. Repeat step 1.
3. Dispense ultrapure water in a cleaned beaker, fill the syringe through the needle with the water from the beaker, and inject out all the water to a wasting container.
4. Repeat step 3 twice.
5. Unscrew the needle from the syringe and fill the syringe with the ultrapure water that is going to be used to test the wettability of the biosurface (see Note 1).
6. Screw the needle back on the syringe and take all the air out of the syringe (see Note 9).
7. Introduce the needle of the syringe into the end of the capillary opposite to the tip all the way to the other end (Fig. 4).
8. Inject the fluid gradually with a slow motion by pushing the piston of the syringe while pulling the needle out of the capillary to avoid that bubbles form inside the liquid dispensed into the capillary (Fig. 4) (see Note 10).
9. Loosen the screw at the very end of the dispenser and introduce the end of the capillary opposite to the tip until slight resistance is felt. This resistance is felt because the capillary touches the silicone seal inside the dispenser (Fig. 3).

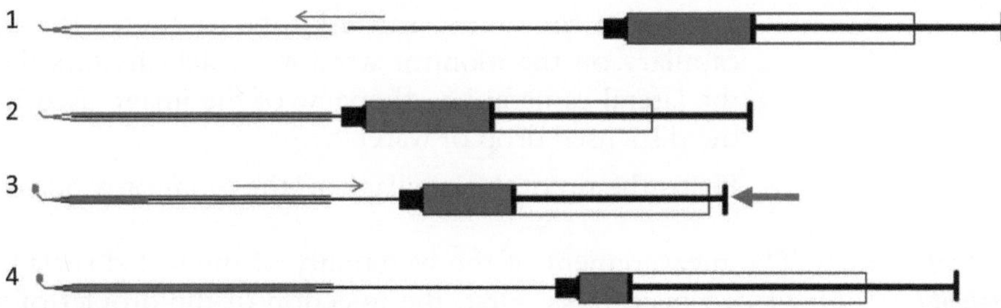

Fig. 4. Schematics showing the steps to follow for filling the capillary with ultrapure water. This corresponds to steps 7 and 8 of the "Subheading 3.2".

10. Force the capillary in until the capillary is inserted into the holder additional 5 mm (Fig. 3).
11. Check that the capillary cannot be moved by applying mild force and tight the screw at the end of the dispenser (Fig. 3).

3.3. Dispenser Adjustment and Focus of the Tip of the Capillary

This task aims to locate the tip of the capillary in front of the cameras and its focusing to clearly visualize the tip on the computer screen before starting the experiments. This task requires patience and some special notes to help overcome common difficulties due to the small dimensions of the tips of the capillaries and thus the use of significant zooming.

1. Set a flat plate of a material with a highly reflective surface (see Note 11) on the sample-holder stage and move the stage to its lowest position using the "up/down" knob.
2. Insert the dispenser with the capillary filled with ultrapure water in the dispenser-holder.
3. Adjust the position of the dispenser until the tip of the capillary is in both the zenithal and the lateral camera field (see Note 12). The visualization of the tip of the capillary in the two screens is not needed in this step. The tip of the capillary should be as close as possible to the middle of the light path of the two cameras, though.
4. Move the dispenser using the *x*(front/back)- *y*(left/right)- and *z*(up/down)-knob until an image of the capillary can be visualized on the TV screen showing the image obtained with the zenithal camera. Adjust the focus until a clear image of the tip is obtained (see Notes 12 and 13).
5. Rotate the dispenser with the capillary to position the tip of the capillary parallel to the line of view of the zenithal camera and thus perpendicular to the sample-holder stage.
6. Open FAMAS v3.1.3 software and open the display monitor window.

7. Use again the *x*-, *y*-, and *z*-knobs to center the tip of the capillary on the monitor window, which displays the view of the lateral camera; i.e., the view of the image used to capture the dispensed drop of water.
8. Focus the tip of the capillary on the monitor window.

3.4. Assessment of the Wettability of the Biosurface

The measurement of the wettability of the tested surface consists of two main tasks. First, the injection of the droplet of ultrapure water on the surface is performed. Secondly, the contact angle between the ultrapure water and the biosurface is measured. The former is the main and substantial difference between assessing the wettability with a traditional macroscopic contact angle meter and the novel microscopic contact angle meter.

Other than the size of the drops used to measure the wettability, which has been commented in the introduction of this protocol, the most significant difference is the way in which the drop is formed and put in contact with the biosurface, which in turn determines the volume of the drop of water. In a traditional contact angle meter, the drop of the water is obtained by pushing the liquid out of a syringe and forming a pendant drop from the needle and the droplet is gently dispensed by putting in contact the surface and the pendant drop. Instead, in the microscopic contact angle meter, the drop is injected from the capillary through its tip, which has a micrometer-size internal diameter, and the drop is blasted on the tested surface. Thus, the adjustment of the pressure and time of injection of water is one of the most important methodological parameters to be monitored and controlled while performing this task.

Once the drop is on the surface, the measurement of the contact angle between the water and the biosurface is carried out the same way as for the traditional contact angle meters. This task is not included in this protocol.

1. Position the sample on the sample-holder stage.
2. Slowly lift up the sample-holder stage until the profile of the tested surface can be seen on the monitor window.
3. Open a contact angle measurement sheet in the FAMAS software.
4. Open the "C.A./P.D. Controller" in the FAMAS software and switch the button "inj/bal" to "inj" (see Note 14).
5. Click on "Measure" and "make a droplet."
6. Adjust the parameters of the "inj" function (pressure, on timer, off timer) and repeat the step 5 until a droplet is formed (see Note 15).
7. Very carefully approach the surface of the sample closer enough to the tip of the capillary for the droplet to get in contact and

stay on the tested surface once the drop is injected out of the tip of the capillary (see Note 16).

8. Click on the "measure" icon and the "measure" button. A picture of the water in contact with the biosurface is automatically taken after a predetermined few ms (see Note 17).
9. Move the sample by using the *x* and *y* micrometers precision-adjustment handles of the sample-holder stage to test other areas of the biosurface.
10. Repeat steps 8 and 9 as many times as desired to map the wettability properties of the biosurface.
11. Finally, after performing all the measurements, the capillary should be emptied. Loose the screw that holds the dispenser in the dispenser holder.
12. Switch the "inj/bal" button to "bal" and the "balance on/off" button to "on."
13. Select the highest pressure and let the water being fully expulsed out of the capillary (see Note 18).
14. Pull the capillary out of the dispenser and carefully place it back into its box wearing medical examination gloves.

4. Notes

1. Different qualities and types of water can be used. The use of ultrapure water is ideal for both to determine wettability with the best reproducibility and to work with the microscopic contact angle meter. This is because particles, microorganisms, ionic potential, and more significantly any surfactant in the water can modify the surface tension of the liquid and thus, the contact angle is measured. Moreover, the purer the water, the lower the likelihood for stopping the dispenser-capillaries up during or after testing sessions. Using thoroughly cleaned/sterilized water containers is recommended.
2. Other water-repellent agents can be used. This is the one that the manufacturer of the microscopic contact angle meter provides with the equipment. It is sold as a water-repellent agent for car windshields.
3. Variations of the syringe and needle for loading the water in the dispenser-capillary are possible, but the gauge of the needle has to be 28G, and the length of the needle has to be ≥50 mm.
4. As a general safety precaution when handling the glass capillary and sharp needles common laboratory safety procedures should be followed to avoid from personal injury by sharp

objectives. When filling the capillary, mounting to and demounting the capillary from the dispenser, and emptying the capillary, never point the capillary to yourself or any person around you. Check and reduce the "bal" and "inj" pressures to zero before mounting to or demounting the capillary from the dispenser.

5. The repellent liquid should never get into the capillary. This would drastically change the surface tension of the dispensed water. That is the reason to adjust a constant flow of air that passes through the capillary during all the time the tip of the capillary is prepared. This is accomplished by using the "bal" function of the automation dispenser controller. The equipment is able to provide with a discontinued flow for injecting air through the capillary using the "inj" function, but this function should be avoided in this part of the protocol. Discarding the capillary is recommended in case that the constant flow of air is interrupted while preparing the capillary with the repellent fluid.

6. Wearing medical examination gloves is highly recommended along the whole protocol, but it is not strictly required. The advantage of working without gloves is that the operator has improved precision when handling the small parts – specially the capillaries and some of the samples – involved in the different tasks of this protocol. However, wearing gloves of the right size prevents the contamination of the testing liquid, the tip of the capillaries, and the samples from any dust, grease, etc. that could be on the hands of the operator while having good handling of the parts. The change in the water contact angle can be significant if some of the mentioned parts get contaminated.

7. In the balance mode (see Note 5), the pressure of injection can be tuned between 0 and 100 kPa, but the values stated in this protocol should be enough to start and maintain a constant flow through the capillary while it is coated with the water-repellent liquid.

8. A critical point for appropriately performing the water filling of the capillary is preventing any contamination (surfactants, cells, particles, ...) of the syringe, the needle, and the liquid (see Notes 1 and 6). This is of even more relevance when filling small internal diameter capillaries; i.e., 30 and 5 μm, because of a higher risk to block the capillary tip than with the capillary with a tip of 50 μm in internal diameter. Thus, a thorough cleaning of the syringe and the needle which are used to fill the capillaries with ultrapure water is needed. To avoid uncontrolled cross-contamination of the capillaries, syringe, and needle, it is highly recommended that each user of the microscopic contact angle meter keeps his/her own set of all these parts. Sonication of the syringe and the needle in acetone and ultrapure

water for 15 min before their use is an alternative/additional task for reassuring that the testing liquid is not going to be contaminated.

9. If small bubbles of air are introduced into or are formed inside the capillary while the capillary is filled with water, the water injection process when performing wettability tests can be significantly affected. This is because bubbles are compressible and transient states during injection can be reached that not only can affect the injection process but also may damage the surface of the sample.
10. If a bubble is formed while filling the capillary with water (see Note 9), stop the injection of the water, and without taking the needle out from the capillary, repeat steps 7 and 8 of the protocol.
11. A flat piece of a material with a highly reflective surface should be used to visualize the tip of the capillary with the zenithal camera. Any polished metallic or enamel surface is appropriate for this purpose.
12. The tip of the capillary is very fragile. Careful positioning of the capillary in the equipment is needed to avoid damage of the tip because of any displacements of the dispenser that could result in contact between the tip and the sample-holder stage or any other part of the equipment. The tip of the capillary is not easily visualized (especially in the case of capillaries with tips with an internal diameter of 5 μm), hence the importance of keeping a safety distance between the sample-holder stage and all the visible parts of the capillary.
13. The step of locating and visualizing the tip of the capillary can be a frustrating task again because of the small dimensions of the tip. A simple way to focus the tip quicker and safer consists of four easy steps. (1) Place the thicker part of the body of the capillary in the light path of the two cameras. This is a much bigger object than the tip and to find it on the TV screen is much easier. (2) Focus this part of the capillary using the *x*-, *y*-, and *z*-knobs. (3) Move the capillary up right until the tip is seen on the TV screen. (4) Focus the tip.
14. In the injection mode, several parameters can be tuned, the injection pressure in the range 0–150 kPa, the time of injection "on timer," and the time of delay "off timer."
15. By adjusting the parameters of the dispenser controller in the "inj" mode the drop can be injected out of the capillary through its tip. There is a minimum pressure, time of injection (on timer), and time of delay (off timer) needed to form the drop depending on the size of the internal diameter of the tip of the capillary. Table 1 shows recommended starting settings for forming drops.

Table 1
Starting settings (pressure, on timer, off timer) of the dispenser controller in the "inj" mode to produce a droplet of ultrapure water

ID (μm)	Pressure (kPa)	On timer (ms)	Off timer (ms)
50	10	2	0
30	20	3	0
5	50	4	0

ID is the internal diameter of the tip of the capillary

By appropriately tuning the parameters with settings other than the ones in Table 1, drops with different volumes can be obtained and discharged on the surfaces. Also, drops with the same volume can be obtained with different settings.

Sometimes, when the drop is discharged the water brings up to the outer surface of the tip instead of being injected on the tested surface. This can be because the time of injection is too short or because the distance between the capillary tip and the surface is too long. In both cases, adjusting parameters of the dispenser controller can overcome this problem; i.e., increasing the time of injection and increasing the pressure, respectively.

16. Most of the times the surface is difficult to be visualized on the monitor window. The approach of the surface to the capillary tip has to be performed very carefully to avoid any contact between the two of them. If the tip and the surface get in contact the tip and/or the surface can be damaged and the surface can be contaminated by the water-repellent liquid that coats the outer part of the tip. Accurate adjustment of the luminosity in the area of measurement while approaching the surface to the capillary tip is the key to clearly distinguish the surface.

17. The system is also able to record the evolution of the contact angle of the ultrapure water on the biosurface as a function of time. The lateral camera can be set to acquire up to 60 pictures per second which can be individually analyzed to assess the contact angle. Videos displaying the dynamic wettability response of the biosurfaces can also be built using that set of images.

    The fast performance of the camera is also useful when the smallest drops of water (10^{-4} to 10^{-5} μL) are discharged. This is because those droplets tend to evaporate quickly, much earlier than the stable state of the drop on the surface is attained.

18. Emptying the capillaries after a session of measurements is a maintenance task to prevent the blocking of the tip of the

capillary. In the case of capillaries with tips with the smallest internal diameter is not unusual that cannot be re-used because they have been obturated after the last session, even if you empty the capillary following the steps of this protocol.

References

1. Young, T. (1805) An Essay on the Cohesion of Fluids *Phil Trans Roy Soc London* **95**, 65–87.
2. Good, J.R., and van Oss, C.J. (1992) The Modern Theory of Contact Angles and the Hydrogen Bond Components of Surface Energies. In: Schraeder, M.E., and Loeb, G.I., eds. Modern Approaches to Wettability. Theory and Applications; New York and London: Plenum Press, 1–28.
3. Butt, H.J., Graf, K., and Kappl, M. (2004) Contact angle phenomena and wetting. In: Physics and Chemistry of Interfaces; Weinheim: Wiley-VCH Verlag GmbH & Co. KGaA, 125–52.
4. Ratner, B.D. (2004) Surface Properties and Surface Characterization of Materials. In: Ratber, B.D., Hoffman, A.S., Schoen, F.J., and Lemons, J.E., eds. Biomaterials Science. An Introduction to Materials in Medicine; San Diego, CA: Elsevier Academic Press, 40–58.
5. Michiardi, A., Aparicio, C. Ratner, B., Planell, J.A., and Gil, F.J. (2007) The influence of surface energy on competitive protein adsorption on oxidized NiTi surfaces *Biomaterials* **28**, 586–94.
6. Pegueroles, M., Aparicio, C., Bosio, M., Engel, E., Gil, F.J., Planell, J.A., and Altankov, G. (2010) Spatial Organization of Osteoblast Fibronectin-Matrix on Titanium Surface – Effects of Roughness, Chemical Heterogeneity, and Surface Free Energy *Acta Biomaterialia* **6**, 291–301.
7. Straumann Holding AH, SLActive – the third generation in implant surface technology http://www.straumann.com/ci_ch_presskit_slactive_ids07.pdf (accessed October 2010).
8. Matsuda, N., Shimizu, T., Yamato, M., and Okano, T. (2007) Tissue Engineering Based on Cell Sheet Technology *Adv Mater* **19**, 3089–99.
9. Chu, K.-H., Xiao, R., and Wang, E.N. (2010) Uni-directional liquid spreading on asymmetric nanostructured surfaces *Nature Mater* **9**, 413–7.
10. Chiou, N.-R., Lu, C., Guan, J., Lee, L.J., and Epstein, A.J. (2007) Growth and alignment of polyaniline nanofibres with superhydrophobic, superhydrophilic and other properties *Nature Nanotech* **2**, 354 – 57.
11. Mata, A., Hsu, L., Capito, R., Aparicio, C., Henrikson, K., and Stupp, S.I. (2009) Micropatterning of bioactive self-assembling gels *Soft Matter* **5**, 1228–36.
12. Hannachi, E., Itoga, K., Kumashiro, Y., Kobayashi, J., Yamato, J., and Okano, T. (2009) Fabrication of transferable micropatterned-co-cultured cell sheets with microcontact printing *Biomaterials* **30**, 5427–32.
13. Maritines, E., Seunarine, K., Morgan, H., Gadegaard, N., Wilkinson, C.D.W., and Riehle, M.O. (2005) Superhydrophobicity and superhydrophilicity of regular nanopatterns. *Nano Lett* **5**, 2097–130.
14. Dupuis, A., Leopoldes, J., Bucknall, D. G., and Yeomans, J. M. (2005) Control of drop positioning using chemical patterning *Appl Phys Lett* **87**, 024103.
15. Masahashi, N., Mizukoshi, Y., Semboshi, S., and Ohtsu, N. (1998) Superhydrophilicity of rutile TiO_2 prepared by anodic oxidation in high concentration sulfuric acid electrolyte *Chem Letters* **37**, 1126–7.
16. Taylor, M, Urquhart, A.J., Zelzer, M., Davies, M.C., and Alexander, M.R. (2007) Picoliter Water Contact Angle Measurement on Polymers *Langmuir* **23**, 6875–8.
17. Erbil, H.Y.(1997) Surface Tension of Polymers. In: Birdi, K.S., ed. Handbook of Surface and Colloid Chemistry; Boca Raton, FL: CRC Press, 259–306.

Chapter 12

AFM to Study Bio/Nonbio Interactions

Holger Schönherr

Abstract

This chapter describes a versatile approach to immobilize proteins and other biomolecules on reactive self-assembled monolayers on gold as a means to study interactions (forces) between these biomolecules and nonbiological entities. Biomolecules are either immobilized on the surface of flat substrates or on the surface of gold-coated atomic force microscopy (AFM) probe tips. In addition to the immobilization protocols, the actual AFM experiments in liquid and the quantitative analysis of adhesive forces receive attention. With these procedures and tools at hand, a wide variety of problems in the area of bio/nonbio interactions can be addressed experimentally.

Key words: Atomic force microscopy, Chemical force microscopy (CFM), Tip modification, Tip-sample forces, Bio/nonbio interactions, Self-assembled monolayers

1. Introduction

Biocompatibility, biointegration, and functionality of natural or man-made biomaterials are frequently governed by the underlying surface properties (1, 2). The importance of these properties is also evident from the crucial role they play for biomedical devices and implants, as well as biosensors in various formats (3, 4). In particular, in biomedical devices, such as, e.g., subcutaneously implanted online sensors for glucose, the surface passivation against nonspecific protein adsorption is a major point of concern. Various strategies have been proposed to modify the corresponding surfaces to impart them with the so-called stealth properties.

In general, processes that occur at the newly formed interface between an object that contacts biological systems, such as bacteria, cells, or fluids of biological origin, such as e.g., blood or serum, determine to a significant extent whether or not the object fulfills its function. The interfaces alluded to above, where the (initial)

Melba Navarro and Josep A. Planell (eds.), *Nanotechnology in Regenerative Medicine: Methods and Protocols*, Methods in Molecular Biology, vol. 811, DOI 10.1007/978-1-61779-388-2_12,

contact between biomaterial and biological systems occurs, are commonly called "biointerfaces" (1, 5). Biointerfaces comprise also the location where a substantial fraction of biochemical reactions and noncovalent molecular interactions (molecular recognition processes) in biology take place.

In more general terms the interaction of biomolecules (e.g., polysaccharides or proteins) (6), and bacteria (7, 8) or cells (9), with artificial, man-made surfaces, and materials can be understood on the basis of molecular forces. Besides the omnipresent van-der-Waals interactions, dipolar, electrostatic, and hydrogen bonding interactions play a pivotal role in this context (10, 11). Since any biomolecule–surface interaction is based on these molecular forces, studies of those forces received increasing attention.

As treated in this chapter, bio/nonbio interactions can be addressed by atomic force microscopy (AFM) approaches, among others. As force-based scanning proximity probe microscopy (12, 13), AFM not only provides high-resolution images of topography and other properties, such as elasticity, but also affords direct access to AFM tip–sample interactions (Fig. 1) (14).

Hence bio/nonbio interactions can be studied, as treated in this chapter, (1) by analyzing a nonbio surface in (or after) contact with a biological system or (2) by direct assessment of the interaction forces between biological and nonbiological molecules (or molecules and materials). In the latter case one type of molecule is firmly attached to the AFM tip (Fig. 1b).

The application of AFM in such studies necessitates the use of AFM in liquid environment at controlled medium, e.g., buffer and temperature conditions. In addition, biological and nonbiological molecules must be immobilized to surfaces (either substrate or

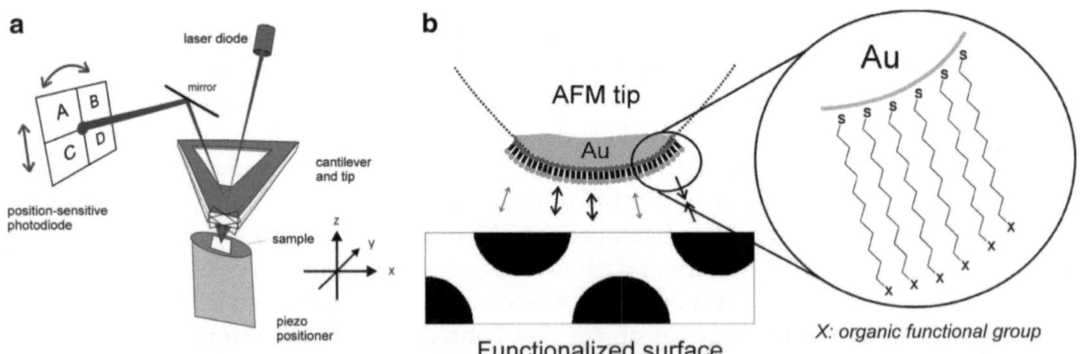

Fig. 1. (**a**) Schematic of AFM. The sample is placed on a piezo transducer, which positions the sample with accuracy <1 nm in all three directions. The interaction forces of the sharp tip mounted on a microcantilever are measured via the detection of the changes in bending angle of the cantilever using an optical deflection system comprising a mirror and a position sensitive photodiode. Topographic images are obtained by maintaining a constant force (i.e., cantilever deflection during scanning). (**b**) Force sensing using an AFM can be rendered chemically specific by modifying the AFM tip with a monolayer exposing a selected functional group functionalized tips (15, 16). This group may be a reactive moiety that enables modification of the tip with biomolecules (17). Adapted with permission from ref. 18.

AFM tip) by appropriate surface chemistry. Due to the wide variety of biological systems of interest, a broad range of different surface modification procedures exists.

The comparison of the specific interaction of cells with (1) covalently attached fibronectin and (2) with a passivated surface comprising a grafted layer of poly(ethylene glycol) serves as an illustrative example (Fig. 2) (21). The interaction of the pancreatic cancer cells (19) shown with two widely different surface chemical compositions was interrogated ex situ. Fixated cells were imaged by intermittent contact mode AFM; the fixation procedure left important subcellular structures as filopodia intact and preserved the cell morphology (20). On fibronectin the cells showed a spread-out appearance, many very long filopodia can be recognized. By contrast, on PEG the cell–film boundary appears to be sharp and the presence of filopodia was not observed.

In this chapter, we focus the attention on a typical *covalent* attachment methodology, which allows one to immobilize proteins

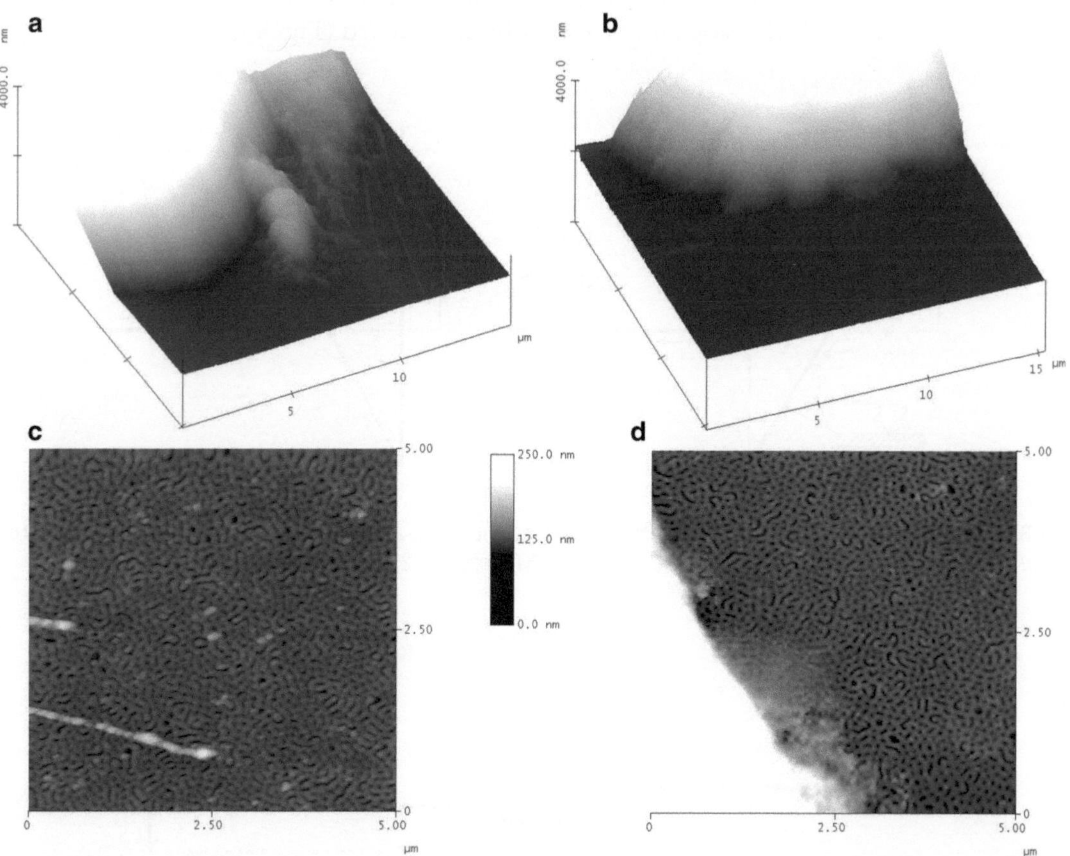

Fig. 2. Tapping mode AFM height images of fixated cancer cells on (**a**), (**c**) fibronectin and (**b**), (**d**) PEG_{5000}-functionalized PS-b-PtBA (data acquired in air). Reproduced with permission from ref. 21. Copyright 2007. Elsevier.

and other biomolecules using carbodiimide chemistry (22–25). In this well-established and broadly applicable reaction, an *N*-hydroxy succinimide active ester reacts under mild conditions in buffered medium with primary amino groups, e.g., of lysine residues. The resulting amide linkage is robust and tethers the biomolecule to the substrate surface or to the AFM tip. Alternatively, the protein may be attached via pending cystein residues directly to a gold-covered AFM tip.

The basis for the force measurements by AFM is the calibration of the optical deflection detection system (Fig. 1a), and the cantilever normal spring constant. There are various procedures for the reliable calibration of normal forces. These include the thermal fluctuation method (26–28), the reference lever (29), or the added mass technique (30). Here, we describe the reference lever method in some detail.

The *sensitivity of the optical deflection detection system* is calibrated by recording a *f–d* curve on a stiff substrate, e.g., a glass slide or a piece of silicon (Fig. 3a). The slope of this plot (i.e., the hard wall contact region) equals to one. If the *f–d* curve is recorded on the compliant reference cantilever of known spring constant, the slope is correspondingly reduced (Fig. 4).

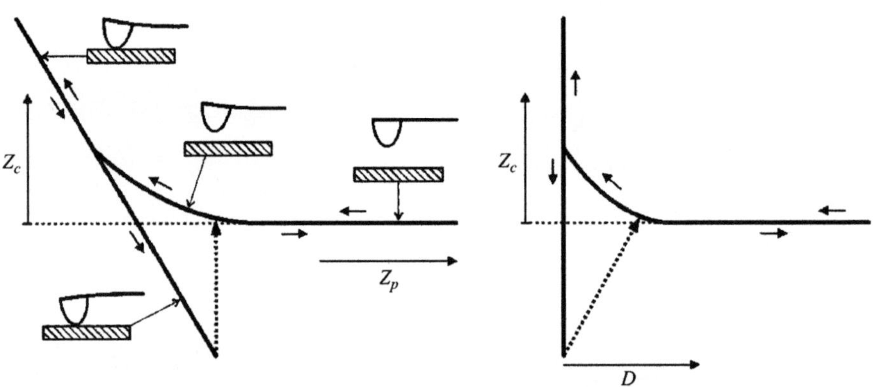

Fig. 3. Schematic of a typical cantilever deflection-vs.-piezo height (Z_c-vs.-Z_p) curve (*left*) and corresponding Z_c-vs.-*D* plot (*right*), with $D = Z_c + Z_p$. Reprinted with permission from ref. 31. Copyright 2005. Elsevier.

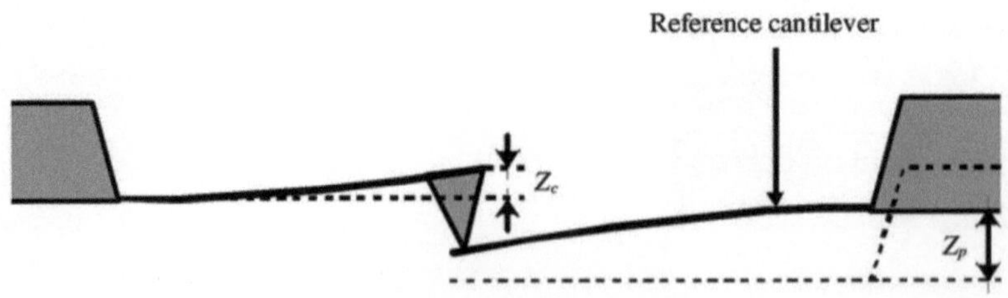

Fig. 4. Schematic of cantilever calibration on reference lever. Reprinted with permission from ref. 31. Copyright 2005. Elsevier.

To ensure adequate sensitivity, the reference lever should have a spring constant close to the cantilever that is calibrated. The spring constant k_N is given by

$$k_N = k_{ref}\frac{Z_p - Z_c}{Z_c} = k_{ref}\frac{1 - Z_c / Z_p}{Z_c / Z_p} \quad (1)$$

where the cantilever deflection is denoted as Z_c and the height of the piezoelectric translator as Z_p (zero is defined for the situation, when the tip just touches the reference cantilever and no deflection has been detected).

Since the spring constant of the reference cantilever k_{ref} is known, k_N can be obtained from the measured slope of the force curve Z_c/Z_p obtained on the reference cantilever in the contact regime.

The surface modification procedures for substrates and AFM tips discussed below, as well as the approaches for AFM imaging and force–displacement measurements serve as illustrative examples and can be expanded to basically any bio/nonbio system of interest to the reader.

2. Materials

1. AFM (Veeco multimode IIIa) with E_v scanner.
2. AFM liquid cell (Veeco).
3. Si_3N_4 AFM cantilevers and tips (Veeco type DNP).
4. Si_3N_4 AFM tips, gold-coated on the tip side (Veeco type NPG).
5. Force calibration cantilever array (Nanoscience Instruments, Inc.).
6. Gold substrates (11×11 mm^2, 250 nm Au on 2 nm Cr on borosilicate glass, e.g., from Arrandee, Werther, Germany).
7. Oxygen plasma cleaner (SPI Supplies, Plasma Prep II).
8. Milli-Q water (Millipore, Milford, MA).
9. 11,11′-Dithio-bis(*N*-hydroxysuccinimidylundecanoate) (NBS Biologicals Ltd.).
10. Bovine serum albumin (BSA) (Sigma, St. Louis, MO).
11. BT-toxin in HEPES buffer.
12. Brush border membrane vesicles in buffer.
13. κ-Carrageenan (Sigma (St. Louis, MO)).
14. Phosphate buffer saline (PBS) tablets (Sigma-Aldrich, Steinheim, Germany) is used to prepare PBS (pH 7.4).

15. Microscopy glass slides.
16. Chloroform (p.a.).
17. Ethanol (p.a.).
18. Argon (gas).
19. Nitrogen (gas).
20. Ferromagnetic AFM sample puck.
21. Double-sided sticky tape.
22. Parafilm (American National Can, Neenah, WI).

3. Methods

3.1. Surface Modification

3.1.1. Capture Layer for Biomolecules

The immobilization of biomolecules with pending primary amino groups on gold substrates that are functionalized with a monolayer of 11,11′-dithio-bis(*N*-hydroxysuccinimidylundecanoate) is described first.

1. Clean glassware and gold substrates by rinsing with chloroform, ethanol, and Milli-Q water (in this order), blow these parts dry in a stream of nitrogen and place them in an oxygen plasma cleaner.
2. Clean precleaned glassware in the oxygen plasma cleaner, e.g., for 5 min, 30 mA, 60 mTorr (refer to manual for operation).
3. Prepare a 10^{-3} M solution of dithio-bis(*N*-hydroxysuccinimidylundecanoate) in ethanol, bubble argon gas for 15 min through solution, store solution in tightly closed beaker in the dark.
4. Clean precleaned gold substrates in the oxygen plasma cleaner (see ref. 2; refer to manual for operation).
5. Immerse plasma cleaned gold substrates into solution of dithio-bis(*N*-hydroxysuccinimidylundecanoate), close beaker tightly, and store in the dark for 12 h.
6. Prepare solution of BSA (1.0×10^{-3} M) in PBS (or alternatively BT toxin in HEPES buffer).
7. Remove gold substrates with tweezers from assembly solution, rinse immediately extensively with chloroform, ethanol, and water to remove any physisorbed material, and immerse immediately into the protein solution for 60 min; then remove sample and rinse with buffer.
8. Transfer of sample to AFM sample puck (see Subheading 3.3.1 and notes).

3.1.2. κ-Carrageenan Layer on Glass

1. Clean microscopy glass slide by rinsing with chloroform, ethanol, and Milli-Q water (in this order), blow dry in a stream of nitrogen and place it in an oxygen plasma cleaner.
2. Clean precleaned glassware in the oxygen plasma cleaner, e.g., for 5 min, 30 mA, 60 mTorr (refer to manual for operation).
3. Dissolve κ-carrageenan in NaCl (0.1 M) in Milli-Q water (conc. ~0.5 g/L).
4. Prepare film by placing a few drops of the κ-carrageenan solution on the cleaned glass slide.
5. Evaporate water in a gentle nitrogen stream; place in vacuum overnight.

3.2. AFM Tip Modification with Capture Layer and Protein

1. Prepare a 10^{-3} M solution of dithio-bis(*N*-hydroxysuccinimidylundecanoate) in ethanol, bubble argon gas for 15 min through solution, store solution in tightly closed vial in the dark.
2. Clean gold-coated AFM tips by rinsing with chloroform, ethanol, and Milli-Q water (in this order), blow these parts dry in a stream of nitrogen and place them in an oxygen plasma cleaner. During rinsing and drying the cantilever-supporting chips must also be held firmly with the tweezers to avoid damage to the cantilever/tip.
3. Immerse plasma cleaned gold substrates into solution of dithio-bis(*N*-hydroxysuccinimidylundecanoate), close vial tightly, and store in the dark for 12 h.
4. Prepare solution of BSA (1.0×10^{-3} M) in PBS.
5. Remove monolayer-coated tips carefully with tweezers from assembly solution, rinse immediately extensively with chloroform, ethanol, and water to remove any physisorbed material, and immerse immediately into the protein solution for 60 min; then remove sample and rinse with buffer.
6. Mount tip in AFM liquid cell.

3.3. AFM Measurements

A NanoScope IIIa Multimode AFM (Veeco) with a liquid cell is used. Prior to the imaging or force experiments, the AFM setup is equilibrated in PBS until the photodiode drift is eliminated. The spring constants of the cantilevers will be calibrated using a reference lever *after* a successful experiment.

3.3.1. AFM Operation in Liquid (Open Cell Configuration, No O-Ring)

A dummy sample comprising an identical substrate like the one to be studied is attached using double side sticky tape to the metallic sample older puck and is placed on the scanner. To protect the scanner from accidentally spilled liquid, the scanner is protected by wrapping it with Parafilm (American National Can, Neenah, WI). For the operation of the multimode AFM without rubber *O*-ring, a conventional cantilever is inserted in the liquid cell, the laser is aligned and the crude approach is carried out in air. The optical head is lowered such that the tip is close to the sample surface, ready for engagement. Subsequently, the optical head is driven briefly upward using the stepper motor. The head is removed vertically and the tip is exchanged against the functionalized tip. The laser is realigned. In addition, the dummy sample is replaced by the actual sample. A drop of buffer solution is placed onto the sample and the optical head is mounted again. The liquid now spans the gap between the sample and the liquid cell. If the drop is too small or does not symmetrically span the gap, additional buffer may be added very carefully through the inlet or from the side using a syringe.

The laser alignment is changed using the mirror to correct for the effect of light refraction. Now the experiment can be started when the differential photodiode signal(s) show stable readings. During the experiment buffer may evaporate, so it should be replenished regularly after withdrawing the tip from the surface.

3.3.2. AFM Imaging Basics in Liquid

The imaging of samples in liquid can be performed at considerably lower forces compared to air. The force setpoint and feedback loop gains are typically adjusted stepwise starting with very small scan sizes (100 nm × 100 nm), to avoid potential damage of the probe tip. After the error (deflection) signal has been minimized, the scan size is increased in small increments, while the gains and scan rate are adjusted correspondingly. Low normal forces result in low lateral forces, thus the minimization of imaging forces is important. The adjustment can be based on acquired force–displacement curves (Fig. 3a); setpoint values close to the out of contact deflection yield minimized normal forces.

3.3.3. AFM Force Measurement Basics in Liquid (Open Cell Configuration, No O-Ring)

Force measurements are performed with BSA-modified tips (Subheading 3.2). To avoid tip damage it is advisable to engage with a scan size of 0 nm^2 and upon engagement to switch to the force data acquisition immediately. Force displacement data is captured using a rate of ~1 Hz. To prevent too large repulsive forces in the tip–sample contact region, a relative trigger (relative to the baseline) is applied. The ramp size is adjusted such that the important features of the force–displacement curve possess enough pixels. For quantitative analysis, 100 or more force curves are collected at various locations. These location are chosen using the manual *x*, *y* offset.

After a successful experiment, the *force constant calibration* is carried out in air. First the sample is replaced by a stiff substrate (e.g., glass) and several *f–d* curves are recorded. From the slope of these curves the cantilever sensitivity (V/nm) is calculated. Subsequently, the sample with the force calibration lever is inserted into the AFM and the tip is engaged on the very end of the chosen lever. Again several *f–d* curves are captured to be able to calculate the spring constant as outlined above.

The deflection measured in nanometer is calculated by dividing the deflection voltage with the cantilever sensitivity; it is converted to force by multiplication with the value of k_N determined above. The extension between sample and cantilever is calculated from the sum of the deflection and the height signals (compare Fig. 3). The observed individual pull-off events can be analyzed individually and plotted in histograms.

3.3.4. Proteins on NHS SAM

Individual BT-toxin molecules can be discerned that were coupled covalently to the NHS esters exposed at the monolayer of dithiobis(*N*-hydroxysuccinimidylhexadecanoate) on gold (Fig. 5). The interaction of individual of these molecules can be probed by AFM approaches. In addition, the interaction with brush border membrane vesicles that carry the corresponding natural receptors can be traced (32). For this experiment, the AFM was operated in intermittent contact (tapping) mode. For TM-AFM operation under liquid, the system is set up similar to contact mode in liquid. However, prior to engaging the cantilevers' resonance frequency is determined in the frequency sweep (cantilever tune).

By AFM the number density of immobilized vesicles and their properties can be directly probed. In addition, the specificity of the adsorption can be tested when different substrates and substrate functionalities are being used.

3.3.5. Forces Between BSA (on AFM Tip) and κ-Carrageenan Films

Force-extension data captured with an unmodified Si_3N_4 AFM tip on a film of κ-carrageenan is shown in Fig. 6. The plot shows a restoring force at extension values of ~80 nm. When the tip–polymer interaction is no longer able to hold the rapidly increasing force exerted by the stretched polymer chain, the contact breaks. The jump-out (pull-off) force provides an estimate of the tip–polymer adhesion force.

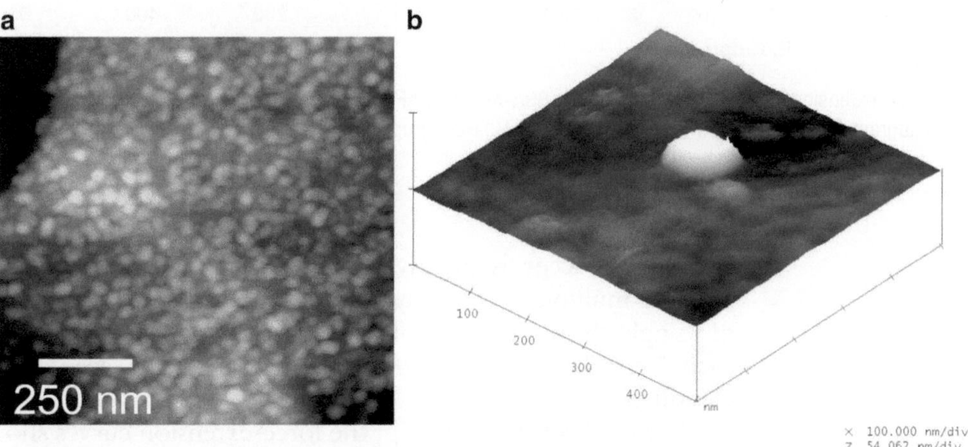

Fig. 5. *Left*: Contact mode AFM height image of BT-toxin covalently coupled to a SAM of a NHS ester disulfide. *Right*: Tapping mode (TM)-AFM image of brush border membrane vesicles attached to BT toxin covalently coupled to a SAM of a NHS ester disulfide (A. T.A. Jenkins; S. Liu; H. Schönherr unpublished data).

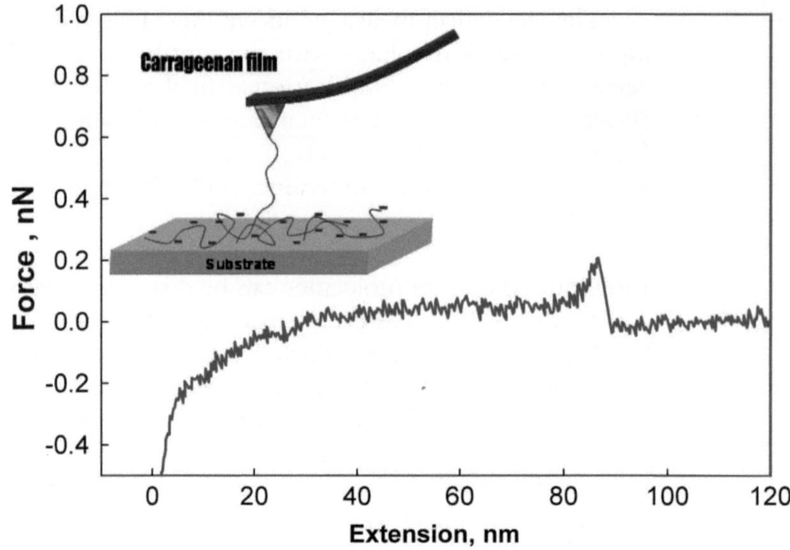

Fig. 6. Force-extension curve measured between silicon nitride AFM tip and κ-carrageenan in 0.1 M NaCl solution. (Lee, J.; Schönherr, H.; Ruengruglikit, C; Huang, Q. unpublished data).

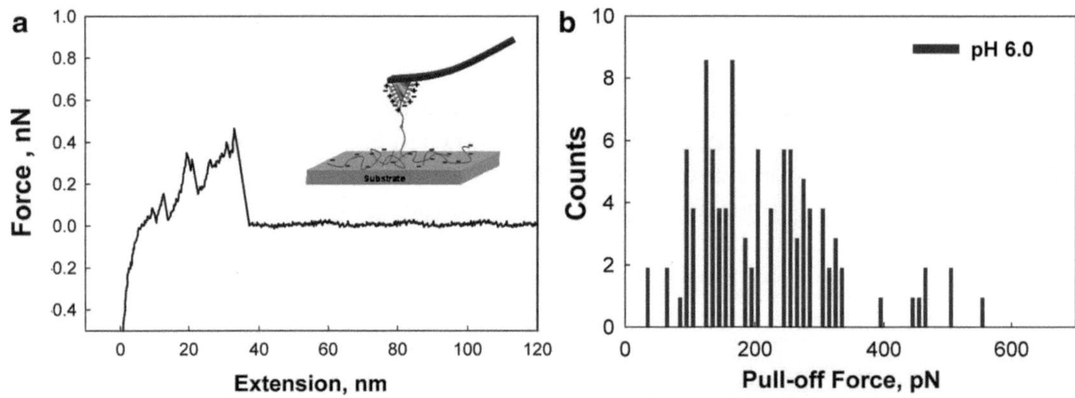

Fig. 7. (**a**) Force-extension curve measured between BSA-modified AFM tip and κ-carrageenan in 0.1 M NaCl solution at pH 6. (**b**) Histogram of pull-off forces. (Lee, J.; Schönherr, H.; Ruengruglikit, C; Huang, Q. unpublished data).

Such curves occur frequently, however, curves with no chain pulling event or multiple events may also occur. Since the elastic properties of a single chain can be obtained by fitting the data to the worm-like chain model (or other models), single molecule events can be identified (33).

For the BSA modified tip, the force-extension curves shown in Fig. 7 display a much richer structure. Several local maxima are observed until the tip–sample contact breaks. These maxima contain contributions of forces from the stretching of the polymer, the protein, or both, and also result from force-induced dissociation of locally strongly interacting bonds. These interactions characterize in parts the specific binding between the BSA immobilized on the AFM tip and κ-carrageenan. The unfolding of proteins is of course a topic of widespread interest and can be addressed on substrates prepared according to section using uncoated AFM tips (34). Likewise the interaction between a given protein and a nonbiological surface can be tackled using the procedure in Subheading 3.2.

4. Notes

1. Surface modification and monolayers: Extremely clean substrates are a prerequisite for successful work. Minute traces of contaminants may adhere strongly to the surface and hence lead to ill-defined surfaces. In particular, the cleaned gold should be hydrophilic during rinsing with water.
2. Contact angle measurement are good measures to assure quality control. The monolayer of NHS-C10 possesses a water contact angle of 60°.
3. The handling of cantilever chips must occur with utmost care; only touch the support chips on the side and avoid contact of the tip or cantilever with any glass vial wall or else.
4. To introduce a sample, which should be kept covered by buffer at all times, into the AFM a precleaned rubber *O*-ring is pressed onto the sample supported on, e.g., a glass slide immersed under buffer. The slide and the rubber ring are then taken out of solution and placed onto a sheet of adsorbent paper to dry the bottom side of the slide. If necessary, buffer is refilled into the rubber ring using a syringe. Next the sample is transferred onto the sample holder which is covered with pressure sensitive adhesive. The holder is mounted onto the protected scanner with minimal delay. Then the rubber ring is removed with tweezers and additional buffer is deposited at the same time onto the sample ensuring that the water spans the distance between the sample and the liquid cell of the AFM after the AFM optical head is put onto the scanner. After a brief equilibration period of typically several minutes in order to obtain a stable photodiode reading, the experiments can be started.
5. The tip is protected by engaging with a scan size of zero, slowly increasing the scan size while adjusting the feedback parameters.
6. During AFM measurements with modified AFM tips lateral forces must be minimized by minimizing normal force and using small scan rates.

7. Normal forces can also be minimized by lowering the setpoint during imaging until the tip does no longer tracks the surface properly. This effect is seen by the appearance of a halo behind elevated objects in the direction of the data collection. To ensure appropriate imaging the scan rate may be reduced, the gains increased and the force/setpoint slightly increased to achieve proper tracking of the surface.
8. AFM in liquid: Care must be taken not to spill any liquid onto the piezo transducer since this can cause a local short circuit and thus terminal damage to the piezo. In order to protect the piezo from accidental liquid spillage, the top of the scanner tube can be wrapped into a very thin polymer film (e.g., Parafilm, American National Can, Neenah, WI). For this purpose, the film is prestretched and then gently wrapped over the tube end.
9. For the force constant calibration one must ensure that the data is acquired at the very end of the force calibration cantilever. For this purpose, the tip is step-wise moved toward the end of this lever. Only data captured at the very end is used in the calculations of k_N.
10. Adhesive interactions (forces) have been shown to depend on the pulling rate (35). This rate is in most cases not equal to the piezo retraction rate because the force builds up nonlinearly due to the elastic properties of soft (usually polymeric) linkers.
11. For TM-AFM in liquid it is necessary to locate the cantilever resonance frequency in the *cantilever tune* menu of the software. The tuning of the cantilever is performed after the crude approach has been carried out. Due to the damping of the forced oscillation by the imaging medium the resonance curves in liquid typically look rather broad. The operation (excitation) frequency is tuned to a side of a peak, where the gradient is steep and near constant. The amplitude is typically lower in liquid than in air, however, care must be taken that it is sufficiently high that the adhesive interactions experienced by the probe tip are overcome during each cycle of oscillation.

Acknowledgments

The author gratefully acknowledges Dr. A.T.A. Jenkins and Prof. Dr. Q. Huang for granting their kind permission to use unpublished collaborative data in this chapter.

References

1. Kasemo, B. (2002) Biological surface science *Surf. Sci.* **500**, 656–677.
2. Kumar, C. S. S. R. (Ed.) (2006) *Tissue, Cell and Organ Engineering*, Wiley-VCH.
3. Schena, M. (2003) *Microarray Analysis*, Wiley-Liss, Wiley & Sons, NJ.
4. Shankaran, D. R. and Miura, N. (2007) Trends in interfacial design for surface plasmon resonance based immunoassays *J. Phys. D: Appl. Phys.* **40**, 7187–7200 and references herein.
5. Castner, D. G. and Ratner, B. D. (2002) Biomedical Surface Science: Foundations to Frontiers *Surf. Sci.* **500**, 28–60.
6. Denis, F.A., Hanarp, P., Sutherland, D.S., Gold, J., Mustin, C., Rouxhet, P.G., and Dufrene, Y.F. (2002) Protein adsorption on model surfaces with controlled nanotopography and chemistry *Langmuir* **18**, 819–828.
7. Vadillo-Rodríguez, V., Busscher, H. J., Norde, W., de Vries, J., and van der Mei, H. C. (2004) Atomic force microscopic corroboration of bond aging for adhesion of Streptococcus thermophilus to solid substrata *J. Coll. Interf. Sci.* **278**, 251–254.
8. van der Mei, H. C., de Vries, J., and Busscher, H. J. (2010) Weibull analyses of bacterial interaction forces measured using AFM *Coll. Surf. B: Biointerfaces*, doi:10.1016/j.colsurfb.2010.03.018.
9. Hoh; J. H. and Schoenenberger, C. A. (1994) Surface-morphology and mechanical-properties of MDCK monolayers by atomic force microscopy *J. Cell Sci.* **107**, 1105–1114.
10. Israelachvili, J. N. (1991) *Intermolecular and Surface Forces*, 2nd edn., Academic Press, London.
11. Israelachvili, J. (2005) Differences between non-specific and bio-specific, and between equilibrium and non-equilibrium, interactions in biological systems *Quarterly Rev. Biophys.* **38**, 331–337.
12. Binnig, G., Quate, C. F., and Gerber, C. (1986) Atomic Force Microscope *Phys. Rev. Lett.* **56**, 930–3.
13. Schönherr, H. and Vancso, G. J. (2010) *Scanning Force Microscopy of Polymers*, Springer, Vienna.
14. Colton, R. J., Engel, A., Frommer, J. E., Gaub, H. E., Gewirth, A. A., Guckenberger, R., Rabe, J., Heckl, W. M., and Parkinson, B., (1998) *Procedures in Scanning Probe Microscopies*, Wiley, New York.
15. Frisbie, C. D., Rozsnyai, L. F., Noy, A., Wrighton, M. S., and Lieber, C. M. (1994) Functional Group Imaging by Chemical Force Microscopy *Science* **265**, 2071–4.
16. Schönherr, H. and Vancso, G. J. (2006) Chemical Force Microscopy: Nanometer Scale Surface Analysis with Chemical Sensitivity *in SPM beyond imaging: Manipulation of Molecules and Nanostructures*, Samori, P. (Ed.), Wiley-VCH, 275–314.
17. Lee, G. U., Kidwell, D. A., Colton, R. J. (1994) Sensing discrete streptavidion biotin interactions with atomic force microscopy *Langmuir* **10**, 354–357.
18. Schönherr, H. (1999) From Functional Group Ensembles to Single Molecules: Scanning Force Microscopy of Supramolecular and Polymeric Systems *Ph.D. thesis*, University of Twente, The Netherlands.
19. Elsässer, H.P., Lehr, U., Agricola, B., Kern, H.F. (1992) Establishment and Characterization of two Cell Lines with Different Grades of Differentiation Derived from one Primary Human Pancreatic Adenocarcinoma *Virchows Arch. B Cell Pathol. Incl. Mol. Pathol.* **61**, 295–306.
20. Kemper, B., Carl, D., Schnekenburger, J., Bredebusch, I., Schäfer, M., Domschke, W., and von Bally, G. (2006) Investigations on Living Pancreas Tumor Cells by Digital Holographic Microscopy *J. Biomed. Opt.* **11**, 34005.
21. Feng, C. L., Embrechts, A., Bredebusch, I., Bouma, A., Schnekenburger, J., García-Parajó, M., Domschke, W., Vancso, G. J., and Schönherr, H. (2007) Tailored Interfaces for Biosensors and Cell-Surface Interaction Studies via Activation and Derivatization of Polystyrene-block-Poly(tert-butyl acrylate) Thin Films *Europ. Polym. J.* **43**, 2177–2190.
22. Schönherr, H. (2009) Coupling Chemistries for the Modification and Functionalization of Surfaces to Create Advanced Biointerfaces *in Surface Design: Applications in Bioscience and Nanotechnology*, Förch, R., Schönherr, H., and Jenkins, A. T. A. (Eds.), WILEY-VCH, pp. 3.
23. Staros, J. V., Wright, R. W., and Swingle, D. M. Enhancement by N-Hydroxysulfosuccinimide of Water-Soluble Carbodiimide-Mediated Coupling Reactions (1986) *Anal. Biochem.* **156**, 220–2.
24. Lahiri, J., Isaacs, L., Tien, J., and Whitesides, G. M. A Strategy for the Generation of Surfaces Presenting Ligands for Studies of Binding Based on Active Ester as a Common Reactive Intermediate: A Surface Plasmon Resonance Study (1999) *Anal. Chem.* **71**, 777–790.

25. Wojtyk, J. T. C., Morin, K. A., Boukherroub, R., and Wayner, D. D. M. Modification of Porous Silicon Surfaces with Active Ester Monolayers (2002) *Langmuir* **18**, 6081–7.
26. Levy, R. and Maaloum, M. (2002) Measuring the Spring Constant of Atomic Force Microscopy Cantilevers: Thermal Fluctuation and Other Methods *Nanotechnology* **13**, 33–37.
27. Ma, H. L., Jimenez, J., and Rajagopalan, R. (2000) Brownian Fluctuation Spectroscopy Using Atomic Force Microscopes *Langmuir* **16**, 2254–2261.
28. Butt, H. J. and Jaschke, M. (1995) Thermal Noise in Atomic Force Microscopy *Nanotechnology* **6**, 1–7.
29. Tortonese, M. and Kirk, M. (1997) Characterization of Application-Specific Probes for SPMs *Proc. SPIE* **3009**, 53–60.
30. Cleveland, J. P., Manne, S., Bocek, D., and Hansma, P. K. (1993) A Non-Destructive Method for Determining the Spring Constants of Cantilevers for Scanning Force Microscopy *Rev. Sci. Instrum.* **64**, 403–405.
31. Butt, H. J., Cappella, B., and Kappl, M. (2005) Force Measurements with the Atomic Force Microscope: Technique, Interpretation and Applications. *Surf. Sci. Rep.* **59**, 1–152.
32. Siqueira, H. A. A., González-Cabrera, J., Ferré, J., Flannagan, R., and Siegfried, B. D. (2006) Analyses of Cry1Ab Binding in Resistant and Susceptible Strains of the European Corn Borer *Appl. Environ. Microbiol.* **72**, 5318–5324.
33. Janshoff, A., Neitzert, M., Oberdörfer, Y., and Fuchs, H. (2000) Force Spectroscopy of Molecular Systems - Single Molecule Spectroscopy of Polymers and Biomolecules *Angew. Chem. Int. Ed.* **39**, 3212–3237.
34. Rief, M., Gautel, M., Oesterhelt, F., Fernandez, J.M., and Gaub, H.E. (1997) Reversible unfolding of individual titin immunoglobulin domains by AFM *Science* **276**, 1109–1112.
35. Ludwig, M., Rief, M., Schmidt, L., Li, H., Oesterhelt, F., Gautel, M., and Gaub, H.E. (1999) AFM, a tool for single-molecule experiments *Appl. Phys. A.* **68**, 173–176.

Chapter 13

Evaluation of Cytocompatibility and Cell Response to Boron Nitride Nanotubes

Gianni Ciofani and Serena Danti

Abstract

While in the last years applications of carbon nanotubes in the field of biotechnology have been largely proposed, so far biomedical applications of boron nitride nanotubes (BNNTs) are still totally unexplored. BNNTs show very interesting physical properties that might be exploited in the nanomedicine field. To fill up the lack of biocompatibility studies on BNNTs, our group has recently begun a rational investigation on interactions between BNNTs and different cell lines. This chapter reports on preliminary cytocompatibility studies carried out on human neuroblastoma (SH-SY5Y) and on mouse myoblast (C2C12) cell lines as model of neural and skeletal muscle cells, respectively, highlighting the methods that allowed us to evaluate the main parameters of interests for cytocompatibility, such as cell viability, metabolism, apoptosis, and differentiation.

Key words: Boron nitride nanotubes, Cytocompatibility, Cellular uptake, SH-SY5Y cell line, C2C12 cell line

1. Introduction

Boron nitride nanotubes (BNNTs) appear of significant interest for the scientific community because of their potentially unique and important properties for structural and electronic applications (1). A BNNT is a structural analogue of a carbon nanotube (CNT) in nature: alternating B and N atoms entirely substitute for C atoms in a graphitic-like sheet with almost no change in atomic spacing. Despite this, CNTs differ from BNNTs for plenty of chemo-physical properties.

In the latest years, applications of CNTs in the field of biotechnology have been largely proposed (e.g., biosensors, DNA chip, and nanovectors for drug, protein, and gene delivery, etc. (2)), but to date biomedical applications of BNNTs are totally unexplored.

Melba Navarro and Josep A. Planell (eds.), *Nanotechnology in Regenerative Medicine: Methods and Protocols*, Methods in Molecular Biology, vol. 811, DOI 10.1007/978-1-61779-388-2_13, © Springer Science+Business Media, LLC 2012

The higher chemical stability of BNNTs results in the problem of their poor dispersibility in the solvents commonly used for biological applications. Several attempts were performed by our group with different kinds of polymers, solvents, and surfactants normally employed for CNTs, but only positively charged agents (e.g., chitosan, poly(lysine), polyethyleneimine) have finally shown satisfactory results. We have shown, for the first time, a method based on BNNT biopolymeric wrapping in order to obtain homogeneous BNNT aqueous dispersions, that allows cell internalization to occur and studies about interactions and effects on living matter to be efficiently carried out (3).

BNNT dispersions were investigated by focused ion beam (FIB) microscopy and transmission electron microscopy (TEM) imaging; size distribution, *Z*-potential, and UV/Vis spectrum were measured. Moreover, we have proposed a useful method for quantification of BNNTs via spectrophotometric analysis, determining the mass extinction coefficient of the BNNT dispersions at a suitable wavelength (4).

We have reported studies on the cytocompatibility of BNNTs on human neuroblastoma cells and demonstrated, for the first time, that they do not have adverse effects on viability, metabolism, and cellular replication of this cell line. These results represent absolutely the first in vitro investigation of interaction between BNNTs and living cells. The fluorescent labeling of BNNTs with quantum dots enabled, moreover, their tracking in the cellular uptake and the investigation of the uptake mechanism (5).

Concerning physical properties, our group has investigated magnetic properties of BNNTs. In a recent work we reported, for the first time, on the superparamagnetic properties of BNNTs and we explored such properties to pursue magnetic-driven drug delivery (6).

Interesting perspectives in the field of boron neutron capture therapy are also under consideration (7).

Collectively, these experimental observations confirm that BNNTs are suitable for the development of novel nanovectors for cell therapy, drug, and gene delivery, and indeed for other biomedical and clinical applications. Such applications may range from sensors and transducers for the detection of biomolecules, to "sensitive" substrates for tissue engineering, or as delivery systems for a variety of diagnostic o therapeutic agents. In addition, the exploitation of their piezoelectric properties (8) could provide the basis for clinical electro-stimulation therapies for various cardiac, skeletal and visceral muscular, and neurogenic disorders (9).

In this chapter, we have summarized the methods employed and the main results achieved in our laboratories concerning BNNT cytocompatibility studies. Because of their potential applications at the level of the nervous system, we have focused our attention on a cell line that could provide a good model of neural cell.

Moreover, due to the piezoelectricity owned by BNNTs, a positive interaction between muscle cells and BNNTs would potentially allow their exploitation in the future as intracellular nanotransducers, conveying mechano-electric stimulation to sensitive cells, such as myoblasts (10).

The in vitro response of human neuroblastoma (SH-SY5Y) and of mouse myoblast (C2C12) cell lines as model of neural and skeletal muscle cells, respectively, was therefore investigated, following up cell exposure to different concentrations of BNNTs, in order to evaluate their effects on cell viability, metabolism, apoptosis, and differentiation.

2. Materials

2.1. Preparation of the BNNT Dispersions

1. BNNTs provided by the Australian National University, Canberra, Australia, are produced by using a ball milling and annealing method (11, 12). Details of sample purity and composition (provided by the supplier) are as follows: yield >80%, boron nitride >97 wt%, metal catalysts (Fe and Cr) derived from the milling process ~1.5 wt% and absorbed O_2 ~1.5 wt%.
2. PBS 10× stock solution (phosphate-buffered saline solution) cell culture tested (P5493 from Sigma). Prepare working solution by dilution of one part with nine parts water and autoclave before storage at room temperature.
3. Polyethyleneimine (PEI, P3143 from Sigma Aldrich) MW 750,000, soluble in water (>50%). Use as 0.1% solution in distilled water.
4. Poly-L-lysine (PLL, 81339 from Fluka, MW 70,000–150,000). Use as 0.1% solution in distilled water.
5. Carboxyl quantum dots (Invitrogen Qdot® 605 ITK™).
6. *N*-ethyl-*N′*-dimethylaminopropyl-carbodiimide (EDC, 03450 from Fluka) as activator.
7. Branson sonicator 2510 (Bransonic) is used for the preparation of the dispersions.
8. UV–Vis spectra are evaluated with a LIBRA S12 Spectrophotometer (Biochrom).
9. Ultracentrifuge Allegra 64R (Beckman) is used for BNNT dispersion purification.

2.2. SH-SY5Y and C2C12 Cell Line Culture

1. Human neuroblastoma cell line (SH-SY5Y) is obtained from American Type Culture Collection (ATCC CRL-2266).
2. C2C12 cell line is obtained from ATCC (CRL-1772). This line, widely used as muscle cells model, differentiates rapidly,

forming contractile myotubes and producing characteristic muscle proteins (13).

3. Dulbecco's Modified Eagle's Medium (DMEM) and Hams F-12 medium (from Lonza).
4. Fetal bovine serum (FBS), l-glutamine, and penicillin/streptomycin solution from Lonza.
5. Solution of trypsin (0.25%) and ethylenediamine tetraacetic acid (EDTA, 1 mM) from Lonza.
6. Insulin–transferrin–selenium (ITS), liquid 100×, from Sigma Aldrich (I3146).
7. T25 flask for cell culture, 6-well, 24-well, and 96-well cell culture plate treated in polystyrene (BD Biosciences).

2.3. Viability and MTT Metabolic Assay

1. Trypan blue 0.4% solution (T8154, Sigma Aldrich).
2. Burker chamber for cell counting.
3. MTT powder (1-(4,5-dimethylthiazol-2-yl)-3,5-diphenylformazan thiazolyl blue formazan; M2003 from Sigma). Store the MTT as powder at −20°C. Once aliquoted in PBS, at −20°C.
4. Dimethyl sulfoxide (DMSO, D8418, Sigma Aldrich).
5. The absorbance is measured on a Versamax microplate reader (Molecular Devices).

2.4. Apoptosis Assays

1. ApoAlert Kit (Clonetech Laboratories, Inc.) is used to evaluate differences between normal and apoptotic cells after treatments. The kit, to be stored at 4°C, contains:
   (a) Annexin V-FITC (20 μg/ml in Tris–NaCl)
   (b) 1× Binding buffer
   (c) Propidium iodide (50 μg/ml in 1× binding buffer)
2. Positive control performed with sodium dodecyl sulfate (SDS, L4390 from Sigma Aldrich) and exposition to UV radiation.

2.5. Internalization Assays and Fluorescence Microscopy

1. LysoTracker dye (L12491, Invitrogen).
2. Sodium azide (NaN_3, 478484, Carlo Erba).
3. Optical and fluorescence microscopy is performed with a Nikon TE2000U fluorescent microscope equipped with Nikon DS-5MC USB2 cooled CCD camera.

2.6. Protein and DNA Quantification on Differentiated C2C12 Cells

2.6.1. Double-Stranded DNA Quantification

1. PicoGreen kit (from Molecular Probes (14)) comprising PicoGreen Dye, buffer, and DNA standards.
2. Double-distilled water (dd-H_2O).
3. Black 96-well plates (PBI).
4. Fluorescence intensity is measured on a plate reader (Victor3, PerkinElmer Inc.) equipped with appropriate filters (excitation/emission at 485/535 nm).

2.6.2. Total Protein Quantification

1. Bicinchoninic acid (BCA) method (Pierce, (15)), comprising reagents A, B, and bovine serum albumin (BSA) standard solution.
2. Double-distilled water (dd-H_2O).
3. Tissue culture flat-bottom 96-well plates (Sarsted).
4. Absorbance is read at 562 nm on a plate reader (BioRad, Hercules).

3. Methods

3.1. BNNT Dispersion Preparation

1. Prepare the samples of BNNTs by weighing a small amount of BNNTs (5 mg) and then mixing this with 10 ml of a 0.1% PEI or PLL solution in a polystyrene tube.
2. Stir the samples over a hot plate at 70°C for 6 h (check the temperature by a thermocouple).
3. After this treatment, sonicate the sample for 12 h with an output power of 20 W.
4. Finally, centrifuge the sample at 1,100 ×*g* for 10 min to remove nondispersed residuals and impurities.
5. Before use in biological assays, ultracentrifuge PEI or PLL wrapped BNNT dispersions three times at 30,000 ×*g*, for 30 min at 4°C in order to remove polymer in excess from the suspension.
6. A stable dispersion of BNNTs without excess of polymer is obtained by removing the supernatant and resuspending the coated BNNTs in appropriate volume of PBS (Fig. 1, see Note 1).
7. Check the concentration of BNNTs monitoring the absorbance at 500 nm (4).
8. After coating, BNNTs can be covalently labeled with carboxyl derivatized quantum dots for cellular tracking studies. Conjugation is performed between the amino-groups of PEI or PLL and carboxylic-groups of quantum dots, the reaction procedure being carried out as specified by the supplier.
9. Briefly, mix 4 ml of BNNT dispersion of 50 μg/ml with 4 μl of Qdots (8 μM) and 60 μl of *N*-ethyl-*N*′-dimethylaminopropyl-carbodiimide (EDC, 10 mg/ml) as activator see Note 2.
10. Stir the solution gently for 90 min at room temperature for optimal conjugation and, thereafter, centrifuge it (1,000 ×*g*, 10 min) to remove large aggregates.
11. Finally, perform ultracentrifugation (twice at 30,000 ×*g* for 30 min at 4°C) and wash in water to remove unbounded quantum dots and side products.

Fig. 1. Focused ion beam (FIB) microscopy allows high-contrast imaging, showing well-dispersed PLL-BNNTs. For this purpose, a small drop (10 μl) of PLL-BNNT dispersion (5 μg/ml) is placed onto a plasma-cleaned silicon wafer and air dried. The FIB system used is a FEI 200 (FIB localized milling and deposition) delivering a 30 keV beam of gallium ions (Ga^{+}). Reproduced from ref. 7, copyright to the authors.

3.2. Cell Culture Preparation

1. SH-SY5Y complete culture medium is a mixture (1:1) of Hams F12 and DMEM supplemented with 2 mMl-glutamine, 100 IU/ml penicillin, 100 μg/ml streptomycin, and 10% heat-inactivated FBS.
2. C2C12 complete culture medium is a mixture of DMEM supplemented with 2 mM l-glutamine, 100 IU/ml penicillin, 100 μg/ml streptomycin, and 10% heat-inactivated fetal bovine serum.
3. Culture cells at 37°C in a saturated humidity atmosphere containing 95% air/5% CO_2 for proliferation in T25 flasks.
4. Split cells when 90% of confluence is reached. After the medium is removed (for SH-SY5Y collect the medium and centrifuge it in order to recover floating cells), rinse the cells with PBS and incubate at 37°C with 0.25% trypsin, 0.53 mM EDTA solution until they detach. After addition of complete medium to inhibit trypsin, centrifuge the cells, dilute in appropriate volume of medium, and seed them (see Note 3).

5. C2C12 differentiating medium is a mixture of DMEM supplemented with 2 mMl-glutamine, 100 IU/ml penicillin, 100 μg/ml streptomycin, ITS 1×, and 1% heat-inactivated FBS.
6. The BNNT-modified culture medium is obtained by mixing BNNT solutions of appropriate concentration and the culture medium at a ratio 1:10 v/v, obtaining final BNNT concentration of 0, 5, 10, and 15 μg/ml.

3.3. MTT and Viability Assay

1. To determine the effect of BNNTs on cell viability and metabolism, Trypan blue exclusion test, followed by the MTT cell proliferation assay, is used.
2. Culture the cells in six-well plate chambers for 24, 48, and 72 h in a BNNT-modified medium at different concentrations (e.g., 0, 5, 10, and 15 μg/ml).
3. Thereafter, detach the cells and carry out the Trypan blue exclusion test. Trypan Blue is the stain most commonly used to distinguish viable from nonviable cells. Viable cells exclude the dye, while nonviable cells absorb the dye and appear blue.
4. Prepare and dilute, as required, a sample of the cell suspension in PBS. Thereafter, dilute the sample 1:4 in 0.4% Trypan blue see Note 4.
5. After counting with a Burker chamber, split each cell culture and seed 30,000 cells in 96-well plate chambers ($n=6$) see Note 5.
6. Once the adhesion is verified (via optical microscopy, about 6 h after seeding), incubate the cells with MTT 0.5 mg/ml for 3 h.
7. Mitochondrial dehydrogenases of viable cells reduce yellow water-soluble MTT to water-insoluble formazan crystals. Remove the MTT-containing medium and replace it with 100 μl of DMSO and let shaking for 10 min on a platform shaker to solubilize the formazan crystals.
8. Finally, read the absorbance at a wavelength of 570 nm.
9. Results are shown as the mean ± standard error, $n=6$. Analysis of data is performed by analysis of variance (ANOVA) followed by Student's t-test to test for significance, set at 5%.
10. After 24, 48, and 72 h of incubation at different PLL-BNNT concentration (5, 10, and 15 μg/ml), C2C12 densities do not differ from the respective controls. In all cases, cell viability assessed with Trypan Blue is higher than 90%.
11. C2C12 exhibit a slight decrease in metabolic activity following incubation with PLL-BNNTs. Cells incubated with 5 μg/ml of nanotubes sustain a decrease of about 10% after 24, 48, and 72 h of incubation, but all these results are not statistically different from the respective controls (in all cases $p>0.05$).

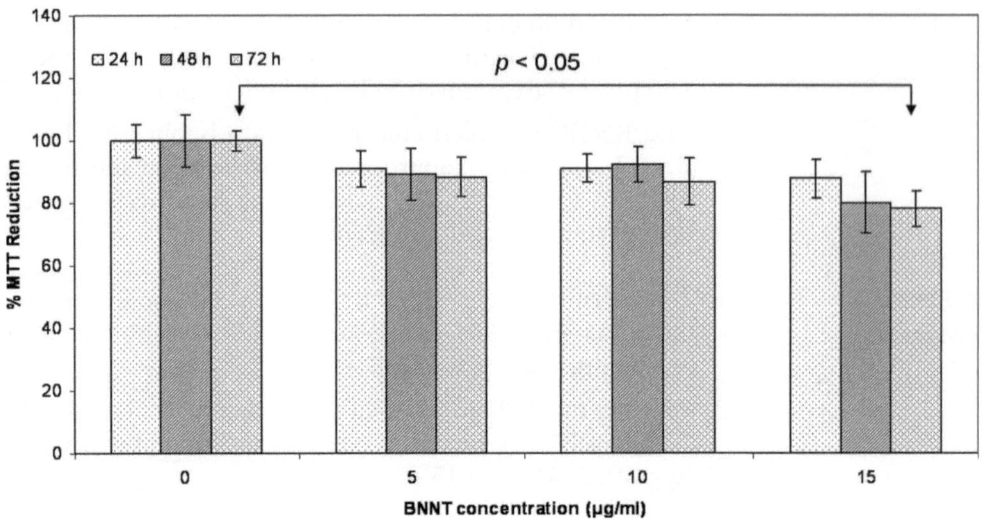

Fig. 2. MTT assay results after 24, 48, and 72 h of incubation of C2C12 cells with 0 (control), 5, 10, and 15 µg/ml of PLL-BNNTs ($n = 6$).

Similar results are obtained by culturing C2C12 with 10 µg/ml of PLL-BNNTs. Incubation with 15 µg/ml of nanotubes provides a nonsignificant MTT reduction after 24 and 48 h (about 15%, $p > 0.05$), but the decrease in metabolic activity becomes significant (about 20%, $p < 0.05$) by the third day (Fig. 2).

12. Similar results are achieved with SH-SY5Y cells, with a slightly higher toxicity mainly due to the different coating experimented (PEI in place of PLL).

3.4. Apoptosis Assay

1. Apoptosis plays a fundamental role in many normal biological processes as well as in several disease states, and it can be induced by various stimuli that all produce systematic and deliberate cell death. One method for studying apoptosis detects changes in the position of phosphatidylserine (PS) in the cell membrane. In nonapoptotic cells, most PS molecules are localized at the inner layer of the plasma membrane, but soon after the initiation of apoptosis, PS redistributes to the outer layer of the membrane, and becomes exposed to the extracellular environment (16). Exposed PS can be easily detected with annexin V-FITC, a fluorescent labeled 35.8 kDa protein that has a strong affinity for PS. Propidium iodide (PI) is a DNA red fluorescent intercalating agent. PI is membrane impermeant and generally excluded by viable cells. It can be therefore used jointly with annexin V-FITC to differentiate necrotic (red stained), early apoptotic (green stained), apoptotic (green and red stained), and normal cells (unstained).

2. For early apoptosis detection, seed 20,000 cells in 24-well plate chambers ($n = 3$) and treat them with 10 μg/ml of PLL-BNNTs for 24 h.
3. Perform a positive control by irradiating a cell culture with UV light for 60 min (to induce apoptosis) and with an SDS solution 0.1% (to induce necrosis); a negative control without PLL-BNNTs has also to be carried out.
4. Rinse cells with 1× binding buffer and add 200 μl of 1× binding buffer to the cells.
5. Add 5 μl of annexin V-FITC and 10 μl of propidium iodide to the wells see Note 6.
6. Incubate cells at room temperature for 15–20 min in the dark.
7. Carry out observation under fluorescent microscope with appropriate filters (i.e., FITC for labeled annexin, TRITC for propidium iodide).
8. C2C12 cells incubated for 24 h with 10 μg/ml of PLL-BNNTs show no evidence of apoptosis, with only a few dead cells comparable to the untreated cultures.

3.5. BNNT Uptake Investigation

1. Incubate cells at a density of 50,000/well in a 24-well plate chamber.
2. Add quantum dot labeled BNNTs to the culture medium at ratio 1:10 for a final concentration of 5.0–10.0 μg/ml of BNNTs.
3. Perform cell internalization studies observing the cultures by fluorescent microscopy at different times of incubation (e.g., after 0.5, 1, 2, 3, 6, and 12 h of incubation).
4. Perform lysosome tracking assays with LysoTracker dye, a fluorescent acidotropic probe for labeling acidic organelles in live cells. The probe fluoresces when it accumulates in cellular compartments with low pH.
5. For these studies, incubate cells 2 h in a growth medium containing LysoTracker in a dilution of 1:2,500 after exposure to labeled BNNTs see Note 7.
6. Observe the cells under fluorescent microscope with appropriate filters (i.e., FITC for LysoTracker, TRITC for quantum dot labeled BNNTs).
7. Tests with sodium azide, an endocytosis inhibitor (17), enable investigation of the mechanism responsible for internalization of the BNNTs. NaN_3 acts by blocking the production of ATP in cells and thus the energy required for the endocytotic pathway.
8. Treat the cell culture for 1 h with an NaN_3 modified medium (in a concentration of 5 mg/ml) and then incubate for 1 h in a medium containing 5 μg/ml of labeled BNNTs.

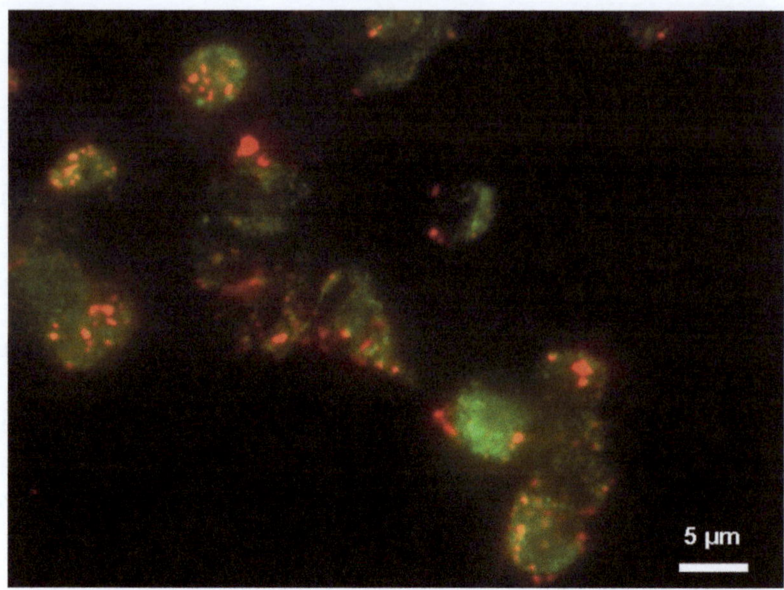

Fig. 3. Merged channel (FITC and TRITC) image following internalization of labeled BNNTs and lysosomal tracking assays on SH-S5SY cells. Adapted from ref. 5 with the permission of John Wiley & Sons Ltd.

9. Thereafter, acquire fluorescent images and compare them with a control culture incubated for 1 h with the same amount of labeled BNNTs, but without NaN_3 pretreatment.
10. The internalization of BNNTs is clearly demonstrated in both the cell lines (SH-SY5Y and C2C12). The high lysosomal activity corresponds with the BNNT internalization and, at higher magnification, the fluorescent spots are seen as small aggregates in well-defined compartments of the cells, rather than being uniformly distributed in the cytoplasm (Fig. 3).
11. Tests carried out in the presence of sodium azide show an inhibition of internalization, suggesting that uptake mechanism of BNNTs is energy-dependent.

3.6. DNA and Protein Quantification on Differentiated C2C12 Cells

1. For DNA and protein quantifications on differentiated C2C12, seed 100,000 cells in six-well plate chambers and when 90% of confluence is reached, change the growth medium (10% FBS) to the differentiating medium (1% FBS and ITS) see Note 8.
2. Perform all assays ($n=3$) after 3 days of incubation with standard differentiating medium (control cultures) and differentiating medium containing 5 μg/ml of BNNTs. Observe cell morphology by inverted light microscopy, for qualitative verification of cellular differentiation.
3. Perform both assays in cascade on the same samples. Additionally, run individual samples in triplicate to minimize operator's error.

4. Carefully remove the culture medium from the cell samples and add dd-H_2O; freeze the samples at −20°C and store them for subsequent assays see Note 9.
5. Perform two freeze/thaw cycles of the samples in order to obtain cell lysates: freeze overnight at −20°C and thaw 10 min at 37°C in a water bath with stirring for 15 s to enable the DNA and the proteins to go in solution.
6. Measure double-stranded (ds) DNA content in cell lysates by using the PicoGreen kit. The PicoGreen dye binds to ds-DNA and the resulting fluorescence intensity is directly proportional to the ds-DNA concentration in solution see Note 10.
7. Prepare standard solutions of DNA in dd-H_2O at concentrations ranging from 0 to 6 μg/ml; load 50 μl of standard or sample in a 96-well black microplate for quantification.
8. Prepare working buffer (5% vol buffer reagent in dd-H_2O-water) and PicoGreen dye solution (4.95% vol buffer and 0.05% vol Picogreen dye in dd-H_2O) fresh before use, and add 100 and 150 μl/well, respectively.
9. After 10-min incubation in the dark at room temperature, measure the fluorescence intensity using an excitation wavelength of 485 nm and an emission wavelength of 535 nm.
10. Total protein concentration is determined with the BCA method following the microplate procedure.
11. The working range of this assay is 20–2,000 μg/ml, therefore prepare standards using BSA ranging from 0 to 2,000 μg/ml.
12. Load cell lysates and working reagent (obtained by mixing reagents A:B at a ratio of 50:1 v/v) inside a 96-well microplate at 25 and 200 μl/well, respectively.
13. Shake the microplate gently and then incubate at 37°C for 30 min.
14. Cool down samples to room temperature and read absorbance at 562 nm on a plate reader.
15. Protein concentrations of cellular specimens are then obtained by reference to BSA standards and, finally, total protein content is normalized by DNA content as determined by the PicoGreen assay.
16. BNNT-treated cultures versus controls exhibit a nonsignificant drop in ds-DNA amount (about 2%), but corresponding to an increase in protein content (about 14%). Consequently, the total proteins, when normalized by ds-DNA (and, therefore, by cell density), result in a rise (about 16%), statistically different respect to the control ($p < 0.05$). Quantitative results show that the normalized protein content is about 92 μg/μg-DNA, in BNNT-treated samples, whereas it is about 78 μg/μg-DNA in controls.

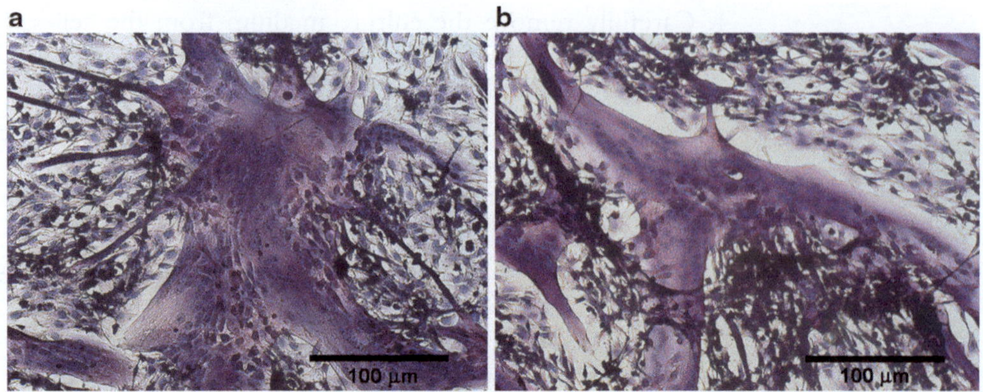

Fig. 4. Qualitative differentiation is assessed by phase contrast microscopy. Images of myotubes after 4 days of incubation in differentiating medium with 0 (**a**) and 5 (**b**) μg/ml of BNNTs. Samples were fixed with 1% buffered-formalin and stained with hematoxylin and eosin (H&E) according to standard histological procedures.

These findings confirm the optimal synthesis capacity by C2C12 differentiated cells in treated cultures.

17. We can therefore conclude that BNNTs only show a modest effect on proliferation of differentiating C2C12 myoblasts; moreover, they affect neither the cellular protein synthesis nor the qualitative development of myotubes in vitro (Fig. 4).

4. Notes

1. At this step, BNNT pellet can be resuspended in the appropriate volume of PBS up to the desired BNNT concentration. It is important to take in account that, because of the dilution with the growth medium, the final concentration in the cell cultures will be scaled 1:10.
2. It is recommended to prepare the appropriate fresh EDAC solution just before its use for the conjugation, in order to achieve a better reaction efficiency.
3. C2C12 cultures must not be allowed to reach confluence, as this will deplete the myoblastic population in the culture, with a consequent decrement of differentiation ability.
4. Since Trypan blue has a higher affinity for serum protein than for cellular proteins, cell suspensions in serum-containing medium will result in a dark background. It is thus recommended to perform Trypan blue staining on cells suspended either in PBS or in serum-free medium.

5. The Burker chamber (an hemocytometer) consists of a thick glass microscope slide with a rectangular indentation that creates a chamber. This chamber is engraved with a laser-etched grid of perpendicular lines. The device is carefully crafted so that the area bounded by the lines is known, and the depth of the chamber is also known. It is therefore possible to count the number of cells or particles in a specific volume of fluid, and thereby calculate the concentration of cells in the fluid overall.
6. Up to the cell type under investigation, it is useful to check against the positive control for the optimal amount of annexin V-FITC to use.
7. Also for lysosomal tracking, the concentration of Lysotracker that gives valuable fluorescence depends on the cell line. If after 2 h of incubation no fluorescence is detected, it is suggested either to increase the dye concentration or to extend its incubation time.
8. Serum depletion is known itself to promote C2C12 differentiation in vitro, however, in our experience it additionally reduces cell metabolic activity and causes cellular suffering. The concomitant addition of ITS enhances C2C12 viability and promotes myotube formation in 36 h.
9. Typically, the addition of 1.5–2 ml dd-H_2O per well of near-confluent six-well plates allows outputs of both DNA and BCA assays to stay within the linear intervals of their standard curves. If 24-well plates are used, the quantity of dd-H_2O to add can be scaled down proportionally to the well area.
10. The ds-DNA content is also directly proportional to the sample cellularity. An estimation of the cell number can be thus obtained according to the DNA weight of a single (diploid) cell that is species-specific (e.g., 7.18 pg in humans, 6.51 pg in rats, and 5.86 pg in mice). As a consequence, the protein content of cellular specimens can be plotted following normalization either per microgram of ds-DNA or per cell.

Acknowledgments

Authors would like to thank Mr. Leonardo Ricotti, Dr. Stefania Moscato, Dr. Delfo D'Alessandro, Mr. Carlo Filippeschi, and Dr. Vittoria Raffa for their valuable help in the experimental procedures, and Prof. Mario Petrini and Prof. Arianna Menciassi, whose laboratories are hosting the research presented in this chapter.

References

1. Terrones, M., Romo-Herrera, J.M., Cruz-Silva, E., López-Urías, F., Muñoz-Sandoval, E., Velázquez-Salazar, J.J., Terrones, H., Bando, Y., and Golberg, D. (2007) Pure and doped boron nitride nanotubes. *Materials Today* **10**, 30–38.
2. Lacerda, L., Bianco, A., Prato, M., and Kostarelos, K. (2006) Carbon nanotubes as nanomedicines: from toxicology to pharmacology. *Adv. Drug Deliv. Rev.* **58**, 1460–1470.
3. Ciofani, G., Raffa, V., Menciassi, A., and Cuschieri, A. (2009) Boron nitride nanotubes: an innovative tool for nanomedicine. *Nano Today* **4**, 8–10.
4. Ciofani, G., Raffa, V., Menciassi, A., and Dario, P. (2008) Preparation of boron nitride nanotubes aqueous dispersions for biological applications. *J. Nanosci. Nanotechno.* **8**, 6223–6231.
5. Ciofani, G., Raffa, V., Menciassi, A., and Cuschieri, A. (2008) Cytocompatibility, interactions and uptake of polyethyleneimine-coated boron nitride nanotubes by living cells: confirmation of their potential for biomedical applications. *Biotechnol. Bioeng.* **101**, 850–858.
6. Ciofani, G., Raffa, V., Yu, J., Chen, Y., Obata, Y., Takeoka, S., Menciassi, A., and Cuschieri, A. (2009) Boron nitride nanotubes: a novel vector for targeted magnetic drug delivery. *Curr. Nanosci.* **5**, 33–38.
7. Ciofani, G., Raffa, V., Menciassi, A., and Cuschieri, A. (2009) Folate functionalised boron nitride nanotubes and their selective uptake by glioblastoma multiforme cells: implications for their use as boron carriers in clinical boron neutron capture therapy. *Nanoscale Res. Lett.* **4**, 113–121.
8. Nakhmanson, S.M., Calzolari, A., Meunier, V., Bernholc, J., and Buongiorno Nardelli, M. (2003) Spontaneous polarization and piezoelectricity in boron nitride nanotubes. *Phys. Rev.* **67**, 235–406.
9. Ciofani G., Raffa V., Danti S., Menciassi A., Dario P., Petrini M., Cuschieri A. Piezoelectric nanotube - mediated cell stimulation. International patent application PCT/IB2010/051602, April 14th, 2010.
10. Ciofani G., Ricotti L., Danti S., Moscato S., Nesti C., D'Alessandro D., Dinucci D., Chiellini F., Pietrabissa A., Petrini M., Menciassi A. Investigation of interactions between poly-L-lysine coated boron nitride nanotubes and C2C12 cells: up-take, cytocompatibility and differentiation. International Journal of Nanomedicine, 5: 285-298 (2010).
11. Chen, Y., Fitz Gerald, J., Williams J.S., Bulcock, S. (1999) Synthesis of boron nitride nanotubes at low temperatures using reactive ball milling. *Chem. Phys. Lett.* **299**, 260–264.
12. Yu, J., Chen, Y., Wuhrer, R., Liu, Z., and Ringer, S.P. (2005) In situ formation of BN nanotubes during nitriding reactions. *Chem. Mater.* **17**, 5172–5176.
13. Yaffe, D., and Saxel, O. (1977) A myogenic cell line with altered serum requirements for differentiation. *Differentiation* **7**, 159–166.
14. Singer, V.L., Jones, L.J., Yue, S.T., and Haugland, R.P. (1997) Characterization of PicoGreen reagent and development of a fluorescence-based solution assay for double-stranded DNA quantitation. *Anal. Biochem.* **249**, 228–238.
15. Stoscheck, C.M. (1990) Quantitation of protein. *Method. Enzymol.* **182**, 50–69.
16. Martin S.J., Reutelingsperger, C.P., McGahon, A.J., Rader, J.A., van Schie, R.C., LaFace D.M., and Green, D.R. (1995) Early redistribution of plasma membrane phophatidylserine is a general feature of apoptosis regardless of the initiating stimulus: inhibition by overexpression of Bcl-2 and Abl. *J. Exp. Med.* **182**, 1545–1556.
17. Shi Kam, N.W., Jessop, T.C., Wender, P.A., and Dai, H. (2004) Nanotube molecular transporters: internalization of carbon nanotube-protein conjugates into mammalian cells. *J. Am. Chem. Soc.* **126**, 6850–6851.

Chapter 14

Nanobiosensors for In Vitro and In Vivo Analysis of Biomolecules

J.-Pablo Salvador, Mark P. Kreuzer, Romain Quidant, Gonçal Badenes, and M.-Pilar Marco

Abstract

This chapter presents as a proof of concept the development of a nanosensor based on the localized surface plasmon resonance for the analysis of biomolecules. The method presented take advantage of the plasmon generated in the surrounding of gold nanoparticles (i.e., 100 nm) for the specific interaction between antigen and antibody. The procedure for the optimization of an assay for the determination of biomolecules consisted mainly of four steps. First, the immobilization of gold nanoparticles over the glass surface using the appropriate ratio, concentration and time-contact of amino-sylilating agent, and nonreactive sylilating agent. Next, the suitable concentration of coating antigen in order to obtain the maximum signal LSPR. Following this step, the interaction between antigen and antibody (specific antibody) is evaluated by measuring the signal LSPR. Finally, a calibration curve was obtained for the detection of a small organic molecule such as stanozolol using this nanobiosensor. As a proof of concept, the use of a model is performed that in this case is for the detection of an anabolic androgenic steroid, such as stanozolol which is banned for the European Commission (EC) as a growth promoter and for the World Anti-Doping Agency (WADA) as a doping agent. The nanosensor developed demonstrates its feasibility for screening purposes due to the limit of detection achieved (0.7 μg/L) is under the MRPL required for both organizations (10 μg/L). A protocol such as that presented here may be generally applied for the analysis of other pollutant such as pesticides or antibiotics, or for biomedical applications for the analysis of biomarkers using the LSPR principle using gold nanoparticles (i.e., 30–120 nm).

Key words: Antibodies, Plasmon, LSPR, Biosensor, Stanozolol, Gold nanoparticles

1. Introduction

Antibodies are polypeptides belonging to immunoglobulin family (Ig). Specifically, the most employed as biorecognition element are Ig type G (IgG, MW ~ 150,000 Da) produced as a consequence from immune system response for an antigen of strange substance. The fact that IgG could recognize selectively a certain antigen has

Melba Navarro and Josep A. Planell (eds.), *Nanotechnology in Regenerative Medicine: Methods and Protocols*, Methods in Molecular Biology, vol. 811, DOI 10.1007/978-1-61779-388-2_14,

been employed as receptors for the detection of pollutants such as proteins or small organic molecules. Immunochemical techniques are based on the affinity of the antibody against an antigen. The formed complex has a high-affinity constant (Ka) that can reach values around 10^{-10} M^{-1}. This interaction is specific between the antigen and the corresponding antibody. The immunochemical techniques use this characteristic as a powerful tool for the detection of pollutants at low concentrations. Theoretically, it is possible to produce IgGs with high affinity and selectivity against any kind of analyte. Moreover, the structural homogeneity of these molecules represents a great potential to standardize procedures. Several immunochemical techniques have been developed for the determination of small molecules (1). IAs have found wide application in forensic, clinic, and veterinary analysis. In immunoassays for small organic molecules such as pharmaceuticals, the reaction antigen–antibody (Ag–Ab) is quantified under competitive conditions.

Biosensors are integrated analytical devices, usually small in size, consisting of a biological component in intimate contact with a physical transducer that converts the biorecognition process into measurable signal. Therefore, nanosensors are devices that used nanostructured systems as transducer for biorecognition events. Within biosensors one of the most employed transducing systems are the optical biosensors, as an example we can found SPR (Surface Plasmon Resonance) biosensors (2, 3). SPR is a physical phenomenon that can occur when plane-polarized light hits a metal film under total internal reflection (TIR) conditions (the incoming light is reflected on the interface of a half circular prism). When the prism is coated with a thin film of a noble metal (gold) on the reflection site, the energy of the photon electrical field can interact with the free electron constellations of the gold surface. The incident light photons are absorbed and converted into surface plasmons. A SPR is an evanescent electromagnetic field generated at the surface of a metal conductor when excited by the impact of light of an appropriate wavelength at a particular angle.

The field of plasmonic biosensing has mainly been governed by extended, flat (thin) films of noble metals known as surface plasmon polaritons (SPPs) and in more recent times by the three-dimensional alternative called localized surface plasmons (LSPs) (4–7). The former has been well documented by the commercial and bench-mark Biacore system, whereas LSP takes advantage of nanometric structures of varying shape (sphere, triangle, rod) and materials (gold or silver) to tune the resonance plasmon peak. Superficial plasmons are very sensitive to the optical properties of the dielectric media that surrounds the noble metal nanoparticle (MNP). Both methods take advantage of the local change of the dielectric function (effective refractive index) at the surface to which the biological compounds have been immobilized.

Localized surface plasmon resonance (LSPR) is elucidated by changes in the intimate environment of the nanoparticle, produced as consequence of the molecular recognition that induces a shift in the resonance absorption wavelength and is recorded as a shift in the resonant peak. Their principal advantages with respect to SPR are the greater amplification of the electromagnetic field that may allow reaching lower detection limits, detection of low molecular weight binding events (i.e., not bulk refractive index changes), the simplicity of the excitation/detection system (there is no need to use prisms or defined angles), and the possibility to confine measurement to a few micrometers. In contrast, SPR is based on delocalized wavelengths difficult to confine.

The use of metallic nanoparticles allows the exclusion for the surface plasmon polarization mode to highlight an alternative approach to plasmon sensing, of the localized kind. Through the three-dimensional arrangement of colloid and/or lithography prepared nanostructures, the main advantages of LSPR-based sensors rely on the fact that the gold plasmon resonant particles with diameters between 30 and 120 nm efficiently scatter light in the visible region and that there is also no need for temperature stabilization as LSPR is not/less prone to such fluctuation as in extended SPP.

The use of colloidal type gold or silver MNP LSP sensors is an emerging non-labeling method due to their unique optical, chemical and electrical properties. MNPs exhibit a strong absorption band in the visible spectrum that is not found for the bulk metal (8). The absorption is induced when the frequency of the incident light is resonant with the collective oscillations of the conduction electrons in the MNP. Since this excitation is localized, it is called LSPR.

Noble metals, especially gold are biocompatible, relatively easy to synthesize and chemically functionalize. Gold MNP surface modification for protein immobilization is straight forward, using amine or thiol chemistry and has been shown to bind peptides, proteins, and DNA ligands. Due to their aforementioned unique properties, spherical gold MNPs, which can be synthesized by solution chemistry have been attractive as sensors. Sensors have been fabricated using colloidal gold MNPs immobilized on glass (9) or in solution for antibody–antigen binding kinetics (10).

LSPR is very suitable for nanoscale detection, because a single MNP can act as a sensor (11–13). However, this sensitivity is directly related to the complexity of the optical set-up. But rather than measuring single MNPs one can get the cumulative effect from arrays of such structures (14, 15), and monolayers of surface binding molecules have been detected with high sensitivity. Although the sensitivity of a sensor is often determined by the equilibrium constant Ka of the affinity ligand, LSPR sensors can measure very small absolute quantities of ligands in a very small sample volume.

Therefore, changes near surface of nanoparticles (in the dielectric function of metal and surrounding medium) could obtain, for instance, a signal coming from the specific recognition between antigen–antibody interactions. LSPR can also show improvements for small molecule detection as the near field is confined to within a few tens of nanometers from the metal surface, more suitable as the sensing volume is smaller and does not extend into bulk solution as in the case of SPP. This fact is advantageous in residue analysis using immunoreagents specific for certain pollutant. As a proof of concept, to demonstrate applicability of this sensing principle for residue analysis, we have employed immunoreagents specifically developed for the detection of stanozolol, an androgenic hormone used illegally as a growth promoter in farm animals and to increase athletic performance. The European Union (EU) has completely prohibited the use of these steroidal hormones for meat production since 1981 (Directive 81/602/EEC (16)), which applies to Member States and imports from third-party countries alike. Regarding as a doping agent, stanozolol is banned by the WADA (17) within S1 subcategory. In both organizations, they established a minimum required performance level (MRPL) for stanozolol at 10 μg/L. Previously to this work, the immunoreagents employed were produced and characterized using ELISA (18), achieving the requirements established from EC and WADA.

Here, the steps followed to develop a LSPR biosensor are described. We have endeavored to highlight the advantages of using these noble metal colloids for the development, through analytical means, of a relatively simple system that rapidly, specifically, and with sufficient sensitivity detects anabolic steroids. The key developments are in the relative ease of preparing the sensors (through the use of colloids and mixed surface-assembled monolayers) both from a fabrication and (bio)chemical viewpoint.

2. Materials

2.1. Chemicals, Standards, and Quality Control

1. Stanozolol (St, Sequoia Research Products, UK).
2. Aminopropyltrimethoxysilane (APTMS, see Note 1; Sigma-Aldrich).
3. Ethyltrimethoxysilane (ETMS, Sigma-Aldrich).
4. Dimethylsulfoxide (DMSO, Sigma-Aldrich).
5. Ethanol (EtOH, Carlo-Erba, Spain).
6. Gold colloid (100 nm, see Note 2, BB International, UK).
   a. Immunoreagents (see Note 3).

1. Polyclonal antisera As147 against St A working aliquot is stored at 4°C and remains stable for about 6 months.
2. Coating antigens St-BSA. A stock solution of 1 mg/mL is prepared in 10 mM PBS and kept at 4°C. The solution is stable for 6 months.

2.2. Buffers and Solutions

1. PBS buffer, 10 mM phosphate buffer, 0.8% saline solution, pH 7.5 (see Note 4).
2. PBST 0.001% buffer, PBS with 0.001% Tween 20.
3. NaOH (2N):EtOH, 1:1, equivolumetric mixture of EtOH and aqueous NaOH (2N).

2.3. General Material

1. Petri dishes used for immersions for the chemical modification of glass substrates (see Subheading 4.2), i.e., Ø 11 cm.
2. Pipettes set from 1 to 10, 10–100, 100–1,000, 1,000–10,000 μL (Eppendorf), and their corresponding tips.
3. Eppendorf tubes, 1.5 mL.
4. PP Plastic tubes, 10 mL.
5. BK7 glass 75 mm × 25 mm (Marienfield, Germany).
6. Silicone isolators eight well, *D* × diam. 2.0 mm × 9 mm (Sigma-Aldrich).

2.4. Optical Arrangement Setup

The optical setup consisted of a standard BX5 microscope (Olympus, Germany) as observed in Fig. 1. The extinction spectra of the gold colloids were measured by conventional dark-field (DF)

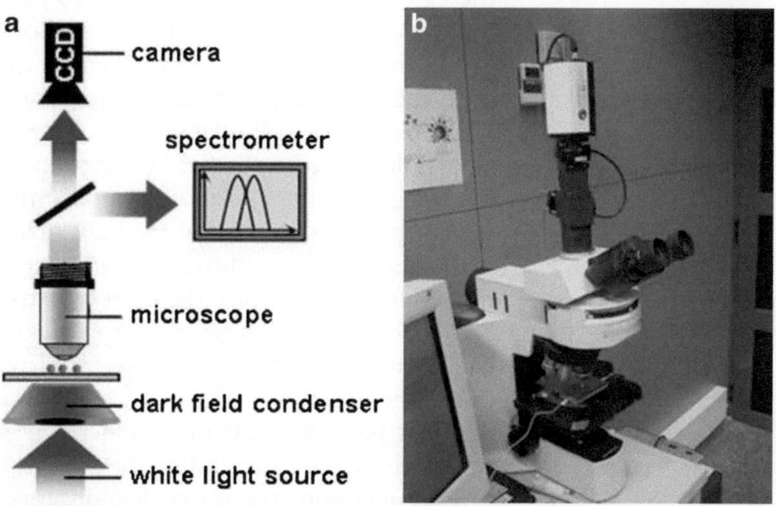

Fig. 1. Schematic setup of the system (**a**) and (**b**) real image representation (with kind permission of Springer Science + Business Media: (19)).

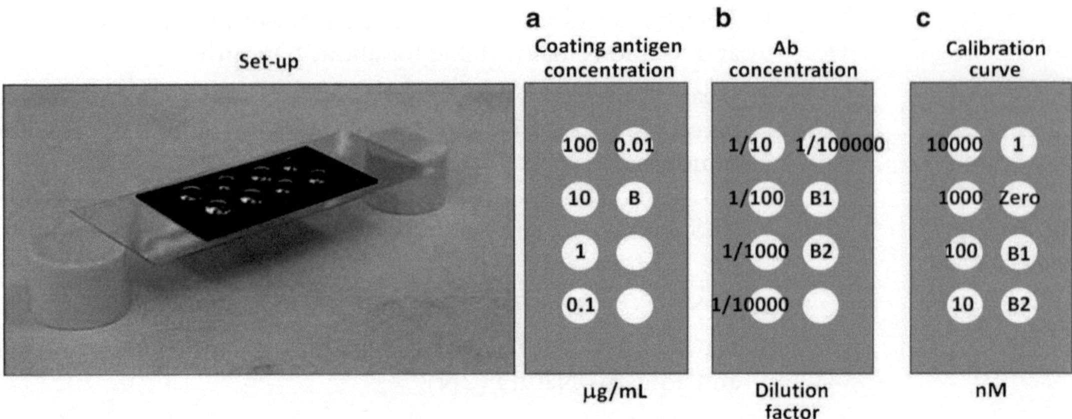

Fig. 2. Experimental set-up for LSPR experiments. (**a**) Schema for different coating antigen concentrations and the blank. (**b**) Schema for different antibody dilution where the coating antigen is immobilized at 10 µg/mL. (**c**) Calibration curve at different concentration for St in nM, at 10 µg/mL of coating antigen and an antibody dilution of 1/50.

spectroscopy. The sample illumination was performed with a halogen lamp (100 W) using an immersion-oil dark-field condenser with high numerical aperture (NA = 1.4 – 1.2). The light scattered by the particles was collected by a long working distance objective lens (×50, NA = 0.5) and sent toward both an Ocean Optics microspectrometer (model HR 2000) and a CCD camera (Q-imaging, Spain, see Note 5). An adjustable diaphragm located at the microscope output was used to reduce the detection area to approximately 100 µm^2. Considering the bandwidth of the particles' extinction spectra, the setup allows for the determination of the central LSPR wavelength with a resolution of ca. 0.3 nm. In all the cases, the volume added was 20 µL/well (experimental setup for glass slides using silicon mask is shown in Fig. 2).

From the viewpoint of the optical detection, it is important to note the limitations of the setup. From experience, we used DF transmission for colloidal samples but alternatives are also available (DF reflection, bright-field extinction), which largely depend on the sample quality and more importantly colloid density. Generally speaking, DF is used when the light, from a wide angle, interacting with the particles is deviated (or scattered) into the objective. In the absence of particles, the light by-passes this collecting objective and the reading is negligible apart from impurities on the glass itself, which may scatter some light. However in the determination of the spectra, this is taken into consideration. It is also worth noting that particles <20 nm inefficiently scatter light. This can be circumvented by increasing the density of colloids per area with care to maintain an interparticle distance not sufficient (>500 nm) to cause broadening of the plasmon peak due to near-field coupling.

If this is the case, the sensitivity or resolution of the peak is compromised.

For the spectral elucidation herein, we considered the following setup points:

1. Use an illumination lamp that covers the spectral range corresponding to the plasmon of the MNP of choice, i.e., gold spheres in the green part of the visible spectrum and gold rods from the red to the near infrared region (depending on aspect ratio, they can extend to the infrared).
2. The DF condenser should have a high NA to provide rays of light approaching the sample at high angles.
3. The collecting objective should have a lower NA than the condenser and a sufficient magnification (50× or 100×) to measure the signal from a large selection of colloids.
4. The absolute signal derived from the illumination area, can be adjusted by changing the fiber core from 50 to 600 μm, thus encompassing more MNPs. The fiber is used to carry the scattered signal from the collecting objective to the spectrometer.

3. Methods

3.1. Preparation of Standards

1. The stock solution of St needed for LSPR standard curve: 10 mM in DMSO, and it should be stored at 4°C.
2. The standard for the calibration curve is prepared at 10,000, 1,000, 100, 10, and 1 nM concentrations in PBST 0.001% Tween 20 prepared as follows:
3. Prepare a dilution of 5 μL of stock solution 10 mM of St in 995 μL of PBST 0.001%.
4. Prepare the 10,000 nM concentration by diluting 200 μL of 50 μM in 800 μL of PBST 0.001%.
5. Prepare the 1,000, 100, 10, and 1 concentrations by serial dilution of 100 μL of the previous concentration in 900 μL of PBST 0.001%.

3.2. Preparation of Solutions

1. The APTMS/ETMS mixture in anhydrous EtOH at 1.13 mM in a 1:4 molar ratio is prepared as follows: 4 μL of APTMS and 14.5 μL of EPTMS are diluted in 100 mL of anhydrous EtOH.
2. The solution of NaOH (2N):EtOH 1:1 (v:v) is prepared by mixing 50 mL EtOH and 50 mL of NaOH (2N). The solution of NaOH (2N) is prepared dissolving 8 g of NaOH in 100 mL of MilliQ water.

3. The stock solution needed for the preparation of coating antigen concentrations is 1 mg/mL.
4. Prepare a dilution of 5 μL of 1 mg/mL stock solution and 495 μL of PBST 0.001% to obtain the desired concentration of 100 μg/mL.
5. Prepare the rest of the coating antigen concentrations 10, 1, 0.1, and 0.01 μg/mL by serial dilution of 50 μL of the previous concentration in 450 μL of PBST 0.001%.
6. Prepare the antiserum dilutions as follows: The dilution of 1/10 of As147 is diluting of 10 μL of As147 in 90 μL of PBST 0.001%. The dilution 1/100, 1/1,000, 1/10,000, and 1/100,000 are prepared by serial dilution from 1/10 as explained before 10 μL of previous concentration in 90 μL of PBST 0.001%.

3.3. Chemical Modification of Glass Substrates (see Note 6)

1. Sonicate the glass substrate for 1 min in 20 mL of ethanol (EtOH) in a Petri dish and dry the slide under a N_2 stream.
2. Immerse the substrate in 20 mL of a solution containing NaOH (2N):EtOH 1:1 for 1 h at room temperature (see Note 7) in a Petri dish covered with a sealing tape. Wash gently with water and EtOH, and dry it under a N_2 stream.
3. A mixture of APTMS/ETMS in 20 mL of anhydrous EtOH at 1.13 mM in a 1:4 molar ratio is applied to the substrate for 2 min at room temperature (the optimal concentration was previously evaluated as it can be observed in Fig. 3) in a Petri dish covered with a sealing tape. Wash gently with EtOH and dry under N_2 stream. The silicon mask was subsequently bonded over the glass slide by self adhesion.
4. Add 20 μL/well of gold nanoparticles 100 nm of particle size for 10, 20, 30, and 40 min at room temperature. Wash gently with MilliQ water and dry under stream of N_2 (see Fig. 4).

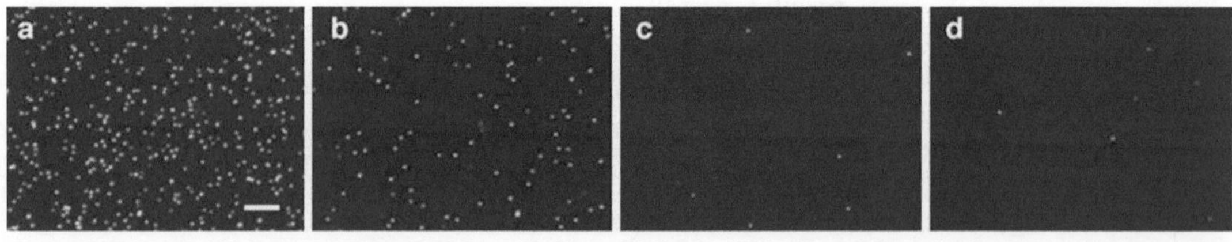

Fig. 3. Dark-field images of gold-modified glass substrates using various ratios of APTMS/ETMS, (**a**) 1:4, (**b**) 1:8, (**c**) 1:16, (**d**) 1:32. Total silanization was 1.13 mM. Scale bar 5 μm (with kind permission of Springer Science + Business Media: (19)).

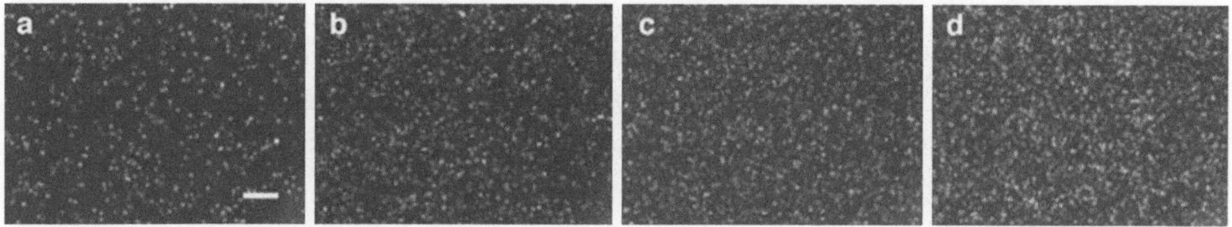

Fig. 4. Dark-field images of the optimization of gold colloid contact time with a 1.13-mM solution APTMS/ETMS (1:4) modified glass substrate for (**a**) 10 min, (**b**) 20 min, (**c**) 30 min, (**d**) 40 min at RT. Scale bar 5 μm (with kind permission of Springer Science + Business Media: (19)).

3.4. Determination of Optimal Concentration of Immunoreagents

3.4.1. Determination of Coating Antigen Concentration

1. Coat 20 μL St-BSA for 30 min at RT in 10 mM PBS pH = 7.4 at 0.01, 0.1, 1, 10, and 100 μg/mL, using the silicon mask as described in Fig. 2.
2. Wash gently with MilliQ water, dry under N_2 stream.
3. Finally, measure the scattered spectrum for each concentration per duplicate (see Note 8).

3.4.2. Determination of Specific Antibody Concentration

1. Coat St-BSA for 30 min at RT in 10 mM pH = 7.4 at 10 μg/mL over the activated glass substrates.
2. Wash gently with MilliQ water and dry under N_2 stream. Measure the scattered spectrum per well. This constitutes the reference spectrum.
3. Prepare antibody dilutions from 1/10, 1/100, 1/1000, 1/10,000, and 1/100,000 in PBST 0.001%. Add 20 μL of each dilution per duplicated in the activated glass with the silicon mask as described in Fig. 2. Let 30 min of binding time at RT.
4. Wash gently with MilliQ water, dry under N_2 stream and measured the scattered spectrum for each well (see Note 9).

3.5. Quantitative Determination of St

1. Coat the silanized glass slides with the help of the silicon mask coating antigen St-BSA at 10 μg/mL.
2. Wash gently with MilliQ water and dry under N_2 stream. Measure the scattered spectrum per well. This constituted the reference spectrum.
3. Prepare the standard curve of St (1–10,000 nM) in PBST 0.001% in a separate mixing plate.
4. Prepare the As147 dilution 1/50 in PBST 0.001%.
5. Preincubate for 15 min at RT, 10 μL of each standard St and 10 μL of As147 1/50.
6. Add 20 μL of each preincubated mixture As147/Standard over each corresponding well and incubate for 30 min at RT as described in Fig. 2.

7. Wash gently with MilliQ water and dry under N_2 stream. Measure the scattered spectrum per well.
8. The standard curve is fitted to a four-parameter logistic equation according to the following formula:

$$y = \left(A - B\left[1 - \left(\frac{x}{C} \right)^{D} \right] \right) + B .$$

where A is the maximum signal, B is the minimum signal, C is the concentration producing 50% of the maximum signal, and D is the slope at the inflection point of the sigmoidal curve.

4. Discussion

4.1. Optimization of Colloid Density in Glass Substrates

This work takes advantage of the characteristic resonance spectrum of gold nanoparticles of 100 nm diameter, in which the typical resonance can be seen around 550 nm.

The APTMS was selected as silanization agent in place of mercaptopropyltrimethoxysilane (MPTMS) due to the similar binding of amino groups to gold surface, but avoiding any oxidation processes of thiols over time (which was evident in sample reproducibility when using MPTMS). A mixture of APTMS with an spacer silanization agent such as ETMS was used in order to obtain an homogenous distribution of gold nanoparticles over the glass surface and also, to block the sites on the glass available for unwanted protein binding. A final concentration of silanization agent was 1.13 mM, and the ratio APTMS/EPTMS was studied. As it can be observed in Fig. 3, the optimum condition to produce consistent binding was found to be at 1:4 ratio. In the figure is showed the decline in gold population when APTMS was less than 1:4. The desired concentration of gold nanoparticles was studied also studying time contact with the mixture of APTMS/ETMS 1:4 at 1.13 mM concentration as shown in Fig. 4. The results showed that the optimum preincubation time was 10 min (see Note 10).

4.2. Biofunctionalization and Optimization of Immunoreagents

As a proof of concept the immunoreagents used for LSPR biosensor development that were studied were the same than the ones described for immunoassay previously for the detection of stanozolol (18). The system utilizing polyclonal serum (As147) and the stanozolol coating conjugate (St-BSA) was assessed. The biofunctionalization of antigen conjugate St-BSA and their specific recognition by its corresponding antiserum As147 should give a measurable wavelength shift. The optimal concentrations of immunoreagents for the determination of stanozolol in LSPR format was first tested evaluating the affinity of As147 to St-BSA immobilized

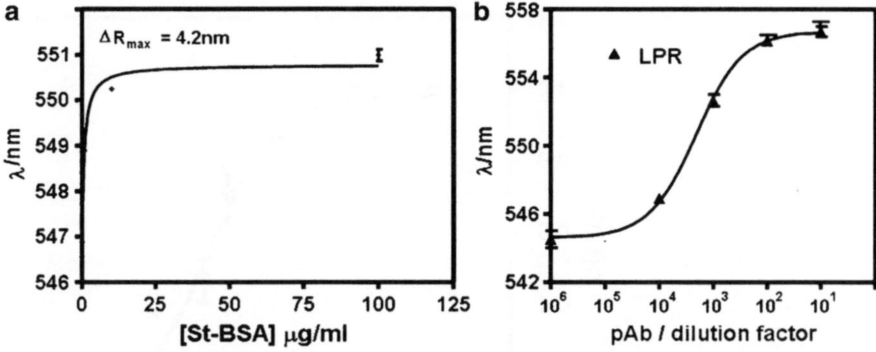

Fig. 5. Dose-dependent curves for the LSPR shift ($\Delta\lambda$), for the binding of (**a**) St-BSA to bare gold colloid and (**b**) dilution of α-St pAb stock to gold colloid coated in 10 μg/mL St-BSA (corrected for St-BSA response as determined in part **a**) in resonance (filled triangle) (with kind permission of Springer Science + Business Media: (19)).

over the gold nanoparticle and evaluating the shift in LSPR signal depending on the concentrations of immunoreagents.

Initial experiments centered on optimizing the coating conjugate St-BSA toward the prepared gold-modified glass substrates. As it can be observed in Fig. 5, concentrations greater than 10 μg/mL did not increase in terms of resonance shift was noted. The maximum shift recorded for St-BSA was 4.2 nm at 10 μg/mL. Therefore, the chosen concentration for the biofunctionalization of gold nanoparticles was 10 μg/mL.

To optimize the concentration of As147, a dose-dependent curve was constructed in order to quantify the shift in the resonance peak depending on the concentration of the specific receptor As147 (see Fig. 5). The data obtained have been fitted to a hyperbolic one-site binding equation. This suggests that the resonance mode is truly measuring the binding event close to the gold surface. The curve is the average on separate days (N=2). Considering this results the dilution factor applied for As147 for further experiments was 1/100. The maximum change in wavelength here was noted at 13.2 nm.

4.3. Calibration Curve for St in LSPR Format

Considering the results obtained from the immunoreagents optimization, stanozolol was determined as described in Subheading 3.5. A preincubation time between the As147 and the stanozolol for 15 min was required. Once the LSPR signal was recorded, the data obtained fit to the nonlinear equation as shown in Fig. 6. The data shown was the averaged result over three individual experiments on separate days (N=3). Data have been normalized to account for any day-to-day variations in the control values (samples B1 and B2) but resonance signals were of the order of those shown during the optimization steps ($\Delta\lambda_{max} = 10 \pm 2$ nm). Table 1 shows the features of the assay obtained. The IC_{50} of this assay,

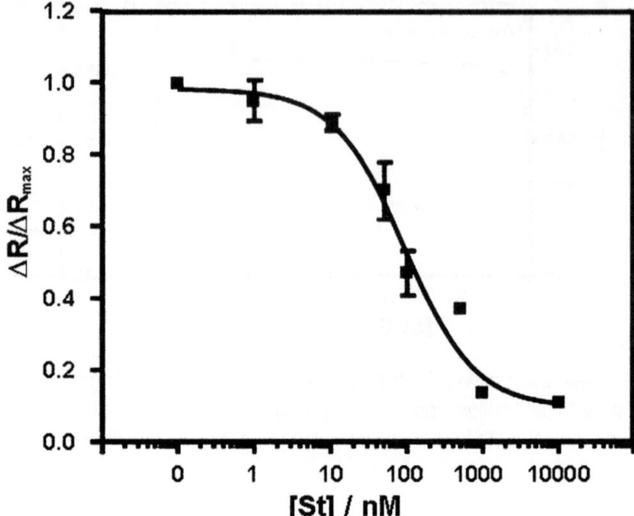

Fig. 6. Quantitation of stanozolol via a preincubation step using gold colloid-coated glass substrates by (filled square) LSPR (with kind permission of Springer Science+Business Media: (19)).

Table 1
LSPR biosensor features

$\Delta\lambda_{max}$	10±2 nm
IC_{50}	103 nM
Slope	-0.9954
LOD	2.4 nM (0.7 μg/L)
r^2	0.991

feature that corresponds to the 50% of total inhibition value, was 103 nM. The limit of detection (LOD) was calculated using the following formula:

$$\mathrm{LOD} = \mathrm{IC}_{50}\left(\frac{\mathrm{Max}-\mathrm{Min}}{\mathrm{Max}-\mathrm{Min}-3\mathrm{SDmax}}-1\right)^{-(1/k)}.$$

The detection limit of the system was calculated as 2.4 nM or 0.7 μg/L St, a value below the MRPL used by the WADA for banned substances.

5. Notes

1. Further studies concluded that other silanization agent such as MPTMS could also be used but one must consider the relative instability of thiols over time.

2. LSPR signal can be obtained from 20 to 120 nm particle size regarding spherical nanoparticles. Sizes smaller than this limit the scattering cross-section of the sphere and in this case alternative geometries like rods are recommended (larger scattering cross-section for equal volume of material). One may also change the material. The shelf-life or stability for commercial gold colloid of 100 nm is limited. In our experience, batches opened more than 1 month aggregated and were not suitable for further use. This is largely due to the capping agent used for aqueous colloids. There are ways to extend the shelf-life by ligand exchange to create more stable analog of the original colloid.

3. The antisera is raised against hapten type B of stanozolol (hB_St or compound 8), synthesized as described (18) covalently coupled by the mixed anhydride method to keyhole limpet hemocyanin. The indirect ELISA uses a homologous coating antigen prepared by conjugation of the same compound 8 to BSA using the active ester method. The assay performs well between pH 7.5 and 9.5. The immunoassay detectabilities do not change significantly when the ionic strength of the media is in the range 5–50 mM in terms of concentration of PBS. The ELISA for St is quite specific, but some cross-reactivity with other steroids, such as NorStanozolol (10%) and Methylboldenone (11%) is observed.

4. Composition for PBS buffer is 8 g of NaCl, 0.2 g of KCl, 1.14 g of Na_2HPO_4, and 0.2 g of KH_2PO_4. Dissolve in 600 mL of MilliQ water and adjust pH at 7.5 using HCl 1N or NaOH 1N. complete the volume to 1 L with MilliQ water.

5. The CCD of the spectrometer should be cooled to −50°C or lower (if possible) to minimize dark counts, which can help to improve signal- to-noise ratio. The grating used (# lines /mm) can also influence the spectral precision by allowing more measurement points/range. Some consideration should be given to the exposure time and whether to use data accumulation, all related to the size of the colloidal scatterer(s) and the density.

6. The detailed optimization, evaluation of the chemical modification of the glass substrates was previously described in our work (19) such as time, concentration, and ratio APTMS/EPTMS.

7. The washing step for the glass surfaces was adopted following the protocol described by Tätte et al. (20) with slight modifications.

8. The selected concentration of coating antigen St-BSA for further studies was 10 μg/mL. This value was chosen considering that a tenfold increase in concentration did not yield an equivalent shift in resonance wavelength.

9. The selected dilution for As147 was 1/100 for the calibration curve experiment.
10. More than 10 min we obtained aggregates and broad spectrum with poor resolution. The final concentration of gold nanoparticles was estimated at 3.3 nanoparticles/μm^2.

Acknowledgments

This work also has been supported by the Ministry of Science and Education (contract number DEP2007-73224-C03-01). The AMR group is a consolidated Grup *de Recerca de la Generalitat de Catalunya* and has support from the Departament d'Universitats, Recerca i Societat de la Informació la Generalitat de Catalunya (expedient 2009 SGR 1343). CIBER-BBN is an initiative funded by the VI National R&D&i Plan 2008-2011, *Iniciativa Ingenio 2010, Consolider Program, CIBER Actions* and financed by the Instituto de Salud Carlos III with assistance from the European Regional Development Fund.

References

1. Salvador, J. P., J. Adrian, R. Galve, D. G. Pinacho, M. Kreuzer, F. Sánchez-Baeza and M. P. Marco (2007). Chapter 2.8 Application of bioassays/biosensors for the analysis of pharmaceuticals in environmental samples. Comprehensive Analytical Chemistry. M. Petrovic and D. Barceló, Elsevier. *Volume* **50**: 279–334.
2. Homola, J. (2008). Surface Plasmon Resonance Sensors for Detection of Chemical and Biological Species. *Chemical Reviews* **108**(2): 462–493.
3. Petz, M. (2009). Recent applications of surface plasmon resonance biosensors for analyzing residues and contaminants in food. *Monatshefte für Chemie / Chemical Monthly* **140**(8): 953–964.
4. Boozer, C., G. Kim, S. Cong, H. Guan and T. Londergan (2006). Looking towards label-free biomolecular interaction analysis in a high-throughput format: a review of new surface plasmon resonance technologies. *Current Opinion in Biotechnology* **17**(4): 400–405.
5. Liu, X., D. Song, Q. Zhang, Y. Tian, L. Ding and H. Zhang (2005). Wavelength-modulation surface plasmon resonance sensor. *TrAC Trends in Analytical Chemistry* **24**(10): 887–893.
6. Pattnaik, P. (2005). Surface plasmon resonance. *Applied Biochemistry and Biotechnology* **126**(2): 79–92.
7. Sepúlveda, B., P. C. Angelomé, L. M. Lechuga and L. M. Liz-Marzán (2009). LSPR-based nanobiosensors. *Nano Today* **4**(3): 244–251.
8. Kreibig, U. and M. Vollmer (1995). Optical Properties of Metal Clusters. Berlin, Springer.
9. Nath, N. and A. Chilkoti (2004). Label-Free Biosensing by Surface Plasmon Resonance of Nanoparticles on Glass: Optimization of Nanoparticle Size. *Analytical Chemistry* **76**(18): 5370–5378.
10. Yu, C. and J. Irudayaraj (2006). Multiplex Biosensor Using Gold Nanorods. *Analytical Chemistry* **79**(2): 572–579.
11. Mock, J. J., M. Barbic, D. R. Smith, D. A. Schultz and S. Schultz (2002). Shape effects in plasmon resonance of individual colloidal silver nanoparticles. *The Journal of Chemical Physics* **116**(15): 6755–6759.
12. Raschke, G., S. Kowarik, T. Franzl, C. Sonnichsen, T. A. Klar, J. Feldmann, A. Nichtl and K. Kurzinger (2003). Biomolecular Recognition Based on Single Gold Nanoparticle Light Scattering. *Nano Letters* **3**(7): 935–938.

13. Sherry, L. J., S.-H. Chang, G. C. Schatz, R. P. Van Duyne, B. J. Wiley and Y. Xia (2005). Localized Surface Plasmon Resonance Spectroscopy of Single Silver Nanocubes. *Nano Letters* **5**(10): 2034–2038.
14. Acimovic, S. S., M. P. Kreuzer, M. U. Gonzalez and R. Quidant (2009). Plasmon Near-Field Coupling in Metal Dimers as a Step toward Single-Molecule Sensing. *ACS Nano* **3**(5): 1231–1237.
15. Barbillon, G., J. L. Bijeon, J. Plain, M. L. de la Chapelle, P. M. Adam and P. Royer (2007). Electron beam lithography designed chemical nanosensors based on localized surface plasmon resonance. *Surface Science* **601**(21): 5057–5061.
16. EC (1996). Directive 96/23/EC. *Off. J. Eur. Commun.* **L 125**: 3.
17. WADA The 2009 Prohibited list. *http://www.wada-ama.org/.*
18. Salvador, J. P., F. Sánchez-Baeza and M. P. Marco (2008). Simultaneous immunochemical detection of stanozolol and the main human metabolite, 3'-hydroxy-stanozolol, in urine and serum samples. *Analytical Biochemistry* **376**(2): 221–228.
19. Kreuzer, M., R. Quidant, J. P. Salvador, M. P. Marco and G. Badenes (2008). Colloidal-based localized surface plasmon resonance (LSPR) biosensor for the quantitative determination of stanozolol. *Analytical and Bioanalytical Chemistry* **391**(5): 1813–1820.
20. Tätte, T., K. Saal, I. Kink, A. Kurg, R. Lõhmus, U. Mäeorg, M. Rahi, A. Rinken and A. Lõhmus (2003). Preparation of smooth siloxane surfaces for AFM visualization of immobilized biomolecules. *Surface Science* **532–535**: 1085–1091.

Chapter 15

Novel Strategies to Engineering Biological Tissue In Vitro

Francesco Urciuolo, Giorgia Imparato, Angela Guaccio, Benedetto Mele, and Paolo A. Netti

Abstract

Tissue engineering creates biological tissues that aim to improve the function of diseased or damaged tissues. In this chapter, we examine the promise and shortcomings of "top-down" and "bottom-up" approaches for creating engineered biological tissues. In top-down approaches, the cells are expected to populate the scaffold and create the appropriate extracellular matrix and microarchitecture often with the aid of a bioreactor that furnish the set of stimuli required for an optimal cellular viability. Specifically, we survey the role of cell material interaction on oxygen metabolism in three-dimensional (3D) in vitro cultures as well as the time and space evolution of the transport and biophysical properties during the development of de novo synthesized tissue-engineered constructs. We show how to monitor and control the evolution of these parameters that is of crucial importance to process biohybrid constructs in vitro as well as to elaborate reliable mathematical model to forecast tissue growth under specific culture conditions. Furthermore, novel strategies such as bottom-up approaches to build tissue constructs in vitro are examined. In this fashion, tissue building blocks with specific microarchitectural features are used as modular units to engineer biological tissues from the bottom up. In particular, the attention will be focused on the use of cell seeded microbeads as functional building blocks to realize 3D complex tissue. Finally, a challenge will be the potential integration of bottom-up techniques with more traditional top-down approaches to create more complex tissues than are currently achievable using either technique alone by optimizing the advantages of each technique.

Key words: Tissue engineering, Bio-fabrication, Cell material interaction, Cell metabolism, Modelling, Bioreactor

1. Introduction

Tissue engineering (TE) is a multidisciplinary field pioneered more than 20 years ago by Langer and Vacanti (1) aiming at the in vitro fabrication of biological substitutes. It was demonstrated that cells of human or animal origin, after being explanted and seeded on a specific substrate (biomaterial), are able to synthesize de novo tissue

Melba Navarro and Josep A. Planell (eds.), *Nanotechnology in Regenerative Medicine: Methods and Protocols*, Methods in Molecular Biology, vol. 811, DOI 10.1007/978-1-61779-388-2_15, © Springer Science+Business Media, LLC 2012

components. Thereby, TE has emerged as a potential alternative to tissue or organ transplantation, and tissue loss or organ failure may be treated either by implantation of an engineered biological substitute. The great success of skin engineering (2) has generated optimism and support for the expansion of tissue-engineering projects. In fact during the last two decades the TE scientific community has addressed numerous studies relaying the in vitro production of several tissues such as liver (3, 4), bone (5, 6), cartilage (7, 8), cardiac muscle (9, 10), tendon and ligament (11, 12), and nerve (13, 14). Nevertheless, the fabrication of functional, tridimensional (3D), and heterotypic tissue as well as the development of a process able to tailor and to control the tissue formation and assembly has not been achieved yet. In this chapter, the strategies used in TE that can be divided in two main branches: top-down and bottom-up approach are reviewed. The classical approach (top-down) is based on seeding cells in a porous scaffold generating a cellular construct, followed by a maturation step in a bioreactor. In the first section, the scenario concerning to this approach is illustrated then the attention is posed on the optimization of the process conditions of the bioreactors. To do this, it is shown that how both metabolic (15) and physic microenvironment (16) remodelling during culture time play a crucial role in tissue development. Finally, the need of a mathematical model able to predict tissue growth by tacking in to account physical and metabolic evolutions is highlighted. In the second section, other emerging strategies aiming at building-up biological tissues in vitro based on modular assembly of small building block that mimic at micrometric scale, the tissue units (17) are described. The discussion will concern on self-assembly properties of cell seeded microbeads. Indeed, it has recently been demonstrated that microbeads surrounded by a micrometric layer of tissue are able to aggregate and generate a 3D tissue equivalent (18).

2. Top-Down Approach

The scaffold design/bioreactor-based strategies relay upon the development of biomaterials with specific biochemical and physical cues (porosity, pore architecture, chemical composition) inserted in a bioreactor that provide mechanical and hydrodynamic stimuli. In this fashion, it is possible to recreate the 3D biological environment and to tailor and to control tissue development within a porous or structure. Indeed, as stated by Martin et al. (19), the bioreactors can be defined "as devices in which biological and/or biochemical processes develop under closely monitored and tightly controlled environmental and operating conditions." In addition, compared with static culture conditions, the bioreactor provides optimized process conditions in order to produce larger and thicker 3D tissue constructs (19–21). The mostly used bioreactor systems are *Mechanical Stimulation, Direct Perfusion, Spinner Flask, and Rotating Wall Bioreactor*. However, the optimized process conditions vary with tissue type, scaffold composition, medium viscosity, and bioreactor configuration. Several kinds of bioreactors have been developed and used for different cell types, among them osteoblasts (22), chondrocytes (23–25), mesenchymal stem cells (26) and cardiac cells as well (9, 27). One of the major limitation during the in vitro culture is to guarantee a suitable mass transport to keep cells viable within the space of the scaffold during culture time. In this direction, spinner flask and rotating wall bioreactor provide a convective transport of culture media and nutrients around the surfaces of the scaffold ameliorating the effect due to stagnant film. These systems were used to produce articular cartilage in vitro after seeding chondrocytes in PGA sponges. It was observed a better tissue deposition and assembly compared with static culture (28). A more efficient system based on hydrodynamic stimulation is the perfusion bioreactor. Culture conditions are characterized by a continuous medium flow through and around the constructs and the convective transport take place in the bulk of the scaffold. In this configuration, nutrients are able to overcome transport barriers and larger tissue can be obtained. Additionally, fluid flow generates shear stresses at pore inner wall, providing mechanical stimulation to the developing constructs and aiding the tissue maturation (29). In general, perfusion culture results in a more homogenous cell distribution, greater cell viability and cell proliferation, enhanced extracellular matrix (ECM) synthesis, and more differentiated phenotype of the engineered tissue compared with static culture and other culture conditions. Cardiomyocytes seeded in porous collagen sponge scaffold and cultured in perfusion bioreactors have shown improved cell seeding efficiency as well as cell viability. Moreover, the cells within the sponge were uniformly distributed throughout the cross section and expressed cardiac markers (29). Dynamic culture provided by using perfusion bioreactor has also been observed to enhance the cell content and matrix synthesis by chondrocytes embedded in porous scaffold or gel up to 2-mm thick (23–25). Mechanical stimulation based bioreactor is extensively used in bone and cartilage tissue engineering. With these systems it is possible to provide dynamic compression that mimics the physiological loading conditions. Indeed, by modulating the frequency, amplitude, and the duration of the load history cells are able to activate specific biosynthetic pathways (30). At the same time mechanical dynamic stimulation has been shown to improve protein mass transport through polymeric hydrogel used in tissue engineering (31). Although the benefits of bioreactor systems was achieved for several tissue type including bone, vascular graft, derma, cartilage, and cardiac muscle (7, 8, 29, 32–41), actually the production of functional tissue-engineered constructs is still limited by the lack of understanding of the regulatory role of specific physicochemical culture parameters on tissue development. The realization of a functional 3D tissue equivalent relies on biomaterial properties, but the choice of a suitable bioreactor culture conditions is a crucial aspect. Herein, it will face the same aspects relying upon the design and the optimization of dynamic culture condition. In particular, the attentions are posed on nutrient consumption, growth factor trafficking, ECM biosynthetic rates, and the evolution of construct properties during culture time. Finally, it shows how this information can be implemented in mathematical framework in order to forecast ab initio tissue maturation under specific culture conditions.

2.1. Metabolic Microenvironment

The quantification of the uptake and production rates of basic metabolites is very important in order to evaluate whether a favourable cellular environment exists within constructs and to optimize and design culture protocols. Today, it is well known that oxygen consumption rate (OCR) depends upon a variety of parameters and the oxygen concentration, within cellular constructs, controls cell biosynthetic activity. Depending on the culture conditions, large gradients in cell viability and/or matrix deposition can arise between the periphery and the centre of the construct, leading to inhomogeneous properties in the neo tissue assembly, in terms of stiffness, permeability, and also cell phenotype. To assess the role of nutrient transport in such processes, several experimental and theoretical surveys were addressed to identify oxygen gradients and to predict the profiles during in vitro culture. Malda et al. (42) measured the oxygen profiles within PEGT/PBT porous scaffolds seeded with chondrocytes and demonstrated that oxygen gradients occurred due to relatively slow diffusion time of oxygen in scaffold thicker than 2 mm. These gradients are even further enhanced by increasing cell density, so indicating that OCRs have a fundamental role in gradient formation as well. Oxygen consumption kinetic was well described by the theoretical Michaelis–Menten constitutive equation corrected with a term accounting for cell density; moreover, it has been observed that cells within the constructs adapt their metabolism to the varied oxygen conditions, and therefore, that OCR varies with culture time. Furthermore, it was demonstrated that a close relationship between matrix synthesis and energy metabolism exists, and these relationships are strongly affected by pericellular environment such as pH, O_2 tension, glucose and lactate level (43). In conclusion, the interplay between scaffold properties and metabolic demand modulates the pericellular environment which, in turn controls cellular proliferative and biosynthetic activities. While the effects of material properties in terms of diffusivity and permeability on oxygen consumption is well established, recent studies demonstrated that OCR is affected

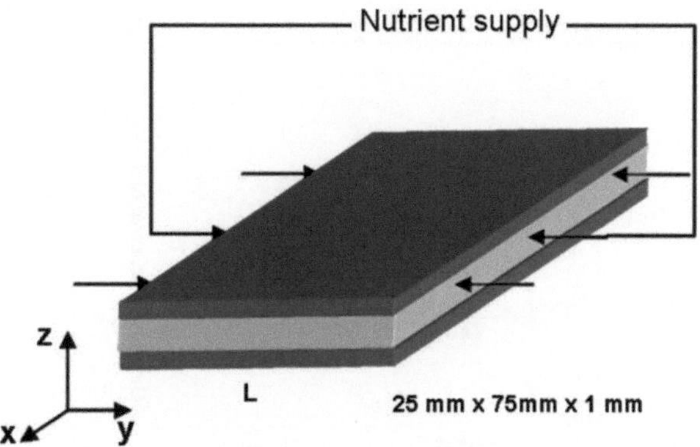

Fig. 1. Sandwich configuration, the oxygen gradient develops mainly along y-direction.

by cell material interaction. Guaccio et al. (15) used two different scaffold having different chemical motifs but very close O_2 diffusivity in order o realize 3D culture of bovine chondrocytes at different densities (400,000 and 4,000,000 cell/ml). Cells were seeded both in agarose and collagen pregel solution and allow to solidify between two glass microscope slides 1 mm spaced (Fig. 1). After gelling, the sandwiches were placed in the culture medium and cultured up to 72 h. In this configuration the transport of basic metabolites occurred only along *y*-horizontal directions. Gel solutions were loaded with a phosphorescent probe Pd-*meso*-tetra (4-carboxylphenyl) porphyrin which is sensitive to oxygen pressure within the cellular constructs. Finally, an optimized experimental setup, based on the principle of phosphorescence quenching microscopy (PQM) (44), was used to monitor the developing oxygen profiles within the construct at different time points along scaffold width.

The systematic investigation of temporal evolution of oxygen profiles along the scaffold was undertaken for both gel-based sandwiches and the results are reported below (Fig. 2). At a cell density of 400,000 cells/ml oxygen profiles within agarose and collagen gels progressively developed during the first 48 h of culture (Fig. 2). The difference in oxygen profiles between the two materials became evident after 12 h of culture. At 24 h of culture in the centre of the agarose gels (Fig. 2a), oxygen tension dropped to values well below 0.05 atm, while it resulted around 0.1 atm in collagen gels (Fig. 2b). To investigate on the effect of cell material interaction the agarose gel was loaded with RGD at two different concentrations 20 and 80 μmol (Fig. 2c, d). The presence of RGD peptides at the lowest concentration used (20 mmol) did not affect oxygen profiles (Fig. 2c). Indeed, oxygen concentration on the edge was found approximately constant around the value of 0.2 atm even after 48 h of culture, while as early as 3 h from the beginning of the culture, in the interior of the scaffolds, cells appear to experience hypoxic

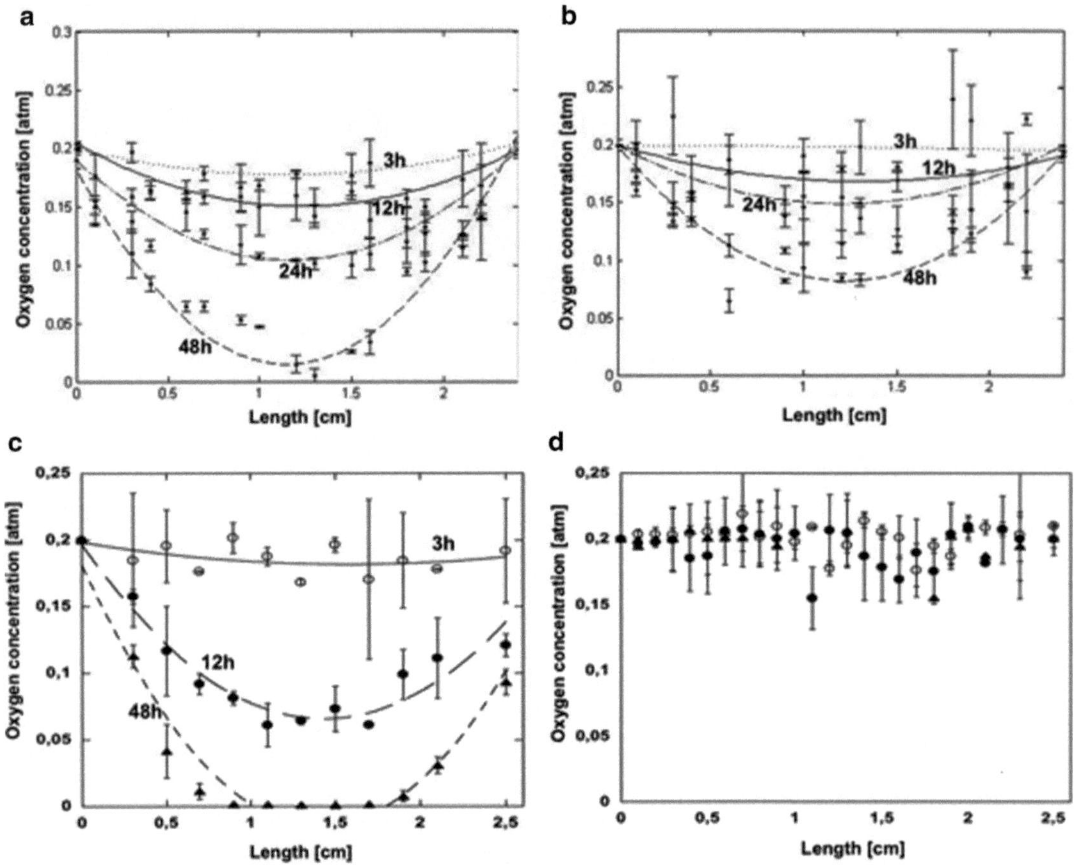

Fig. 2. Oxygen profile within cellular constructs at different time points. Agarose 400,000 cell/ml 3h, 12h, 24h, 48h (**a**); collagen 400,000 cell/ml 3h, 12h, 24h, 48h (**b**); agarose 400,000 cell/ml + RGD 20 μmol 3h, 12h, 48h (**c**), agarose 400,000 cell/ml + RGD 80 μmol at 3h (full circle), 12h (empty circle), 24h (triangle) (**d**).

conditions (Fig. 2c). When the RGD concentration was increased a slight decrease in the oxygen gradient was recorded compared with the lower RGD concentration. Finally, at 80 μmol RGD concentration, the oxygen values recorded laid within the range of 0.2–0.15 atm in the time interval from 3h to 48h (Fig. 2d). The oxygen profiles registered in the case of agarose loaded with RGD 80μmol, match those observed in the collagene gels at same cell density (Fig. 2b). OCR was evaluated using (1) and (2) that describe the mass balance and associated boundary conditions, respectively:

$$\frac{\partial c}{\partial t} = D\frac{\partial^2 c}{\partial y^2} - \frac{\mu_{\max} c}{c_{\mathrm{m}} + c}\rho \tag{1}$$

$$\begin{aligned} &\frac{\partial c}{\partial y}\left(x = \frac{L}{2}, t\right) = 0 \\ &c(x = 0, t) = c_{\mathrm{o}} \\ &c(x, t = 0) = c_{\mathrm{o}} \end{aligned} \tag{2}$$

In (1) c is the oxygen concentration, μ_{max} the maximum OCR, c_m the oxygen concentration at oxygen consumption equal to half the maximum value, c_o is the external concentration, and ρ is the cell density. Evaluating the OCR by means of (1) it was observed that from 1 h to 24 h of culture, μ_{max} values in collagen reduced by a half (from 4.69×10^{-16} to 2.47×10^{-16} g/cell/s), while the c_m reduced from 3.56×10^{-7} to 2.13×10^{-7} g/ml. Analogously, in agarose, from 1 to 24 h of culture μ_{max} reduced by a half (10×10^{-16} to 4.69×10^{-16} g/cell/s), while c_m increased from 1.92×10^{-7} to 3.56×10^{-7} g/ml. Because of the diffusivity of the oxygen within two hydrogel does not vary too much the differences in OCR is mainly due to the cell material interaction. Collagen fibres display molecular motifs that promote cell adhesion via integrin-mediated receptors, while these are not present in the agarose structure. Then, the difference in molecular cues displayed by the two materials would represent a more plausible explanation since the mechanisms of biological recognition occur in a much shorter time frame. This conclusion is supported by the strong reduction of the OCR measured in the chondrocyte–agarose constructs when 80 μmol of soluble RGD was added prior to the agarose gelation (Fig. 2d). Chondrocytes seeded in RGD-enriched agarose gels exhibited an OCR comparable to that measured in collagen. In synthesis, the results indicated that cell OCR in 3D scaffolds is also regulated by cells material molecular interaction and it is inhibited in those scaffolds that display integrin molecular cues. Therefore, to describe the nutrient transport, the knowledge of the kinetics constant is restricting. Finally, the formulation of a constitutive equation able to fully describe the oxygen transport, needs to account also for the level of other metabolites and the effect of material on cell receptors.

2.2. Physical Microenvironment Remodelling

A fundamental tissue engineering design hypothesis is that the scaffold should provide a transient biomimetic physical microenvironment in terms of mechanical and transport properties for initial function, physical cues, cell migration, tissue deposition, and molecular trafficking (45). To reach this aim several techniques were employed to build-up 3D porous structure with controlled and optimized geometry that match the requirements of a specified cell type (46). The initial scaffold properties are required to design the optimal set of physical stimuli. Indeed, in 3D cellular constructs under direct perfusion or mechanical stimulation culture, properties such as porosity, tortuosity, pore architecture, degree of crosslink, modulate the shear stress acting on cell surface, fluid velocity and nutrient supply, and mechanical load/strain transfer on cells (47). As cells synthesize ECM components, tissue growth and remodelling in vitro rely on a complex interplay between cell proliferation, matrix biosynthesis, cell–scaffold interaction, and metabolic microenvironment. From a macroscopic point of view the transport and mechanical properties of the construct change as a consequence of interaction between tissue deposition and scaffold properties. The time and space evolution of such properties makes the metabolic cell requirements variable with culture time and on the other hand the set of stimuli provided by the bioreactor need to be updated in order to guide and tailor the cell fate (48). Matrix deposition, in extreme cases, may impair transport of nutrients and metabolic wastes throughout cellular construct, leading to loss of viable cells and necrotic regions (49). To better design and to control bioreactors, a deep understanding of spatial and temporal evolution of transport properties as a function of culture time and ECM deposition is required. In cartilage tissue engineering it was postulated that transport hindrance to fluid, as well as to small and large molecules within the ECM, depends mainly on glycosaminoglycan (GAG) content (50–52), then a correlation between the total amount of GAGs and the transport hindrance is expected. The time and space evolution of diffusion coefficient of different probes within agarose–chondrocyte cartilage constructs have been evaluated along with fluid and mechanical properties (16). Chondrocytes was seeded in agarose gel and cultured up to 28 days, each week elastic modulus, hydraulic permeability, and diffusion coefficient were measured. At the same time point total amount of GAG and cell number were quantified in order to correlate the evolution of physical properties with biosynthetic activity. The determination of hydraulic permeability (K) and elastic modulus (H) was performed by means of confined compression test (53). Further, diffusion coefficients (D) of two probes, dextran (500 kDa) and bovine serum albumin (BSA), were measured with high spatial and temporal resolution by fluorescent recovery after photobleaching (FRAP). This technique allows measurements of diffusivity with spatial resolution in the order of cell size. In Fig. 3 is shown the value of diffusivity ratio (D/D_0) of two probes in a region very close to a cell imbibed in the agarose gel, along with culture time. Mechanical tests performed on the cellular constructs at different culture times, demonstrated the variation of elastic modulus and hydraulic permeability up to 28 days of culture (Fig. 4). It was observed that elastic modulus increased while permeability decreased.

The results indicated that compression stiffness, hydraulic conductivity, and the diffusion coefficient depend on culture time and finally upon GAG concentration. An empirical relationship between GAG content and diffusion coefficient as well as elastic modulus and hydraulic permeability was proposed ((3)–(5)).

$$\log\frac{k}{k_0} = -A_k C_{\mathrm{GAG}} \qquad (3)$$

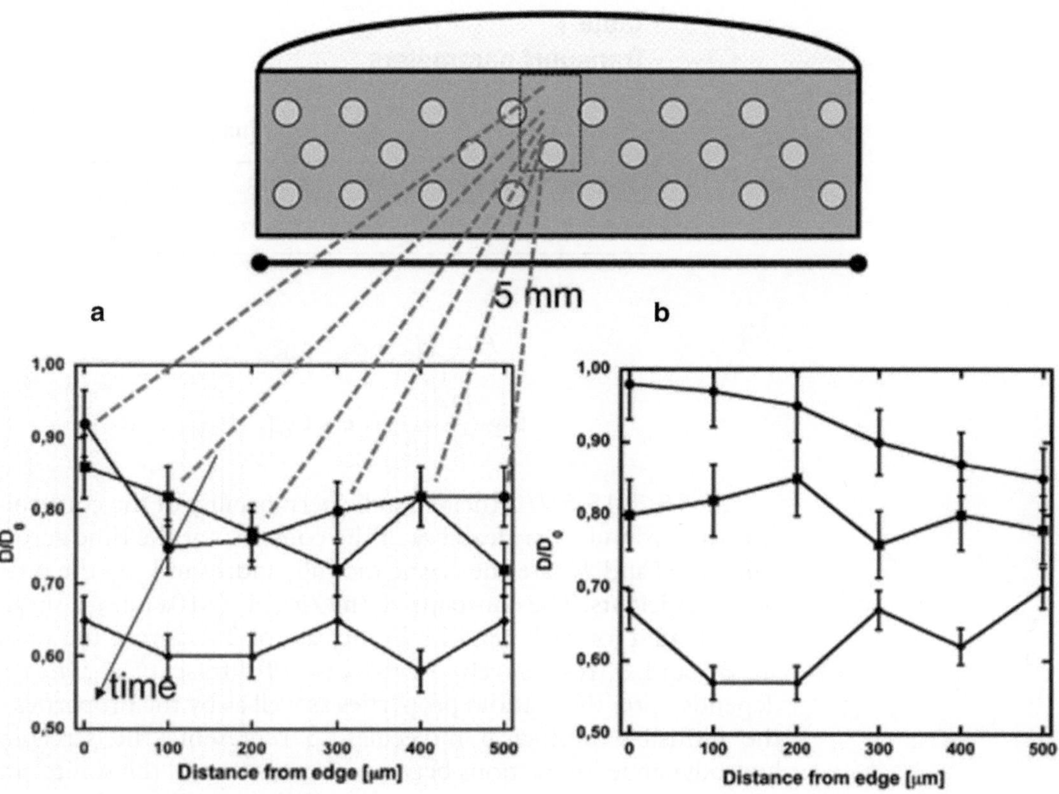

Fig. 3. Drawing of the constructs (top): circles represent the cells and area limited by dot lines represents the region where diffusivity ratio has been measured. Values of diffusivity ratio (bottom) in different regions close to the cell, at different time points (time = 0, 14, 28 days). Dextran 500 kDa (bottom, a), BSA (bottom, b).

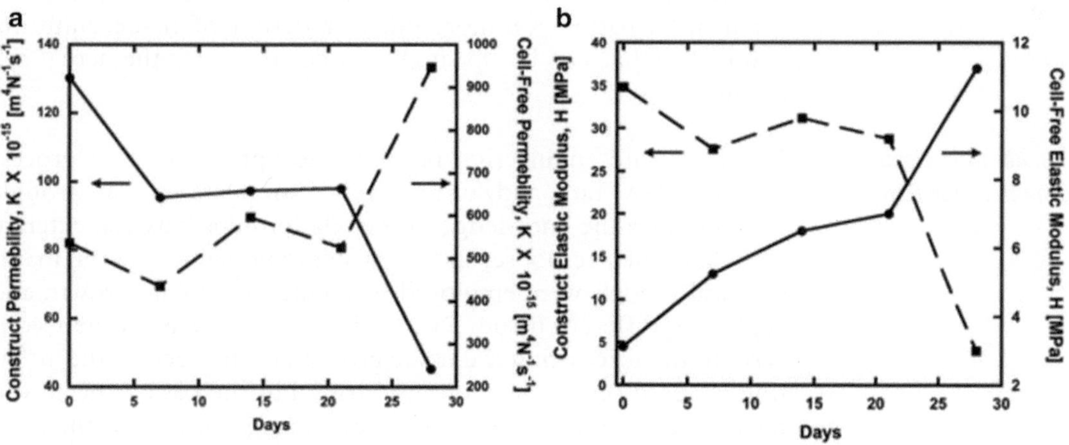

Fig. 4. Time evolution of hydraulic permeability (**a**) and elastic modulus (**b**), of cell-free agarose gel and cell-seeded agarose gel.

Table 1
Transport parameters

	A_D	r (nm)	β
Dextran (500 kDa)	40	15	0.25
BSA (67 kDa)	22	5	0.62

$$H = H_0 + A_H C_{\mathrm{GAG}} \tag{4}$$

$$\log \frac{D}{D_0} = -A_D \mathrm{e}^{-\beta r} C_{\mathrm{GAG}} \tag{5}$$

In (15.3–15.5) k is the hydraulic permeability of the construct, k_0 is the hydraulic permeability of the construct at the time zero of culture, H and H_0 are the elastic moduli, and D and D_0, the diffusion coefficients. The constants A_k (ml/g), A_H (MPa ml/g), should be evaluated for each scaffold, in the case of 2% agarose the values are 50 and 3, respectively. In the case of diffusivity A_D (ml/g), depends upon the scaffold properties as well as by the properties of the diffusible biological molecules. β represents the screening hydrodynamic interactions between the matrix and the solute, and r is the hydrodynamic radius of the diffusing probe. In Table 1 are summarised the values of the coefficient for the diffusivity ratio. This data show that during culture time the initial physical properties of the scaffold are remodelled by EMC deposition, aiding help to understand the mutual influence between biosynthetic activities and transport/mechanical properties. Moreover, they provide a useful information for more efficient design of tissue culturing conditions that can be updated in order to satisfy the actual cell requirement.

2.3. Ab Initio Tissue Growth Prediction

In vitro tissue engineering requires the optimization of process variables (flow rate, and/or mechanical stimulation) that strongly depend upon the knowledge of the relationships between external stimuli, cellular response, and construct remodeling. Several mathematical models were employed to predict neo tissue growth and applied to different tissue. Due to the complexity and variety concerning the forecasting the tissue growth, in this section the attention is posed on cartilaginous construct but the basic concept can be applied to other tissue. Hascall et al. (54) considered the average concentration of proteoglycan (PG) in the tissue as a function of time to depend upon synthesis and degradation rates. This model accounted for overall matrix deposition, but neither transport mechanisms nor space-varying tissue properties were considered. Time and space evolution of neo-synthesized matrix was taken in to account in the work of Obradovic et al. (55). This model is based on the coupling between total GAG synthesis and oxygen consumption. Even if the model was able to predict spatial and temporal evolution of GAG synthesis in PGA scaffold, several aspects were not taken in to account. Firstly, the GAG component was modeled as "total GAG" while it is now accepted that GAGs are present in three states named soluble (s), bound (b), degraded (d), respectively (56). "Soluble" refers to a neo-synthesized GAG that is further able to assemble in the extracellular space to form the "bound" fraction of the tissue (b), followed by the degradation of the bound GAG, to produce the "degraded" GAG (d). Furthermore, the effect of perfusion on oxygen uptake and transport of the soluble species was neglected. The effect of fluid velocity on biosynthesis as well as convective transport of soluble cartilaginous biomolecules was considered in the work of Klein and Sah (57). Results shown that perfusion velocity can affect the distribution of GAG within the scaffold. Indeed, high velocity field leads to a washing out effect of the soluble GAG effecting at same time the concentration of the bound and degraded molecules, but the effect of fluid velocity on nutrient supply was not provided for. In the previous sections, it has been shown that while tissue is synthesized, variation of diffusion coefficient, hydraulic permeability as well as elastic modulus and OCR, occurs due to the ECM synthesis. These variations lead to time and space evolution of the nondimensional parameters that govern the balance between different transport mechanisms: nutrient consumption, synthesis, assembly, diffusion, and convection. The objective of this section is to show a new concept in modelling tissue growth within polymeric structure by introducing the constitutive equations describing microenvironment remodelling. The scaffold is modelled as homogeneous porous structure described by the Darcy's low (6). The matrix accumulation within the scaffold is regulated by the model proposed by Di Micco et al. (56), thus synthesis, assembly, and degradation of ECM components are taken in to account ((7)–(9)). An extension to this model is adopted by inserting in the equation of the synthesis rate, the product inhibition effect as well as the dependence upon local oxygen concentration (right-hand side of (11)). Additional information adopted rely on the dependence of hydraulic permeability (K) and diffusion coefficient of soluble and degraded matrix (D), upon total GAG content ((3) and (5)). The approach proposed herein is a semicontinuum approach, the cell are modelled as "real," imbibed in a polymeric hydrogel (Fig. 5). Each cell represents a source of soluble tissue and oxygen well. In the extracellular space diffusion, convection, assembly, and degradation take place. The polymeric hydrogel is schematised as a porous medium subjected to a perfusion culture. The set of the equation used is schematised as follows.

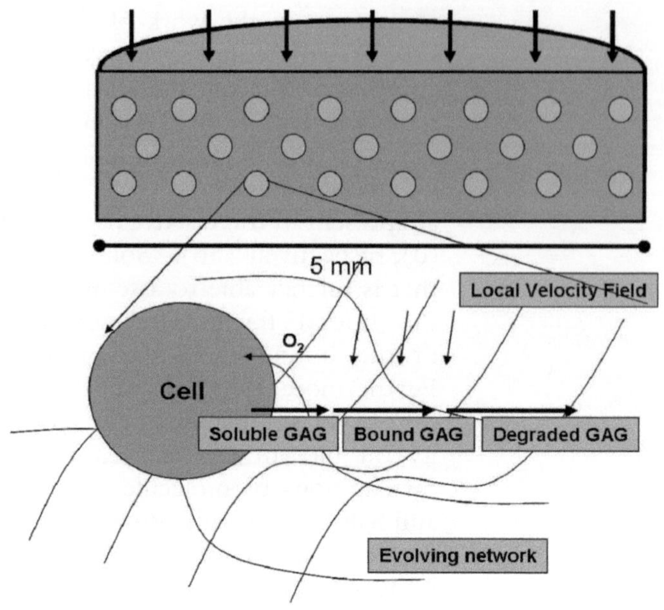

Fig. 5. Schematic representation of the used model.

2.4. Filed Equations

2.4.1. Momentum Balance

The flow in the hydrogel domain was modelled as Brinkmann's low:

$$\underline{\nabla} p = -\frac{\mu}{K}\underline{v} \tag{6}$$

2.4.2. Mass Balance

Evolution of the ECM components and oxygen transport:

$$\frac{\partial C_s}{\partial t} = -\underline{\nabla}\cdot\left[D_s \underline{\nabla} C_s - C_s \underline{v}\right] + r_f - r_b \tag{7}$$

$$\frac{\partial C_b}{\partial t} = +r_b - r_d \tag{8}$$

$$\frac{\partial C_d}{\partial t} = -\underline{\nabla}\cdot\left[D_d \underline{\nabla} C_d - C_d \underline{v}\right] - r_d \tag{9}$$

$$\frac{\partial C_{O_2}}{\partial t} = -\underline{\nabla}\cdot\left[D_s \underline{\nabla} C_{O_2} - C_{O_2} \underline{v}\right] + r_{O_2} \tag{10}$$

2.4.3. Kinetics

Kinetics terms, synthesis rate, bund rate degradation rate, OCR:

$$r_f = k_f \rho \left(1 - \frac{C_T}{C_{max}}\right) C_{O_2} \tag{11}$$

$$r_b = K_b C_b \tag{12}$$

$$r_d = K_d C_d \tag{13}$$

$$r_{O_2} = \rho \frac{\mu_{max} C_{O_2}}{c_{max} + C_{O_2}} \tag{14}$$

$$C_{T,GAG} = C_s + C_b + C_d \tag{15}$$

2.4.4. Constitutive Equations for Transport Parameters

$$\log \frac{k}{k_0} = -A_k C_{T,GAG} \tag{3}$$

$$\log \frac{D_i}{D_{0,i}} = -A_D e^{-kr} C_{T,GAG} \quad i = s, d \tag{5}$$

where p = pressure, C_{O_2} = oxygen concentration, C_s = soluble GAG concentration, C_b = bound GAG concentration, C_d = degraded GAG concentration, $C_{T,GAG}$ = total GAG concentration, μ_{max} = OCR (15), c_m = oxygen concentration at which OCR is $\mu_{max/2}$ (15), K = hydraulic permeability (16), D_s = diffusivity of soluble GAG (57), D_d = diffusivity of degraded GAG (57), μ = medium viscosity, K_f = kinetics constant for soluble GAG (57), K_b = kinetics constant for soluble GAG (57), K_d = kinetics constant for soluble GAG (57), and ρ = cell density.

2.5. Model Results

The set of equations (3)–(15) has been used to simulate the growth of cartilage tissue within a polymeric hydrogel both under static and perfusion culture. Agarose hydrogel has been used as "dummy" scaffold. The value for hydraulic permeability at time zero has been extrapolated from Fig. 4a, while the value for diffusivity ratio of soluble and degraded GAG has been taken from Table 1 by considering that GAGs macromolecules are similar to BSA. The initial partial pressure of the oxygen in the perfusion culture media as well as in the scaffold was set to 20%. Figure 6a shows the scheme of the model, the matrix distribution around the cell and in the scaffold. Figure 6b shows the effect of perfusion on total oxygen content. In static conditions, the total oxygen in the scaffold decrease faster compared to the perfusion culture because the fluid velocity sustains higher oxygen values. Figure 6c, d shows the time evolution of total soluble GAG and total bound GAG, respectively. The curves rely upon the basic model with perfusion flow ((7)–(9), neglecting oxygen and scaffold evolution equations) and the improved model under static and dynamic cultures. The predictions of basic model are very different compared with the improved model. In this case, the total amount of matrix is higher at each culture time. This is due to the variations of scaffold properties during culture time. As ECM components are synthesised and assembled, the reduction of diffusivity as well as hydraulic permeability, induce an increasing in the residence time of total matrix

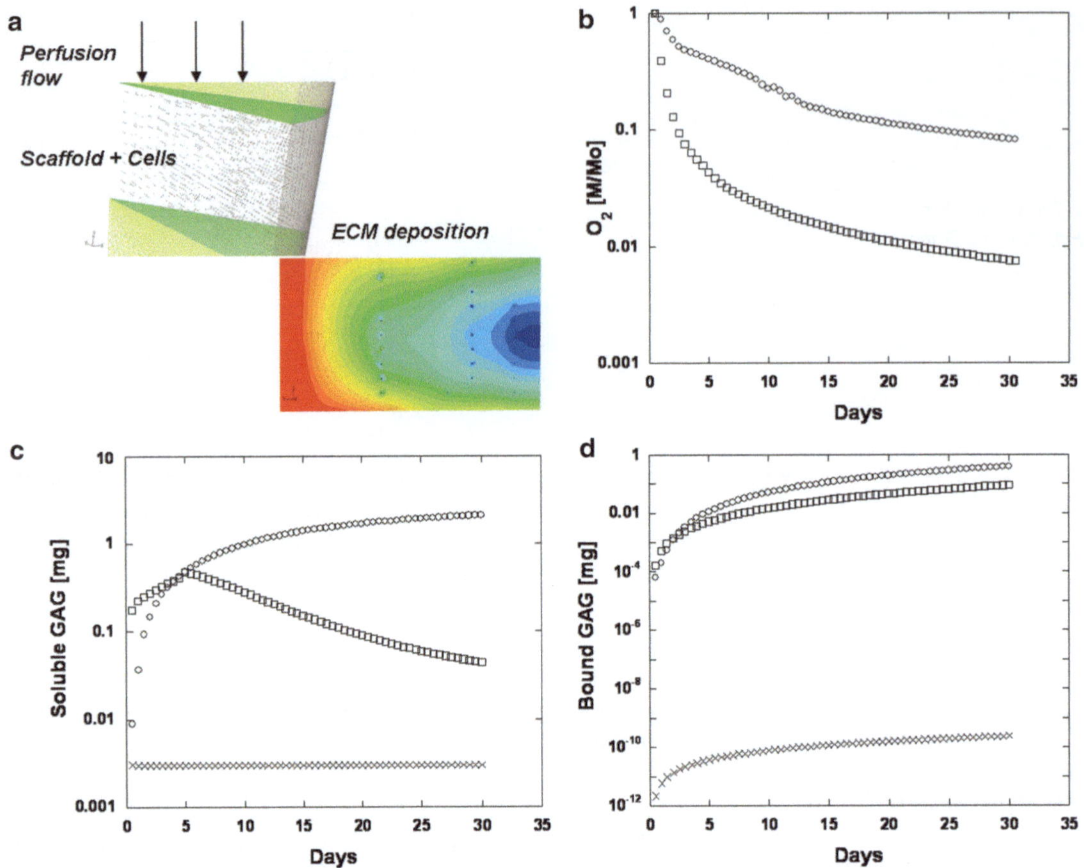

Fig. 6. Schematic representation of the model and ECM deposition in the scaffold (**a**). Total oxygen in the construct, perfusion – *circle*, static – *square* (**b**). Total soluble GAG, basic model – *cross*, static condition – *square*, perfusion – *circle* (**c**). Total bound GAG in the construct, basic model – *cross*, static condition – *square*, perfusion – *circle* (**d**).

components leading to higher content of ECM in the scaffold. In Fig. 6c is also shown the curve of soluble GAG evolution under static conditions. The decrease in oxygen content leads to a reduction in synthesis rate (11). The results presented herein show how the effects of oxygen uptake and time and space evolution of scaffold properties need to be considered in modelling tissue growth in vitro. Indeed, by considering the actual requirements of cells it is possible to predict the best tissue culture condition and to design a suitable bioreactor system in an optimised fashion.

3. Bottom-Up Strategies

The bottom-up strategies is based on the concept to assemble large tissue construct by the biological sintering of smaller building blocks (17) that mimic natural tissue composition and architecture at micrometric scale. These building blocks can be created by much of ways such as through self-assembled cell aggregates (58, 59), microfabrication of cell-laden microgel (60), creation of cell sheet (61), microfabrication of cell seeded microbeads (18). Once obtained, these building blocks can be assembled in larger tissue through a number of methods such as random packing, stacking of layers, or direct assembly. In this context, promising results have been obtained by using cell sheet (61).This technology aims at build up functional thick tissues without the use of biodegradable scaffolds and bioreactor by layering individual cell sheets of few microns thick. It was reported that intact cardiomyocyte sheets show spontaneous pulsation in vitro and can adhere and communicate to each other because of the presence of deposited ECM (61, 62). Promising results by using this approach have been obtained in realizing 3D viable tissue for corneal and myocardial tissue regeneration and it is today in clinical trial in Japan. However, this technology presents some shortcomings concerning the cell sheets handling difficulties, the production of a tissue of a desired shape with a high extra cellular matrix/cells ratio (e.g. derma, bone, cartilage) as well as limited number of cell sheets that can be effectively layered without core ischemia or hypoxia (61, 63, 64). To form building blocks of specific geometries, researchers began mixing cells with natural and synthetic hydrogel creating cell-laden microgel. Microgels are microscale hydrogels fabricated by the merger of microscale technologies and hydrogel chemistry that exhibit many desirable properties for tissue engineering applications, such as tunable geometries mechanical strength and biodegradability, and resemble the natural ECM for cell encapsulation at tissue densities (60). Therefore, cell-laden microgels may be used as building blocks to fabricate 3D tissue constructs that mimic the in vivo tissue structures by containing repeating functional units that make up most tissues (i.e. islet, nephron, or sinusoid) (65). The bottom-up assembly of cell-laden microgels has been gaining increased attention in tissue engineering research, with numerous approaches developed including random assembly (66), manual manipulation (67), multi-layer photo-patterning (68), microfluidic-directed assembly (69), and hydrophobic interactions (70). The microscale tissue constructs can be used as a "3D co-culture system" in which both cell types are physically separated and cultured in a 3D microenvironment tailored for specific requirements. An exemplary co-culture system can comprise hepatocytes in microgels (such as collagen microgels for maintaining the liver-specific functions) as functional cells and fibroblast in the bulk hydrogel (such as PEG hydrogel to provide mechanical support) as supporting cells.

3.1. Microtissue Precursor Assembly

Among the above-mentioned strategies to produce functional biological tissue starting from smaller building block, in this section, a tissue engineering strategy is described which allows the production of viable thick 3D tissues with a compact ECM by using the self-assembly properties of micrometric tissue precursors (μTPs, (18)). μTPs is obtained by means of dynamic cell seeding of bovine fibroblasts on porous gelatine microcarriers using a spinner flask bioreactor. Numerous studies illustrated that the particle cultivation technique is more effective than cell culture on flat substrates such as culture dishes (71, 72). It was observed that under optimal culture conditions cells were able to adhere, proliferate, and in particular synthesize ECM components to form a thin layer of tissue around the microbeads, generating microscale tissue named μTPs (18). The μTP was used as an ideal "material" for bio-fabrication of 3D tissue constructs. Such building blocks can be assembled, in an appropriate assembling chamber, by means of the tissue layers surrounding them, which allow their fusion through cell–cell and cell–matrix interactions. Following this strategy, a 3D functional dermal tissue equivalent has been realized (18). The first step of the process is to build the μTP. To reach this aim a dynamic cell seeding is used by means of spinner flask that was loaded with 10^5 cell/ml and 1 mg/ml of microbeads in 250 ml of culture medium, corresponding to 50 cells per bead. The first 5 h of the seeding phase are characterized by intermittent stirring (5 min at $0.04\times g$, 30 min in static) to improve the cell-to-bead distribution and to obtain a lower proportion of unoccupied beads. The disappearance of free cells from the inoculated spinner cultures was considered to indicate the attachment of cells to the microcarriers. By determining the concentration of the fibroblasts in the culture medium during the intermittent seeding, it was possible to observe that the cell density in the culture medium (cells/ml) decreased by up to 60% after 6 h because of cell attachment to the microbeads and cell death. Figure 7a shows the evolution of μTP during the seeding phase. The number of cell per microbeads increase during culture time (from 20 to 80) indicating that cell attaches and proliferate. The histology reported in Fig. 7b shows that cells are able to synthesize ECM components which act as biological glue generating aggregates of μTPs (Fig. 7c). The spontaneous assembly of μTPs after only 4 days of spinner culture suggests that this promising property can be exploited to realize and assemble 3D biohybrids by placing the μTPs in mutual proximity in an assembling/maturation chamber (Fig. 8).

The assembling chamber is designed to work both under static or perfusion conditions. It was observed that just after 1 week in the maturation chamber under static condition the building blocks were able to assemble leading to a compact tissue equivalent. Figure 8 shows a derma-like tissue (10 mm diameter and 1 mm thick) obtained by μTPs assembly. Histological images show

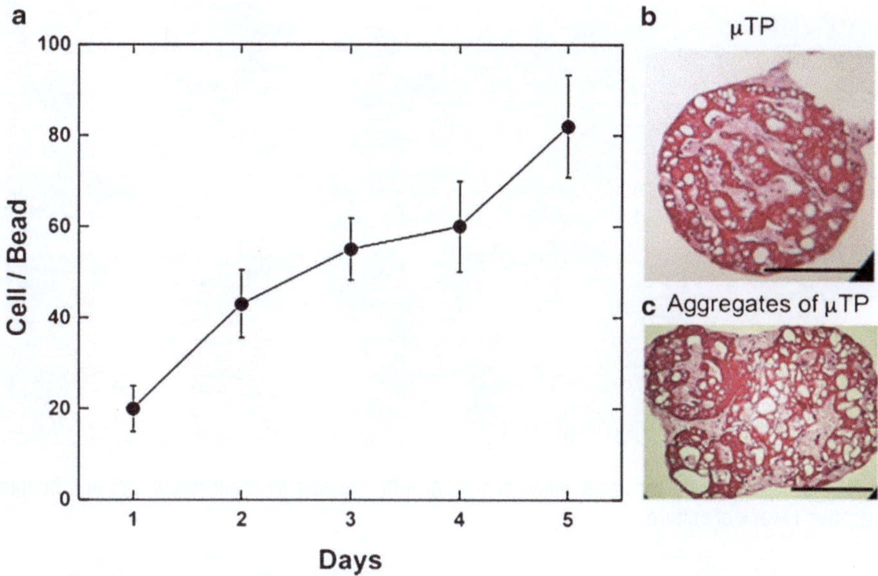

Fig. 7. Time evolution of cell/microbeads ratio in spinner culture up to 4 days (**a**). Cell colonised microbeads after 2 days of spinner culture (**b**). Aggregates of μTPs after 4 days of spinner culture (**c**).

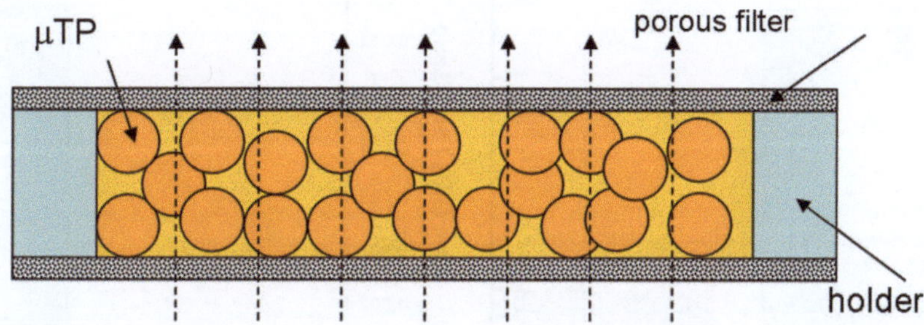

Fig. 8. Maturation chamber used to kept in close proximity the microtissue precursor. The chamber is provided with two porous filters that allow the nutrient and fluid transport and retain that microtissue inside the maturation space.

abundant ECM that connects the microbeads, and Masson tricromic stain shows organised collagen fibres in the matrix (Fig. 9). Experimental investigation carried out in our laboratories showed that although microscale assembly of biological tissue is a promising strategy to build-up tissue or organ in vitro, the maturation of the tissue equivalent can be modulated by culture conditions. Firstly, culture time longer than 1 week lead to a necrotic region in the bulk of the tissue. Furthermore, it was observed that hydrodynamic regimes modulate the assembly of the neo-synthesised tissue and the mechanical properties as well. By inducing a perfusion flow to the maturation chamber it was observed that necrotic region in the centre of the tissue equivalent can be avoided, due to more efficient nutrient transport. Moreover, by using different flow regimes difference in final tissue properties and composition was observed. Continuous perfusion induces a washing out effect of the tissue that can be observed by an abundant release of ECM

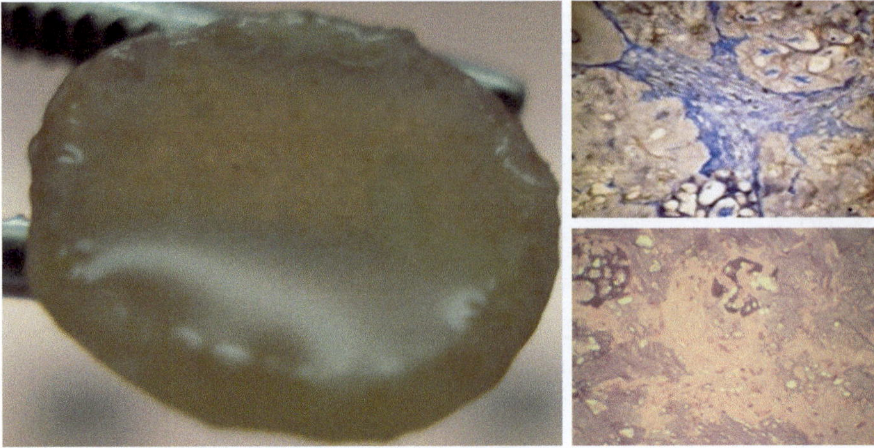

Fig. 9. Image of the tissue equivalent under static condition (*left*), Masson tricromic (*right-top*) and hematossiln/eosin (*right-bottom*), after 1 week of culture.

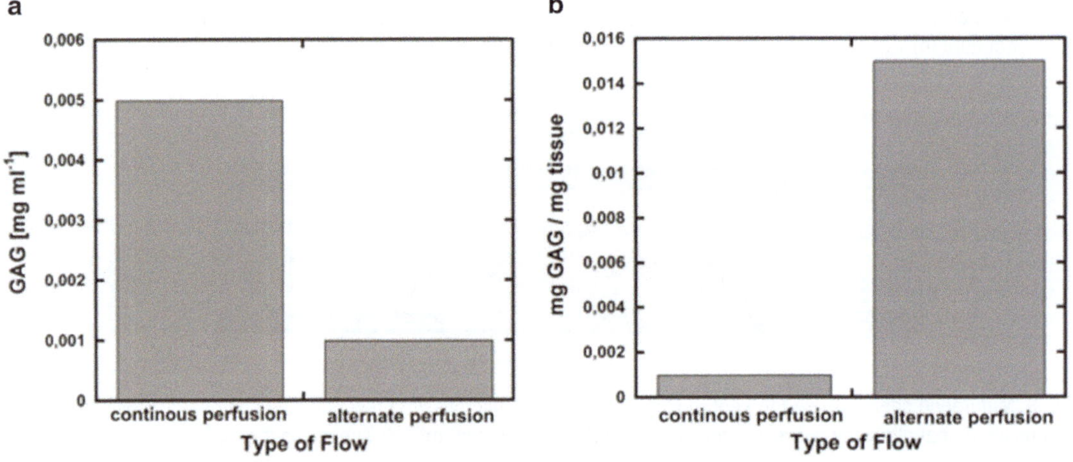

Fig. 10. GAG content released in the culture media (**a**); GAG content in the tissue equivalent (**b**) after 8 weeks of culture.

components in the culture medium (Fig. 10a). Operating with alternate flow, 12 h perfusion and 12 h tangential flow it was observed that content of the ECM released in the medium was smaller (Fig. 10a). In contrast, the amount of GAG within the tissue equivalent was higher in the case of alternate perfusion. Culture conditions affected mechanical properties of the tissue equivalent. After 8 weeks of culture the elastic modulus of the tissue equivalent kept under continuous perfusion, did not show any significant variation ranging from 2.9×10^3 to 3.2×10^3 Pa. Under alternate flow, the elastic modulus was observed to vary from 2.8×10^3 to 5×10^3 Pa. By comparing the results concerning the biochemical composition (Fig. 10b) and the mechanical properties is evident that the tissue equivalent with higher value of GAG show increased mechanical properties. These results suggest that the self-assembly property of micrometric tissue precursor can be used to build up 3D functional tissue. In the to classical top-down approach the onset of the maturation step is characterised by the absence of ECM components implying high OCR. The main advantage by using μTPs assembly is that during the spinner culture the tissue precursor is obtained in a dynamic environment where transport barrier to nutrient and waste are very low. Then, the presence of ECM components in the first stage of the maturation step entails minor metabolic request. The subsequent maturation step can be further improved by tailoring the tissue maturation by using specifics culture protocols.

References

1. Langer, R. and Vacanti, J.P. (1993) Tissue engineering. *Science* **260**, 920–926.
2. Cuono, C.B., Langdon, R., Birchall, N., Barttelbort, S., and McGuire, J. (1987) Composite Autologous-Allogeneic Skin Replacement: Development and Clinical Application. *Plast. Reconstr. Surg.* **80**, 626–635.
3. Bhandari, R.N., Riccalton, L.A., Lewis, A.L., Fry, J.R., Hammond, A.H., Tendler, S.J., and Shakesheff, K.M. (2001) Liver tissue engineering: a role for co-culture systems in modifying hepatocyte function and viability. *Tissue Eng.* **7**, 345–357.
4. Dvir-Ginzberg, M., Gamlieli-Bonshtein, I., Agbaria, R., and Cohen, S. (2003) Liver Tissue Engineering within Alginate Scaffolds: Effects of Cell-Seeding Density on Hepatocyte Viability, Morphology, and Function. *Tissue Eng.* **9**, 757–766.
5. Bancroft, N.G., Sikavitsas, V.I., and Mikos, A.G. (2003) Design of a flow perfusion bioreactor system for bone tissue-engineered applications. *Tissue Eng.* **9**, 549–544.
6. Yu, X., Botchwey, E.A., Levine, E.M., Pollack, S.R., and Laurencin, C.T. (2004) Bioreactor-based bone tissue engineering: the influence of dynamic flow on osteoblast phenotypic expression and matrix mineralization. *PNAS* **101**, 11203–11208.
7. Nettles, D.L, Vail, T.P., Morgan, M.T., Grinstaff, M.W., and Setton, L.A. (2004) Photocrosslinkable Hyaluronan as a Scaffold for Articular Cartilage Repair. *Ann. Biomed. Eng.* **32**, 391–397.
8. Gemmiti, C.V. and Guldberg, R.E. (2006) Fluid flow increases type II collagen deposition and tensile mechanical properties in bioreactor-grown tissue-engineered cartilage *Tissue Eng.* **12**, 469–479.
9. Radisic, M., Marsano, R., Maidhol, R., Wang, Y., and Vunjak-Novakovic, G. (2008) Cardiac tissue engineering using perfusion bioreactor systems. *Nat. Protoc.* **3**, 719–738.
10. Masuda, S., Shimizu, T., Yamato, M., and Okano, T. (2008) Cell sheet engineering for heart tissue repair. *Adv. Drug. Deliv.* **60**, 277–285.
11. Sahoo, S., Ouyang, H., James, C., Goh, H., Tay, T.E., and Toh, S.L. (2006) Characterization of a Novel Polymeric Scaffold for Potential Application in Tendon/Ligament *Tissue Eng.* **12**, 91–99.
12. Woo, S.L.Y., Hildebrand, K., Watanabe, N., Fenwick, J.A., Papageorgiou, C.D., and Wang, J. (1999) Tissue Engineering of Ligament and Tendon Healing *Clin. Orthop. Relat. R.* **367**, S312–S323.
13. Yanga, F., Murugan, R., Ramakrishna, S., Wang, X., Ma, Y.-X., and Wang, S. (2004) Fabrication of nano-structured porous PLLA scaffold intended for nerve tissue engineering *Biomaterials* **25**, 1891–1900.
14. Hadlock, T., Sundback, C., Hunter, D., Cheney, M., and Vacanti, J.P. (2000). A Polymer Foam Conduit Seeded with Schwann Cells Promotes Guided Peripheral Nerve Regeneration *Tissue Eng.* **6**, 119–127.
15. Guaccio, A., Borselli, C., Oliviero, O., and Netti, P.A. (2008) Oxygen consumption of chondrocytes in agarose and collagen gels: A comparative analysis *Biomaterials* **29**, 1484–1493.
16. De Rosa, E., Urciuolo, F., Borselli, C., Gerbasio, D., Imparato, G., and Netti, P.A. (2006) Time and Space Evolution of Transport Properties in Agarose–Chondrocyte Constructs *Tissue Eng.* **12**, 2193–2201.
17. Nichol, J.W. and Khademhosseini, A. (2009) Modular Tissue Engineering: Engineering Biological Tissues from the Bottom Up *Soft Matter.* **5**, 1312–1319.
18. Palmiero, C., Imparato, G., Urciuolo, F., and Netti, P. (2010) Engineered dermal equivalent

tissue in vitro by assembly of microtissue precursors. *Acta Biomater.* **6**, 2548–2553.

19. Martin, I., Wendt D., and Heberer, M. (2004) The role of bioreactor in tissue engineering. *Trends Biotechnol.* **22**, 80–86.
20. Portner, R., Nagel-Heyer, S., Goepfert, C., Adamietz, P., and Meenen, N.M. (2005) Bioreactor Design for tissue engineering. *J. Biosci. Bioeng.* **100**, 235–245.
21. Martin, Y., and Vermette, P., (2005) Bioreactors for tissue mass culture: design, characterization, and recent advances. *Biomaterials* **26**, 7481–7503.
22. Yu, X., Botchwey, E. A., Levine, E. M., Pollack, S.R. and Laurencin, C.T. (2004) Bioreactor-based bone tissue engineering: The influence of dynamic flow on osteoblast phenotypic expression and matrix mineralization *PNAS* **101**, 11203–11208.
23. Xu, X., Urban, J.P.G., Tirlapur, U., Wu, M.H., Cui, Z., and Cui, Z. (2006) Influence of perfusion on metabolism and matrix production by bovine articular chondrocytes in hydrogel scaffolds. *Biotechnol. Bioeng.* **93**, 1103–1111.
24. Freyria, A.M., Yang, Y., Chajra, H., Rosseau, C.F., Ronziere, M.C., Herbage, D., and El Haj, A.J. (2005) Optimization of dynamic culture conditions: effects on biosynthetic activities of chondrocytes grown in collagen sponges. *Tissue Eng.* **11**, 674–684.
25. Vunjack-Novakovic, G., Martin, I., Obradovic, B., Treppo, S., Grodzinsky, A., Langer, R., and Freed, L.E. (1999) Bioreactor cultivation conditions modulate the composition and mechanical properties of tissue-engineered cartilage. *J. Orthop. Res.* **17**, 130–138.
26. Zhao, F., and Ma, T. (2005) Perfusion Bioreactor System for Human Mesenchymal Stem Cell Tissue Engineering: Dynamic Cell Seeding and Construct Development. *Biotechnol. Bioeng.* **91**, 482–493.
27. Carrier, R.L., Rupnick, M., Langer, R., Schoen, F.J., Freed, L.E., and Vunjak-Novakovic, G. Perfusion improves tissue architecture of engineered cardiac muscle. (2002) *Tissue Eng.* **8**, 175–188.
28. Pei, M., Solchaga, L.A., Seidel, J., Zeng, L., Vunjak-Novakovic, G., Caplan, A.I. and Freed, L.E. Bioreactors mediate the effectiveness of tissue engineering scaffolds (2002) *FASEB J.* **16**, 1691–1694.
29. Dvir, T., Benishti, N., Shachar, M., and Cohen, S. (2006) A novel perfusion bioreactor providing a homogenous milieu for tissue regeneration. *Tissue Eng.* **12**, 2843–2852.
30. Seidel, J.O., Pei, M., Gray, M.L., Langer, R., Freed, L.E. and Vunjak-Novakovic, G. (2004) Long-term culture of tissue engineered cartilage in a perfused chamber with mechanical stimulation. *Biorheology* **41**, 445–458.
31. Urciuolo, F., Imparato, G. and Netti, P.A. (2008) Effect of Dynamic Loading on Solute Transport in Soft Gels Implication for Drug Delivery. *AIChE J.* **54**, 824–834.
32. Kalyanaraman, B., Supp, D.M., and Boyce, S. (2008) Medium flow rate regulates viability and barrier function of engineered skin substitutes in perfusion culture. *Tissue Eng.* **14**, 583–593.
33. Minuth, W.W., Schumacher, K., Strehl, R., and Kloth, S. (2000) Physiological and cell biological aspects of perfusion culture technique employed to generate differentiated tissues for long term biomaterial testing and tissue engineering. *J. Biomater. Sci. Polym.* **11**, 495–522.
34. Sikavitsas, V. I., Bancroft, G.N., Holtorf, H.L., Jansen, J.A., and Mikos, A.G. (2003) Mineralized matrix deposition by marrow stromal osteoblasts in 3D perfusion culture increases with increasing fluid shear forces. *PNAS.* **100**, 14683–14688.
35. Buttafoco, L., Engbers-Buijtenhuijs, P., Poot, A.A., Dijkstra, P.J., Vermesa, I., and Feijen, J., (2006) Physical characterization of vascular grafts cultured in a bioreactor. *Biomaterials* **27**, 2380–2390.
36. Hoerstrup, S.P., Zünd, G., Sodian, R., Schnell, A.M., Grünenfelder, J., and Turina M.I. Tissue engineering of small caliber vascular grafts. (2001) *Eur. J. Cardiothorac. Surg.* **20**, 164–169.
37. Sun, T., Norton, D., Haycock, J.W., Ryan, A.J., and MacNeil, S. (2005) Development of a closed bioreactor system for culture of tissue-engineered skin at an air-liquid interface. *Tissue Eng.* **11**, 1824–1831.
38. Halberstadt C.R., Hardin, R., Bezverkov, K., Snyder, D., Allen, L., and Landeen, L. (1993) The in vitro growth of a three-dimensional human dermal replacement using a single-pass perfusion system. *Biotechnol. Bioeng.* **43**, 740–746.
39. Yang, Y.I., Seol, D.L., Kim, H.I., Cho, M.H., and Lee, S.J. (2007) Continuous perfusion culture for generation of functional tissue-engineered soft tissues. *Current Applied Physics.* **7 S1**, e80–e84.
40. Mizuno, S., Watanabe S., and Takagi, T. (2004) Hydrostatic fluid pressure promotes cellularity and proliferation of human dermal fibroblast in a three-dimensional collagen gel/sponge. *Biochem. Eng. J.* **20**, 203–208.
41. Raimondi, M.T., Boschetti, F., Falcone, L., Fiore, G.B., Remuzzi, A., Marinoni, E., Marazzi, M., and Pietrabissa, R. (2002) Mechanobiology of engineered cartilage cultured under a

quantified fluid-dynamic environment. *Biomech. Model. Mechanobiol.* **1**, 69–82.

42. Malda, J., Rouwkema, J., Martens, D.E., Le Comte, E.P., Kooy, F.K., Tramper, J., van Blitterswijk, C.A. and Riesle, J. (2004) Oxygen Gradients In Tissue-Engineered Pegt/Pbt Cartilaginous Constructs: Measurement And Modeling. *Biotechnol. Bioeng.* **86**, 9–18.
43. Sengers, B.G., Heywood, H.K., Lee, D.A., Oomens, C.W. and Bader, DL. (2005) Nutrient Utilization by Bovine Articular Chondrocytes: A Combined Experimental and Theoretical Approach. *Biomech. Eng.* **127**, 758–766.
44. Guaccio, A. and Netti, P.A. (2010) Monitoring oxygen uptake in 3D tissue engineering scaffolds by phosphorescence quenching microscopy. *Biotechnol. Progr.* **26**, 1494–1500.
45. Hollister, S.J. (2005) Porous scaffold design for tissue engineering. *Nat. Mater.* **4**, 518–524.
46. Hutmacher, D.W., Sittinger, M., Risbud, M.V. (2004) Scaffold-based tissue engineering: rationale for computer-aided design and solid free-form fabrication systems *Trends Biotechnol.* **22**, 354–362.
47. Lee, D.A., Knight, M.M., Bolton, J.F., Idowu, B.D., Kayser, M.V. and Bader, D.L. (2000) Chondrocyte deformation within compressed agarose constructs at the cellular and sub-cellular levels. *J. Biomech.* **33**, 81–95.
48. Sengers, B.G., van Donkelaar, C.C., Oomems, C.W.J., and Baaijens, F.P.T. (2004) The local matrix distribution and the functional development of tissue engineering cartilage, a finite element study. *Ann. Biomed. Eng.* **32**, 1718–1727.
49. Sutherland, R.M., Sordat, B., Bamat, J., Gabbert, H., Bourrat,B., and Mueller-Klieser, W. (1986) Oxigenation and differentiation in multicellular spheroids of human colon carcinoma. *Cancer Res.* **46**, 5320–5329.
50. Grodzinsky, A.J. (1983) Electromechanical and physicochemical properties of connective tissue. *Crit. Rev. Biomed. Eng.* **9**,133–199.
51. Lorenzo, P., Bayliss,M.T., and Heinegard, D. (1998) A novel cartilageprotein (CILP) present in the mid-zone of human articular cartilage increases with age. *J. Biol. Chem.* **273**, 23463–23468.
52. Mow, V.C., and Ratcliffe, A. (ed.) (1997) *Basic Orthopaedic Biomechanics.* Raven Press, NY
53. Ateshian, G.A., Warden, W.H., Kim, J.J., Grelsamer, R.P.,and Mow, V.C. (1997) Finite deformation biphasic material properties of bovine articular cartilage from confined compression experiment. *J. Biomech.* **30**, 1157–1164.
54. Hascall, V.C., Luyten, F.P., Plaas, A.H.K. And Sandy, J.D. (1990) Steady state metabolism of proteoglycans in bovine articular cartilage, in *Methods in Cartilage Research* (Maroudas, A. and Kuettner, K. ed.). Academic Press, San Diego, pp. 108–112.
55. Obradovic, B., Meldon, J.H., Freed, L.E. and Vunjak-Novakovic, G. (2000) Glycosaminoglycan deposition in engineered cartilage: Experiments and mathematical model. *AIChE J.* **46**, 1860–1871.
56. Di Micco, M.A. and Sah, R.L. (2003) Dependence of Cartilage Matrix Composition on Biosynthesis, Diffusion, and Reaction *Transp. Por. Med.* **50**, 57–73.
57. Klein, T.J. and Sah, R.L. (2007) Modulation of depth-dependent properties in tissue-engineered cartilage with a semi-permeable membrane and perfusion:a continuum model of matrix metabolism and transport. *Biomech. Model. Mechanobiol.* **6**, 21–32.
58. Jacab, K., Norotte, C., Damon, B., Marga, F., Neagu, A., Besch-Williford C.L., Kachurin, A., Church, K.H., Park, H., Mironov, V., Markwald, R., Vunjak-Novakovic, G., and Forgacs, G. (2008) Tissue engineering by self-assembly of cell printed into topologically defined structures. *Tissue Eng.* **14**, 413–421.
59. Kelm, J.M., Djonov, V., Ittner, L.M., Fluri, D., Born, W., Hoerstrup, S.P., and Fussenegger, M. (2006) Design of custom-shaped vascularized tissues using microtissue spheroids as minimal building units. *Tissue Eng.* **12**, 2151–2160.
60. Du, Y., Lo, E., Ali, S. and Khademhosseini, A. (2008) Directed assembly of cell-laden microgels for fabrication of 3D tissue constructs. *PNAS* **105**, 9522–9527.
61. Yang, J., Yamato, M., Shimizu, T., Sekine, H., Ohashi, K., Kanzaki, M., Ohki, T., Nishida, Kohno, C., Nishimoto, A., Sekine, H., Fukai, F., and Okano, T. (2007) Reconstruction of functional tissues with cell sheet engineering. *Biomaterials* **28**, 5033–5043.
62. Yang, J., Yamato ,M.K., Kohno, C., Nishimoto, A., Sekine, H., Fukai, F., and Okano, T. (2005) Cell sheet engineering: recreating tissues without biodegradable scaffolds. *Biomaterials* **26**, 6415–6422.
63. Ng, K.W. and Hutmacher, D.W. (2006) Reduced contraction of skin equivalent engineered using cell sheets cultured in 3-D matrices. *Biomaterials* **27**, 4591–4598.
64. Ng, K.W., Khor, H.L. and Hutmacher, D.W. (2004) In vitro characterization of natural and synthetic dermal matrices cultured with human dermal fibroblasts. *Biomaterials* **25**, 2807–2818.
65. Costanzo, L. (2006) *Physiology.* Saunders, Philadelphia.
66. McGuigan, A.P. and Sefton, M.V. (2006) Vascularized organoid engineered by modular

assembly enables blood perfusion. *PNAS* **103**, 11461–11466.

67. Yeh, J., Ling, Y., Karp, J.M., Gantze, J., Chandawarkard, A., Eng, G., Blumling, J., Langer, R. and Khademhosseini, A. (2006) Micromolding of shape-controlled, harvestable cell-laden hydrogels. *Biomaterials* **27**, 5391–5398.
68. Tsang, V.L., Chen, A.A., Cho, L.M., Jadin, K. D., Sah, R.L., DeLong, S., West, J.L. and Bhatia, S.N. (2007) Fabrication of 3D hepatic tissues by additive photopatterning of cellular hydrogels. *FASEB J.* **21**, 790–801.
69. Chung, S.E, Park, W., Shin, S., Lee, S.A. and Kwon, S. (2008) Guided and fluidic self-assembly of microstructures using railed microfluidic channels. *Nat Mater* 7, 581–587.
70. Du, Y., Lo, E., Ali, S. and Khademhosseini, A. (2008) Directed assembly of cell-laden microgels for fabrication of 3D tissue constructs. *PNAS* **105**, 9522–9527.
71. Himes, V.B. and Hu, W.S. (1987) Attachment and growth of mammalian cells on microcarriers with different ion exchange capacities. *Biotechnol. Bioeng.* **24**, 1155–1163.
72. Huang, S., Deng, T., Wanf, Y., Deng, Z, He, L., and Liu, S. (2008) Multifunctional implantable particles for skin tissue regeneration: preparation, characterization, in vitro and in vivo studies. *Acta Biomater.* **4**,1057–1066.

Chapter 16

Enabling Biomedical Research with Designer Quantum Dots

Nikodem Tomczak, Dominik Jańczewski, Denis Dorokhin, Ming-Yong Han, and G. Julius Vancso

Abstract

Quantum Dots (QDs) are a new class of semiconductor nanoparticulate luminophores, which are actively researched for novel applications in biology and nanomedicine. In this review, the recent progress in the design and applications of QD labels for in vitro and in vivo imaging of cells is presented. Surface chemical engineering of hydrophobic QDs is required to render them water soluble and biocompatible. Further surface modification and attachment of bioactive molecules to the surface of QDs, such as peptides, aptamers, or antibodies are intensively explored for targeted imaging of living cells, and disease states in animals. Specially designed surface coatings can drastically decrease nonspecific interactions between QDs and cells, minimize degradation of QDs under in vivo physiological conditions, reduce the cytotoxicity of QDs, and prolong circulation lifetimes in animals. New generations of QD probes are also promising for imaging cellular processes at the single-molecule level. Ultimately, QDs as components of complex therapeutic nanosystems are poised to contribute significantly to the field of personalized medicine.

Key words: Quantum dots, Nanocrystals, Luminescence, Surface modification, In vivo imaging, Cytotoxicity, Cell membrane, Endocytosis, Living cells

1. Introduction

Nanomedicine is a field of science located at the intersection of nanotechnology and medicine, which gives prominent attention to the application of engineered luminescent nanoparticles for molecular level in vitro and in vivo diagnostics and therapeutics (1–5). In particular, luminescent semiconductor nanocrystals [Quantum Dots (QDs)] are emerging as robust optical labels and versatile optical transducers of molecular level binding events for imaging biologically relevant processes or disease states (6–8). QDs are designed to serve as local probes to monitor biological processes and as molecular level imaging and labeling agents. A QD-based

Melba Navarro and Josep A. Planell (eds.), *Nanotechnology in Regenerative Medicine: Methods and Protocols*, Methods in Molecular Biology, vol. 811, DOI 10.1007/978-1-61779-388-2_16,

"diagnostic tool-box" requires the development of nanoparticles made from functional materials with proper and adjustable physicochemical properties, tailored surface composition for local molecular level targeting, diagnostics, and drug release, all integrated into one nanoscale system that can be tracked in live cells (9, 10), tissues of living organisms (11), and ultimately in human body (2), with high spatial and temporal resolutions. These highly complex and functional nanoparticles are expected to be of particular benefit in genomic and proteomic technologies, and in personalized medicine. It is anticipated that nanoparticle-based therapeutics (12) will be useful for stem cell studies (13), tissue engineering to treat organs, retina regeneration, and neuron repair.

QDs are gradually replacing conventional labels in many biological applications. Compared to fluorescent dyes, QDs are much brighter, less prone to photodegradation, have broad absorption spectra and narrow emission lines. These features are extremely useful in multiplexed imaging and for long-term monitoring and detection of targets at low concentrations. Good photophysical parameters allow one, for instance, to visualize single QDs diffusing in cells or interacting with the biochemical machinery of the cell.

In this review, we present examples of QD applications in labeling and imaging of cells, tissues, and animals. QDs are shown to be excellent contrast agents for cell labeling and monitoring of the progress of therapies. Issues related to QD surface functionalization and conjugation to biomolecules for cell targeting are presented. Advances in QD applications in biological imaging have been recently the subject of Volume 374 of Methods in Molecular Biology series (14). We therefore tend to focus our attention on more recent literature examples of QD applications.

2. QDs as Luminescent Biolabels

QDs are made of elements from the II–VI and III–V groups of the periodic table, e.g., CdSe, InP, or ZnS (15). The size of the nanocrystals, within the 1–10 nm range, determines the absorption and luminescent properties of QDs, with the emission wavelength shifting to lower values with decreasing QD size. Colloidal stability of the nanocrystals is usually achieved by surface ligands, which are usually hydrophobic, such as trioctylphosphine oxide (TOPO). Compared to other luminescent or fluorescent probes used in biology, like organic chromophores, QDs offer some important advantages, including size-tunable optical emission, large molar absorption coefficients, and relatively large quantum yields. (In an excellent recent review, Resch-Genger et al. make extensive comparison between QDs and standard organic fluorophores used in biomedical imaging (16)). The brightness of QDs is higher than that of most

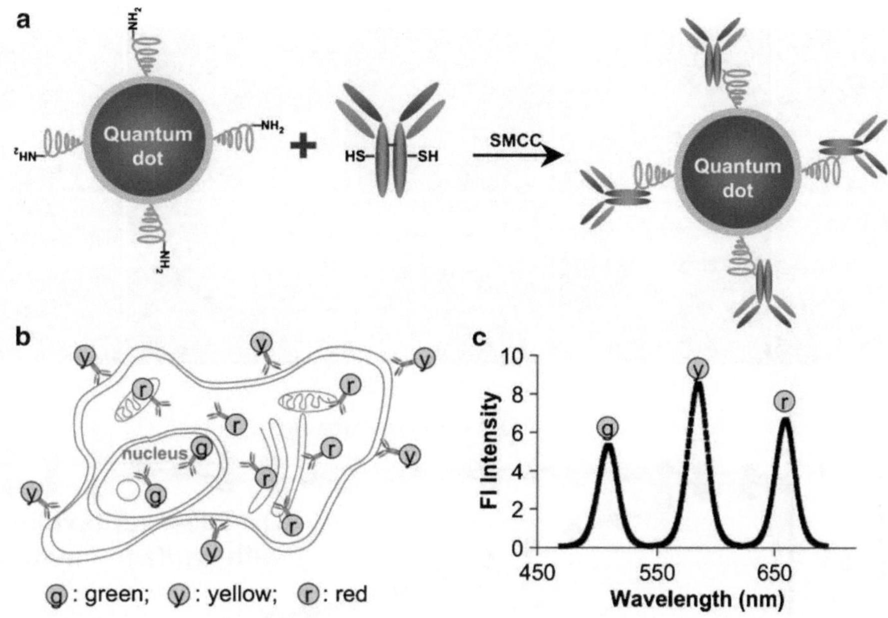

Fig. 1. Scheme illustrating the multicolour labeling of various cell components with QDs emitting in the *green* (g), *yellow* (y), and *red* (r). (**a**) The QDs are bioconjugated to different antibodies, (**b**) which can target cell membrane, internal cell components, and cell nucleus. (**c**) The multicolour labeling allows for quantitative comparison of the relative expression levels of the biomarkers. Copyright Wiley-VCH Verlag GmbH & Co. Reproduced with permission from (20).

common chromophores. The needed amount of the labels for cell imaging can be therefore significantly reduced. Chromophores exhibit narrow absorption spectra and excitation of many different dyes requires the use of many different excitation light sources. The emission spectra of chromophores exhibit significant "red tails," causing a cross talk between different molecules in multiplexed fluorescence detection. In contrast, the absorption spectrum of QDs is broad and it increases toward higher energies allowing for the excitation of many different QDs with one excitation wavelength. The narrow emission lines furthermore minimize the cross talk between different detection channels (17, 18). There is an increasing need to study many cell components (biological targets) simultaneously to understand the complexity and dynamics of biological interactions in live cells. Due to their spectral characteristics, QDs are excellent labels for multicolor imaging of multiple targets within a cell using a single excitation wavelength (Fig. 1) (18, 19). Imaging of multiple biomarkers in tumor cells and tissues (20) multiple cell surface proteins (21), colocalization of cell surface receptors (22), multicolor coding of beads (23) and cells (24, 25), and multicolor imaging of tumor cells in vivo (26) was realized.

QDs have high resistance to photobleaching and retain their luminescence orders of magnitude longer than most chromophores (Fig. 2) (8, 19, 27, 28). This allows performing single cell monitoring over extended periods of time. The low photobleaching

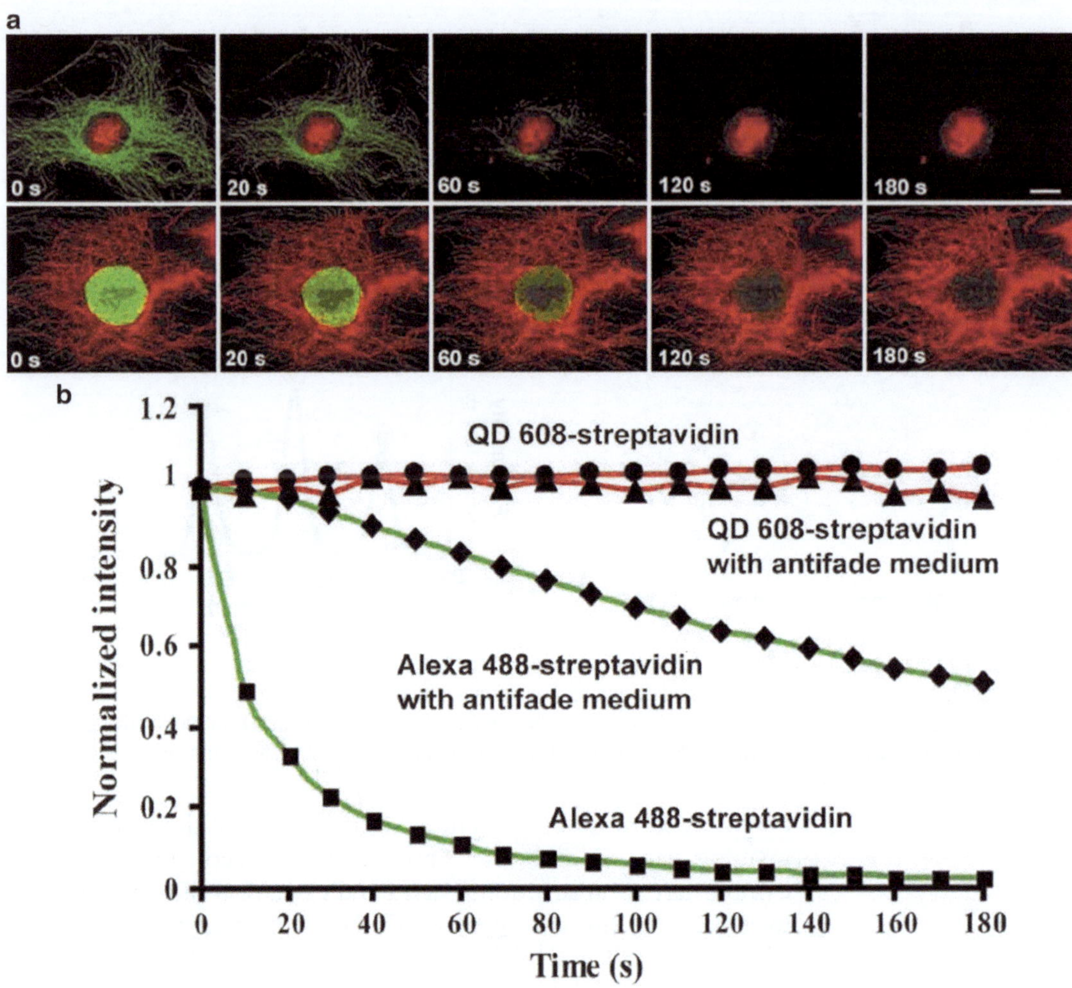

Fig. 2. Comparative analysis of the photostability of QD and a chromophore. (**a**) *Top row*: the nucleus is labeled with QDs (*red*), and the microtubules are labeled with Alexa 488 (*green*). *Bottom row*: Microtubules are labeled with QDs (*red*) and the nucleus is stained with Alexa 488 (*green*). Continuous illumination of the cells for longer periods of time (images at 0, 20, 60, 120, and 180 s time intervals are shown) causes a significant decrease in the signal of the dyes while the QD signals showed no change. (**b**) Quantitative analysis of the signal intensity changes of chromophores and dyes used in (**a**) clearly demonstrates the higher stability of the QD probes compared to Alexa 488. Reprinted by permission from Macmillan Publishers Ltd. (18).

thresholds are used in QD based imaging of cell motility and metastatic potential (29). Finally, yet importantly, the synthesis of QDs is relatively simple and one can obtain QDs emitting at different wavelengths using similar chemical procedures. Advances in QD synthesis resulted in nanoparticles emitting in the near-infrared region, a region of the electromagnetic spectrum with low absorbance by living tissue (30). Imaging of organs or tumors through thick layers of tissue, at greater depths than with traditional dyes, is therefore possible by using near infrared probes.

Resistance to photobleaching, large saturation intensities, and high brightness of the QDs allowed for tracking of individual QDs in cells. Single QD imaging is crucial to understand individual mechanisms underlying cell physiology (31, 32). For instance, single QD imaging was realized by coating QDs with small hapten-modified peptides, which targeted cell surface fusion proteins containing the corresponding single chain fragment antibody (33). Similar strategy was used to image individual transmembrane protein integrin (34), or cell surface G-protein coupling receptors (GPCRs) (35). Diffusion processes at the single QD level were also visualized. Dahan et al. (36) obtained trajectories of individual glycine receptors in the neuronal membrane of living cells for a period of time from milliseconds to minutes, much longer than the available duration when using conventional chromophores. Individual microtubule motor steps in living cells were observed by following endocytic vesicles with encapsulated QDs (37); the motion of QD-tagged kinesin motors was followed in living HeLa cells (38); and the lateral diffusion of membrane proteins (39), or multiple individual proteins in living cell (40) were visualized. New developments in single-molecule imaging methods allow for the visualization of a three-dimensional diffusion of individual QDs in cells (41). Single QD detection has however its own shortcomings. The major inconvenience is related to the blinking behavior of a single QD (42, 43). Blinking refers to random switching between "on" and "off" emission states. While the QD is in the "off" state no localization is possible, and the diffusion time-traces are as a result severely shortened. Suppression of blinking has been an active topic of research but better methods applicable to biological environments have yet to be developed (44, 45). On the other hand, blinking can be exploited for super resolution imaging in living cells (46). Super resolution imaging is relatively new and it allows one to obtain optical images with unprecedented resolution far below the diffraction limit. This approach is therefore very attractive to image the crowded cellular environment.

Interesting developments can be observed in the application of nonspherical nanocrystals, the so-called Quantum Rods (47–50). Quantum Rods (QR) offer some advantages compared to QDs. They have larger absorption cross-sections and larger surface areas per particle. The stokes shift can be controlled by their aspect ratio and they display polarized emission. Live cancer cell imaging with QR has been demonstrated recently for two-photon excitation. QRs were also used for multiplexed imaging of pancreatic cancer cells, in in vitro studies of cultured cells and fixed tissue sections (51).

2.1. QD Surface Modification

For applications of QDs in biology, one must carefully consider the structure of the QD organic ligand shell employed for steric stabilization. The chemical composition of the original shell is given by the QD synthetic protocols. For biological applications, it is necessary to make the QDs dispersible in water and therefore to modify the original stabilizing ligands. For this purpose, polymers are often employed (52). Additionally, biological molecules must be coupled to the surface of the QDs to achieve biological specificity.

High quality nanocrystals are currently synthesized via methods based on the decomposition of organometallic precursors in the presence of a coordinating solvent. The resulting nanocrystals are coated with a layer of a hydrophobic ligand, e.g., trioctylphosphine oxide (TOPO), and therefore the QDs are not soluble in water. Primary surface modifications are related to surface passivation with inorganic shells. It was found that coating the nanocrystals with an additional layer of an inorganic semiconducting material is an excellent method to improve their optical properties (53–55). Problems related to water-dispersibility and stability in aqueous buffers can be solved by growing a silica shell onto the QD core (17, 55, 56). The coating procedure is relatively simple and cost-effective. It is based on surface modification with a cross-linked primer silane-coupling agent (for example, 3-(mercaptopropyl) trimethoxysilane (MPS)) and growing the silica shell directly in water or water/ethanol solution. Alternatively, one can use micro-emulsion synthesis resulting in individual QDs embedded at the center of a silica sphere (57, 58). Coupling of biomolecules to the surface of QDs can be realized by proper surface functionalization with bifunctional cross-linkers, e.g., amines, thiols, carboxylic acids, or phosphonic acids (17, 59, 60). In general, functionalized silica shells provide stability in a wider range of buffer solutions and pH conditions. Additionally, the silica coating gives protection against photooxidation and may act as a barrier for diffusion of possibly toxic ions out of the nanocrystals into the environment, reducing therefore the QD toxicity (61).

Several requirements exist regarding the QD dispersibility in biological buffers, maintaining the attractive luminescent properties, and ensuring colloidal, chemical, and photochemical stabilities. Proper tailoring of ligands at the QD surface seems to be half of the success and great attention was given to the choice of the ligand chemistry and attachment strategies. Loss of surface functionality leads to aggregation and precipitation, which ultimately makes the QDs unreliable as luminescent probes.

There are well-established methods available for rendering the QDs compatible with aqueous environment, a primary requirement for the QDs to become useful in biology. Thiols were found to bind strongly to the QDs surface and ligand exchange reactions between TOPO and bifunctional thiols displaying also a hydrophilic group have been routinely performed (62, 63). Currently, detailed and proven protocols can be found in the literature for the preparation of water soluble QDs based on dihydrolipoic acid (DHLA) (64), PEG-based bidentate ligands with higher stability in biological media (65), or for the encapsulation of QD in phospho-

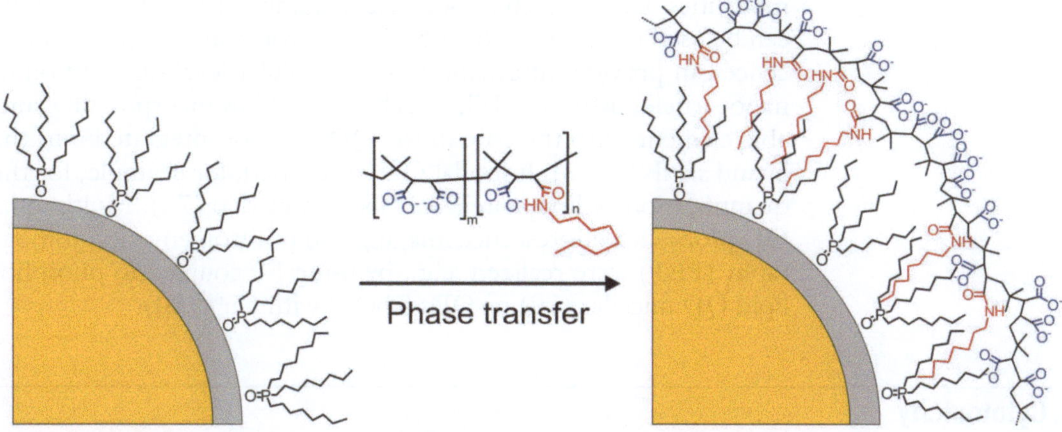

Fig. 3. General scheme of coating hydrophobic, TOPO-coated QDs with an amphiphilic polymer. The polymer has hydrophobic (*red*) and hydrophilic (*blue*) units in its structure. The hydrophobic part of the amphiphile interacts with TOPO via hydrophobic interaction, while the hydrophilic part provides water dispersibility. Using such polymers, the QDs can be efficiently transferred to water without loss of their luminescent properties (63).

lipid micelles for use in cellular and in vivo imaging (66). The latter approach is based on the attachment of an amphiphilic coating on top of the original ligand via hydrophobic interactions. Small organic molecules, as well as specifically tailored amphiphilic polymers were shown to coat effectively the QDs (67–71). Dubertret et al. (72) encapsulated QDs within phospholipid block copolymer micelles. The encapsulated QD showed reduced toxicity and improved stability by including PEG in the micelles. An important advantage of using amphiphilic coatings is that no ligand exchange reactions are needed and the original surface ligands remain unperturbed. The QDs therefore retain their luminescent properties when coated. We have recently developed a novel polymeric coating platform for functionalization and derivatization of hydrophobically coated nanoparticles (Fig. 3). The general structure of the polymer consists of hydrophobic octyl side chains for interaction with TOPO, and the hydrophilic carboxylic groups render the system water-soluble.

Our approach allows for relatively easy introduction of various chemical functionalities on the surface of QDs, including vinyl groups, PEG chains, amide acrylates, amines, and lanthanide complexes with kryptand units, or units designed for "click chemistry" (73). The latter can be used for easy coupling of biomacromolecules. A disadvantage of the amphiphilic polymer coatings is that the size of the QDs increases significantly above >15 nm even without any further modification.

Surface modifications are of interest to obtain multifunctional nanoparticles integrating therapeutic and diagnostic capabilities. In particular, the luminescence detection can be combined with other modalities, like magnetic resonance imaging (MRI) (74, 75). MRI can be used to assess the distribution of probes in vivo, and fluorescence can provide information on the cellular level distribution of nanoparticles. Multimodality can be achieved by incorporating suitable elements into the core of the QDs (76) or integrated into the ligand shell (77). Multimodal QDs were used, for example, for the quantification of lipoprotein metabolism in vivo (78). Multimodal QD probes for fluorescence imaging and positron emission tomography (PET) were realized, e.g., by using F^{18} coupled to phospholipid QD micelles (79) or QDs labeled with Cu^{64} (80).

3. Cytotoxicity of QDs

Cytotoxicity of nanoscale materials is of growing concern and detailed understanding of the interactions between nanoparticles and biological interfaces is important (81–83). Major concerns must be addressed before QD-based therapeutics will be available for clinical applications (84–86). It is believed that the toxicity of the QDs arises mainly from degradation of the QDs leading to the release of toxic ions, like Cd^{2+} (84, 85). Avoiding toxic materials, like cadmium, in QDs seems to be a straightforward solution; however, the photophysical parameters of the labels, e.g., silicon nanocrystals (87) or InP QDs (88, 89) should match those of the currently available compounds. Not only the QD composition, but also the size, shape, chemistry, and the physicochemical properties of the surface coatings were shown to play a significant role in cell death, proliferation, and differentiation (9, 84, 90). Cellular response to nanoparticles, like binding and activation of membrane receptors and subsequent protein expression is strongly size-dependent, although the related mechanisms are still poorly understood (91). Synthesis of more compact coatings to reduce the QD hydrodynamic radius for in vivo applications is needed to understand better the QD in vivo clearance mechanisms (92, 93). Size reduction may be achieved by fabrication of monovalent QDs (94) or by using smaller antibody fragment like single-domain antibodies (95).

Noninternalized QDs as well as internalized into cells can induce damage (86). It is therefore important to distinguish between whether the toxicity arises from the interaction between the surface groups with the cell membrane or from the intracellular uptake of QDs. The chemistry of the surface groups affects endocytosis, so one has to distinguish between intracellular and extracellular exposure concentrations (96). Reduced endocytosis at 4°C as well as coating the QDs with PEG results in improved biocompatibility because of minimized endocytic uptake. After endocytosis, QDs with PEG on the surface showed no different cytotoxic effect than "bare" QDs. Intracellular environment may degrade QDs due to low pH, metabolic degradation, oxidative environment or irradiation and destruction of the ligand shell. All this destruction paths may lead to cell death (84). For example, Lovric et al. observed reduced QD-induced cell-death after addition of an antioxidant indicating that redox reactions inside the cells play a significant role in QD degradation. (86). Overall, the QD cytotoxicity correlates better with intracellular QD levels than with extracellular exposure. Standard tests for cytotoxicity can be accompanied by label free tests, e.g., by studies of cellular micromotility using electrical cell-substrate impedance analysis (97).

Interestingly, the toxic nature of QDs may be used in therapeutics. Photoexcited QDs can participate in energy and electron transfer processes to nearby molecules. These processes can result in the formation of highly reactive species, which may induce damage to the cells (Fig. 4) (98). This is the basis of photodynamic therapy (PDT), a cancer treatment method where the apoptosis of the cancer cells is initiated by reactive oxygen species (ROS) or radicals generated by a photosynthesizing agent (99, 100).

Combined with active and selective targeting of cancer cells by surface engineered QDs PDT becomes a highly local method, i.e., the generation of reactive species happens only where the QDs are located and when illuminated with appropriate light. The tunability of the emission of the QDs up to the infrared region allows one to use them for the treatment of deep-seated tumors. Conventional molecular photosensitizers can also be bound to QDs (101) and their excitation can be realized via FRET from QDs. PDT can then

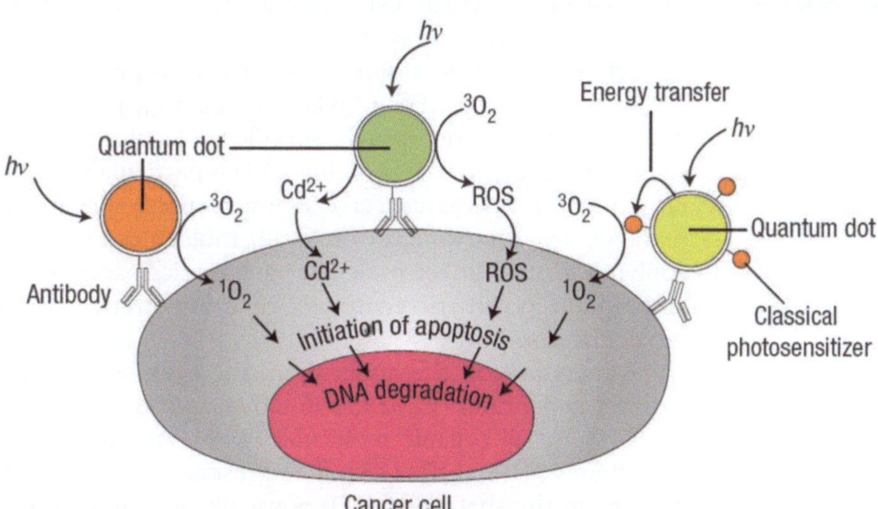

Fig. 4. Schematic presentation of possible routes for the initiation of apoptosis of cancer cells via generation of reactive oxygen species (ROS), highly reactive singlet oxygen, or degradation of the QDs and release of toxic ions by simply illuminating the QDs with short wavelength light. Attachment of QDs to specific cancer cells can be realized by using properly functionalized QDs. Reprinted by permission from Macmillan Publishers Ltd. (8).

be used at excitation wavelengths were the photosensitizer does not absorb. By adjusting the spectra of the QDs to that of the photosensitizer, one could also generate specific ROS depending on the QD type.

4. Cellular Imaging and In Vivo Studies

The design of successful QD labels includes the synthesis of a QD of proper type and size, surface modification with inorganic shells, surface modification for water dispersibility, and bioconjugation to specific biomolecules. The availability of bright QDs dispersed in water opened the possibility to use them as biological labels. Surface functionalization strategies with a layer of biomolecules has been developed, making the QD labels more specific to particular biological targets. Low photobleaching rate and spectral properties make the QDs ideal for long-term multiplexed imaging. After solubilization, the hydrophilic groups on the surface of QDs include amines, carboxylic acids, or alcohols, which can further be modified for coupling to biomolecules of interest, usually through a carbodiimide-mediated amidation, since most of the proteins have amine groups in their structure, or by electrostatic assembly (102–105). Additionally, functionalization of the QD surface with PEG is usually performed, as PEG increases water dispersibility, significantly reduces nonspecific interactions with cells and increases the circulation time in vivo.

4.1. Cellular Imaging

QDs have been widely used for the imaging of biomolecules and biomolecular interactions and processes on the cell surface and in the cell interior. Depending on the targeted process or molecule, the surface coating of the QDs must be adjusted and optimized. The coating may incorporate specific molecules to target specific cell surface receptors or proteins. This is particularly interesting for targeting tumors, as cancer cells overexpress specific cell-surface markers. The attachment of relevant molecules can be realized by applying a small linker molecule, polymeric spacers (e.g., PEG or amphiphilic polymeric coatings), or using streptavidin for supramolecular assembly (18, 36).

Depending on the application, the QDs were functionalized with short peptides (22, 24, 106–109), antibodies (18, 19, 36, 69, 71, 110), small organic molecules (111), aptamers (112), or carbohydrates (105, 113–115). The presence of multiple functional groups on the surface of QDs is usually advantageous. QDs with multiple ligands can in principle act as multivalent ligands and cross-link cell surface receptors (91) increasing the affinity of QDs toward specific cells.

Binding of QDs to cell membranes is highly dependent on the surface coating. Targeting cell surface receptors, ion channels, and transporters was reported and imaging of serotonin receptors in HeLa cells (111), glycine receptors (36), erbB/HER receptors (22), AMPA receptors in neurons (116), $GABA_C$ receptors expressed in Xenopus oocytes (117), TrkA receptors (118), and GPCRs (35) was described, among others, in the literature. Each of these applications required a specific surface coating. For example, integrin labeling (34) with QDs required to include a cyclic Arg-Gly-Asp (RGD) sequence peptide on the surface of the QDs, or development of QD-IgG probes was needed to selectively stain breast cancer cell marker Her2 on the surface of live breast cancer cells after the cells were incubated with a monoclonal anti-Her2 antibody (18).

Intracellular labeling faces more obstacles compared to the labeling of the cell surface, primarily because specific uptake must dominate nonspecific binding and the QDs must be allowed to escape the endosomal compartments (119). Additionally, the unbound labels should be removed from the cytoplasm. Nevertheless, cellular uptake of QDs was demonstrated to be possible via a number of different mechanisms. The specific mechanism is strongly dependent on the molecular functionality on the surface of the QDs. It is still challenging to functionalize the QDs with multiple ligands to decouple, e.g., cell membrane translocation and targeted detection of biomolecules inside the cells. Membrane translocation was realized by electroporation (120), lipid-based transfection (26, 84), microinjection (72, 84), receptor-mediated endocytosis (22, 62, 88, 120), or peptide-mediate endocytosis (24, 108, 121–124). After internalization, the QDs tend to accumulate in vesicular compartments, endosomes. The mechanism of QD endocytosis is not explored in details and it is not known what exact endocytotic route the QDs follow. PEG on the surface of the QDs was shown to inhibit QD uptake by the cells. The linkage chemistry between the PEG and the QD can be designed to be sensitive to enzymatic cleaving. After cleavage, cell-penetrating peptides are exposed on the surface of the QDs, and the QDs may internalize into cells. Light sensitive cleavage can lead to targeted delivery of QDs by enzyme-modulated cell uptake (125, 126).

There are many problems related to endocytosis and to the degradation of QDs entrapped in organelles, such as endosomes and lysosomes (123). For instance, QDs coated with cell-penetrating peptides, such as polyarginine or Tat (a protein transduction domain of HIV-1 Tat peptide) are trapped in endosomes, and are not available for imaging, recognition, or targeting. Overall, the trafficking of QDs after internalization is poorly understood. The trafficking mechanism can be modulated by the QD size, coating type (121) or disease state, and detailed presentation of this subject is outside of the scope of this review. Recent results showed that

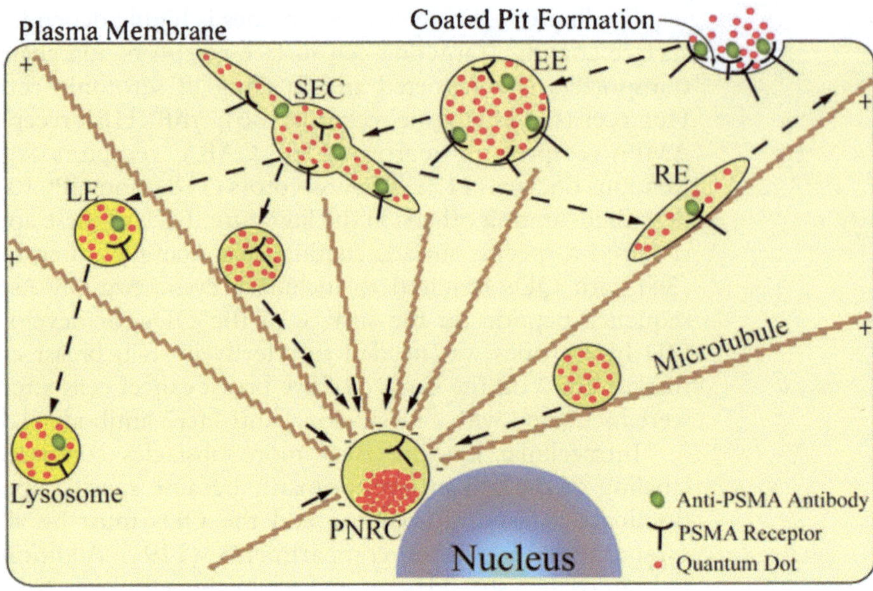

Fig. 5. Scheme showing the possible trafficking and intracellular localization of charged QDs in prostate cancer cells following clathrin-mediated uptake. *EE* early endosomes, *SEC* sorting endosomal complex, *RE* recycling endosomes, *LE* degradative late endosomes and lysosomes, *PNRC* perinuclear recycling compartment, *MTOC* the microtubule organizing center. Copyright Wiley-VCH Verlag GmbH & Co. Reproduced with permission from (127).

trafficking of QDs in cells might also depend on the cancer-cell phenotype, adding to the complexity of the trafficking of QDs inside live cells (Fig. 5) (127).

If the cellular uptake of QDs proceeds via endocytosis, endosomal escape, and internalization in the cytoplasm is needed to label individual cellular components. It was shown that the QDs can escape the endosomes by, e.g., disrupting them by endosomolytic surface coatings, e.g., polyethyleneimine (PEI) (128). The disruption is believed to occur due to the positive charges of the PEI and a proton sponge effect.

A growing number of different cellular components have been successfully labeled with QDs. Specific target recognition ability is needed for effective labeling in the crowded environment of a cell (17). Among others, labeling of subcellular components, like actin filaments (17, 18), nucleus/nuclear proteins (18, 84, 120), mitochondria (84), microtubules (18), or endocytic compartments (121) have been reported.

QD delivery into cells can be optimized by employing cell-penetrating peptides (13, 123, 129). Cationic peptide, polyarginine, and a hydrophobic counterion pyrenebutyrate were used for the delivery of QDs across the plasma membrane to the cytosol of kidney cells bypassing the barrier of endocytic vesicles. The uptake was not affected by decreasing the temperature, indicating that uptake was proceeding via a route independent of endocytosis

(130, 131). QDs labeled with an insect neuropeptide, allatostatin 1, were shown to be able to cross into the nucleus. For nuclear targeting the QDs must escape the endosomes, have a nuclear localization signal to be transported to the nucleus and to interact with nuclear pore complex, and be small enough to enter the nucleus (<20 nm), making the design of QD probes specifically targeting the biomolecules in the nucleus even more challenging.

4.2. In Vivo Animal Imaging

QDs were widely used as contrast agents for imaging of live animals. Despite the large light attenuation in in vivo imaging, the large absorption coefficients and low photobleaching threshold allow one to use higher excitation levels and monitor the QD probes for extended periods of time. In a pioneering paper, the imaging of frog embryos with QDs encapsulated in phospholipid micelles was presented by Dubertret et al. (72). The QDs were injected into individual blastomeres and the embryo was imaged to the tadpole stage, clearly demonstrating the advantage of low photobleaching rates of QD probes. Stability of the QDs under in vivo conditions and long circulation times enabled imaging of QDs injected into tumors and their migration to sentinel lymph nodes. Mapping these lymph nodes was among the first uses of QDs in in vivo imaging (27, 132). The large two-photon absorption cross-section of QDs was utilized in multiphoton microscopy to image the vasculature in live animals. Angiography was performed by imaging the QDs directly through skin after they were intravenously injected into mice (133).

Passive targeting of tumors is based on accumulation of QDs near the tumors due to the enhanced permeability and retention effect (EPR), while active targeting is mediated by antibody, aptamer, or peptide binding to cancer-specific cell surface biomarkers (69, 106). Improvements in the QD-coating designs allowed for more specificity in targeting tumors, for example, prostate cancer (69, 134) or pancreatic cancer (71). Tumor targeting in vivo with high selectivity was demonstrated by Akerman et al. (106). The QDs were coupled to different peptides targeting specific organs, like lungs blood vessels or lymphatic vessels of mice. Coating the QDs with PEG reduced significantly the accumulation of the QDs in liver and spleen. In vitro histological results showed that QDs homed to tumor vessels guided by the peptides and escaped clearance by the reticuloendothelial system. Gao et al. used multifunctional nanoparticles based on amphiphilic polymer coatings for cancer targeting and multicolor imaging of a human prostate cancer growing in mice (69). Multicolor imaging with QDs was used to visualize tumor cells intravenously injected into mice and to follow them until they extravasated into lung tissue (26).

Improved tissue penetration and greater imaging depths in in vivo imaging can be achieved by using near infrared QD probes (30, 135, 136). Sentinel lymph node imaging is an increasing practice

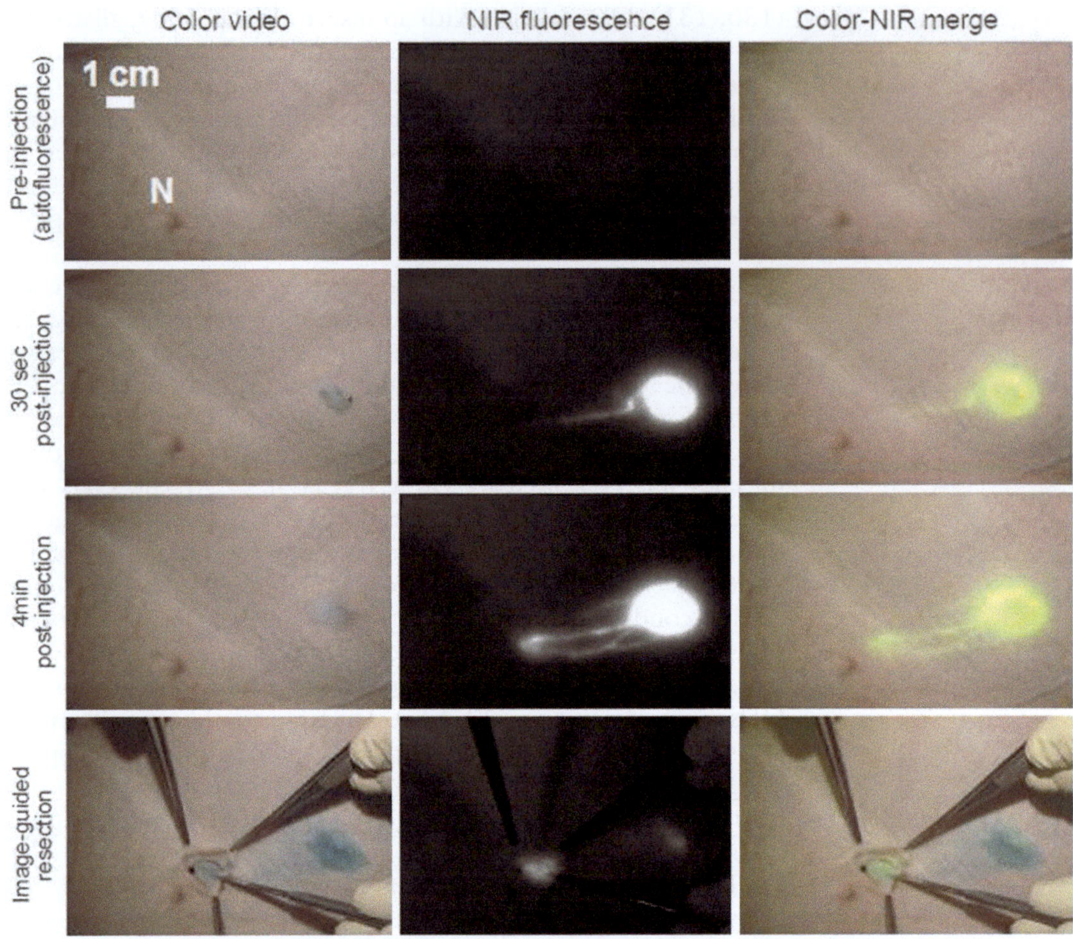

Fig. 6. Images of the surgical field in a pig injected intradermally with NIR QDs at different time points (*vertical labels*). In the columns color video (*left*), NIR fluorescence (*middle*) and color-NIR merged (*right*) images are shown. Reprinted by permission from Macmillan Publishers Ltd. (30).

in cancer treatment. QDs can be used to guide tumor surgery without the need of radioactive labels or dyes and 1 cm depth imaging in large animals, like mouse or pig, in real time was demonstrated after intradermal QD injection (30). The injected QDs were shown to migrate to lymphatic nodes within minutes (Fig. 6).

Background reduction and high signal-to-noise ratios can be achieved by using bioluminescence imaging (137). The so-called self-illuminating QDs luminesce by bioluminescence resonance energy transfer from bioluminescent proteins coupled to the QD surface. The donor energy comes from a chemical reaction catalyzed by donor enzyme. In imaging, this results in strongly reduced fluorescent background arising from endogenous chromophores. Using bioluminescence resonant energy transfer (BRET) small animal imaging with QDs has been performed from 3 mm deep tissue (137).

5. Conclusions and Outlook

QDs were shown to constitute a new class of versatile and robust luminescent labels and imaging agents for biological and biomedical investigations. Synthetic advances in QD research have resulted in a broad spectrum of possible surface chemistries that can be used in cell labeling, in vivo imaging, or guided surgery. These advances, although clearly remarkable, are slowed down by a number of challenging problems. These problems are mainly related to the relatively small size of the nanocrystals and the associated colloidal and toxicological properties. As the QD size gives rise to a desired emission wavelength (depending on the semiconductor type) and is fixed, the colloidal properties are mainly manipulated via chemical engineering of the nanocrystals' surfaces. Still poorly understood are the interactions of QDs with cells on the molecular level. Properly surface engineered QDs display reduced toxicity, increased specific binding, and reduced nonspecific interactions with the cell membrane. The problem of surface chemistry is therefore multifaceted, and a complicated nanoscale system incorporating inorganic semiconductor cores and organic, functional, corona emerges, with only several well-established protocols discussed in details in the literature available for their fabrication. Trial and error methods of achieving proper QD coating integrating biomolecules are not uncommon, and a concerted effort toward designing of proven protocols is needed for the QDs to become more widespread in biology. Going beyond imaging, and the scope of this review, active integration of QDs with cells, especially interfacing with neurons to trigger neuronal functions is a very exciting research direction (109, 138) in neuroscience (10, 36, 139). QDs can also be used as vehicles for drug delivery (122, 140) and monitoring of the therapeutic progress (1, 12). Surely many new developments in these fields will be seen in the coming years.

Acknowledgments

We are grateful to the Institute of Materials Research and Engineering of A*STAR for financial support.

References

1. Peer, D., Karp, J. M., Hong, S., Farokhzad, O. C., Margalit, R., Langer, R. (2007) Nanocarriers as an emerging platform for cancer therapy. *Nature Nanotechnol.* **2**, 751–760.
2. Iga, A. M., Robertson, J. H. P., Winslet, M. C., Seifalian, A. M. (2007) Clinical potential of quantum dots. *J. Biomed. Biotechnol.* **2007**, 76087.
3. Douma, K., Prinzen, L., Slaaf, D. W., Reutelingsperger, C. P. M., Biessen, E. A. L., Hackeng, T. M., Post, M. J., van Zandvoort, M. A. M. J. (2009) Nanoparticles for optical

molecular imaging of atherosclerosis. *Small.* **5**, 544–557.

4. Farokhzad, O. C., Langer, R. (2009) Impact of nanotechnology on drug delivery. *ACS Nano.* **3**, 16–20.
5. Riehemann, K., Schneider, S. W., Luger, T. A., Godin, B., Ferrari, M., Fuchs, H. (2009) Nanomedicine-Challenge and perspectives. *Angew. Chem. Int. Ed.* **48**, 872–897.
6. Alivisatos, P. (2004) The use of nanocrystals in biological detection. *Nature Biotechnol.* **22**, 47–52.
7. Alivisatos, A. P., Gu, W., Larabell, C. (2005) Quantum dots as cellular probes. *Annu. Rev. Biomed. Eng.* 7, 55-76.
8. Medintz, I. L., Uyeda, H. T., Goldman, E. R., Mattoussi, H. (2005) Quantum dot bioconjugates for imaging, labeling and sensing. *Nature Mater.* **4**, 435–446.
9. Maysinger, D., Lovric, J., Eisenberg, A., Savic, R. (2007) Fate of micelles and quantum dots in cells. *Eur. J. Pharma. Biopharma.* **65**, 270–281.
10. Pathak, S., Cao, E., Davidson, M. C., Jin, S., Silva, G. A. (2006) Quantum dots applications to neuroscience: New tools for probing neurons and glia. *J. Neurosc.* **26**, 1893–1895.
11. Gao, X., Yang, L., Petros, J. A., Marshall, F. F., Simons, J. W., Nie, S. (2005) In vivo molecular and cellular imaging with quantum dots. *Curr. Op. Biotechnol.* **16**, 63–72.
12. Derfus, A. M., Chen, A. A., Min, D.-H., Ruoslahti E., Bhatia S. N. (2007) Targeted quantum dot conjugates for siRNA delivery. *Bioconj. Chem.* **18**, 1391–1396.
13. Lei Y., Tang, H., Yao, L., Yu, R., Fen, M., Zou, B. (2008) Applications of mesenchymal stem cells labeled with Tat peptide conjugated quantum dots to cell tracking in mouse body. *Bioconj. Chem.* **19**, 421–427.
14. Hotz, Charles Z., Bruchez, Marcel (Eds.) (2007) Quantum Dots. Applications in Biology. *Methods in Molecular Biology*, **374**, Humana Press.
15. Reiss, P., Protiere, M., Li, L. (2009) Core/shell semiconductor nanocrystals. *Small.* **5**, 154–168.
16. Resch-Genger, U., Grabolle, M., Cavaliere-Jaricot, S., Nitschke, R., Nann T. (2008) Quantum dots versus organic dyes as fluorescent labels. *Nature Meth.* **5**, 763–775.
17. Bruchez, M., Moronne, M., Gin, P., Weiss, S., Alivisatos, A. P. (1998) Semiconductor nanocrystals as fluorescent biological labels. *Science.* **281**, 2013–2016.
18. Wu, X., Liu, H., Liu, J., Haley, K. N., Treadway, J. A., Larson, J. P., Ge, N., Peale, F., Bruchez, M. P. (2003) Immunofluorescent labeling of cancer marker Her2 and other cellular targets with semiconductor quantum dots. *Nature Biotechnol.* **21**, 41–46.
19. Jaiswal, J. K., Mattoussi, H., Mauro, J. M., Simon, S. M. (2003) Long-term multiple color imaging of live cells using quantum dot bioconjugates. *Nature Biotechnol.* **21**, 47–51.
20. Yezhelyev, M. V., Al-Hajj, A., Morris, C., Marcus, A. I., Liu, T., Lewis, M., Cohen, C., Zrazhevskiy, P., Simons, J. W., Rogatko, A., Nie, S., Gao, X., O'Regan, R. M. (2007) In situ molecular profiling of breast cancer biomarkers with multicolor quantum dots. *Adv. Mater.* **19**, 3146–3151.
21. Chen, I., Choi, Y.-A., Ting, A. Y. (2007) Phage display evolution of a peptide substrate for yeast biotin ligase and application to two-color quantum dot labeling of cell surface proteins. *J. Am. Chem. Soc.* **129**, 6619–6625.
22. Lidke, D. S., Nagy, P., Heintzmann, R., Arndt-Jovin, D. J., Post, J. N., Grecco, H. E., Jares-Erijman, E. A., Jovin T. M. (2004) Quantum dot ligands provide new insights into erbB/HER receptor-mediated signal transduction. *Nature Biotechnol.* **22**, 198–203.
23. Han, M., Gao, X., Su, J. Z., Nie, S. (2001) Quantum-dot-tagged microbeads for multiplexed optical coding of biomolecules. *Nature Biotechnol.* **19**, 631–635.
24. Lagerholm, B. C., Wang, M., Ernst, L. A., Ly, D. H., Liu, H., Bruchez, M. P., Waggoner, A. S. (2004) Multicolor coding of cells with cationic peptide coated quantum dots. *Nano Lett.* **4**, 2019–2022.
25. Mattheakis, L. C., Dias, J. M., Choi, Y.-J., Gong, J., Bruchez, M. P., Liu, J., Wang, E. (2004) Optical coding of mammalian cells using semiconductor quantum dots. *Anal. Biochem.* **327**, 200-208.
26. Voura, E. B., Jaiswal, J. K., Mattoussi, H., Simon, S. M. (2004) Tracking metastatic tumor cell extravasation with quantum dot nanocrystals and fluorescence emission-scanning microscopy. *Nature Med.* **10**, 993–998.
27. Ballou, B., Lagerholm, B. C., Ernst, L. A., Bruchez, M. P., Waggoner, A. S. (2004) Noninvasive imaging of quantum dots in mice. *Bioconj. Chem.* **5**, 79–86.
28. Michalet, X., Pinaud, F. F., Bentolila, L. A., Tsay, J. M., Doose, S., Li, J. J., Sundaresan, G., Wu, A. M., Gambhir, S. S., Weiss, S. (2005) Quantum dots for live cells, in vivo imaging, and diagnostics. *Science.* **307**, 538–544.
29. Parak, W. J., Boudreau, R., Le Gros, M., Gerion, D., Zanchet, D., Micheel, C. M., Williams, S. C., Alivisatos, A. P., Larabell, C. (2002) Cell motility and metastatic potential studies based on quantum dot imaging of phagokinetic tracks. *Adv. Mater.* **14**, 882–885.

30. Kim, S., Lim, Y. T., Soltesz, E. G., De Grand, A. M., Lee, J., Nakayama, A., Parker, J. A., Mihaljevic, T., Laurence, R. G., Dor, D. M., Cohn, L. H., Bawendi, M. G., Frangioni, J. V. (2004) Near-infrared fluorescent type II quantum dots for sentinel lymph node mapping. *Nature Biotechnol.* **22**, 93–97.

31. Jaiswal, J. K., Simon, S. M. (2007) Imaging single events at the cell membrane. *Nature Chem. Biol.* **3**, 92–98.

32. Durisic, N., Bachir, A. I., Kolin, D. L., Hebert, B., Lagerholm, B. C., Grutter, P., Wiseman, P. W. (2007) Detection and correction of blinking bias in image correlation transport measurements of quantum dots tagged macromolecules. *Biophys. J.* **93**, 1338–1346.

33. Iyer, G., Michalet, X., Chang, Y.-P., Pinaud, F. F., Matyas, S. E., Payne, G., Weiss, S. (2008) High affinity scFv-hapten pair as a tool for quantum dot labeling and tracking of single proteins in live cells. *Nano Lett.* **8**, 4618–4623.

34. Lieleg, O., Lopez-Garcia, M., Semmrich, C., Auernheimer, J., Kessler, H., Bausch, A. R. (2007) Specific integrin labeling in living cells using functionalized nanocrystals. *Small.* **3**, 1560–1565.

35. Zhou, M., Nakatami, E., Gronenberg, L. S., Tokimoto, T., Wirth, M. J., Hruby, V. J., Roberts, A., Lynch, R. M., Ghosh, I. (2007) Peptide-labeled quantum dots for imaging GPCRs in whole cells and as single molecules. *Bioconj. Chem.* **18**, 323–332.

36. Dahan, M., Levi, S., Luccardini, C., Rostaing, P., Riveau, B., Triller, A. (2003) Diffusion dynamics of glycine receptors revealed by single-quantum dot tracking. *Science.* **302**, 442–445.

37. Nan, X., Sims, P. A., Chen, P., Xie, X. S. (2005) Observation of individual microtubule motor steps in living cells with endocytosed quantum dots. *J. Phys. Chem. B* **109**, 24220–24224.

38. Courty, S., Luccardini, C., Bellaiche, Y., Cappello, G., Dahan, M. (2006) Tracking individual kinesin motors in living cells using single quantum-dot imaging. *Nano Lett.* **6**, 1491–1495.

39. Bannai, H., Levi, S., Schweizer, C., Dahan, M., Triller, A. (2006) Imaging the lateral diffusion of membrane molecules with quantum dots. *Nature Protocols.* **1**, 2628–2634.

40. Roullier, V., Clarke, S., You, C., Pinaud, F., Gouzer, G., Schaible, D., Marchi-Artzner, V., Piehler, J., Dahan, M. (2009) High-affinity labeling and tracking of individual histidine-tagged proteins in live cells using Ni^{2+} tris-nitrilotriacetic acid quantum dot conjugate. *Nano Lett.* **9**, 1228–1234.

41. Holtzer, L., Meckel, T., Schmidt, T. (2007) Nanometric three-dimensional tracking of individual quantum dots in cells. *Appl. Phys. Lett.* **90**, 053902.

42. Nirmal, M., Dabbousi, B. O., Bawendi, M. G., Macklin, J. J., Trautman, J. K., Harris, T. D., Brus, L. E. (1996) Fluorescence intermittency in single cadmium selenide nanocrystals. *Nature.* **383**, 802–804.

43. Empedocles, S. A., Neuhauser, R., Shimizu, K., Bawendi, M. G. (1999) Photoluminescence from single semiconductor nanostructures. *Adv. Mater.* **11**, 1243–1256.

44. Spinicelli, P., Mahler, B., Buil, S., Quelin, X., Dubertret, B., Hermier, J.-P. (2009) Non-blinking semiconductor colloidal quantum dots for biology, optoelectronics and quantum optics. *ChemPhysChem.* **10**, 879–882.

45. Smith, A. M., Nie, S. (2009) Next-generation quantum dots. *Nature Biotechnol.* **27**, 732–733.

46. Fernandez-Suarez, M., Ting, A. Y. (2008) Fluorescent probes for super-resolution imaging in living cells. *Nature Rev. Mol. Cell Biol.* **9**, 929–943.

47. Yong, K.-T., Qian, J., Roy, I., Lee, H. H., Bergey, E. J., Tramposch, K. M., He, S., Swihart, M. T., Maitra, A., Prasad, P. N. (2007) Quantum rod bioconjugates as targeted probes for confocal and two-photon fluorescence imaging of cancer cells. *Nano Lett.* **7**, 761–765.

48. Fu, A., Gu, W., Boussert, B., Koski, K., Gerion, D., Manna, L., Le Gros, M., Larabell, C. A., Alivisatos, A. P. (2007) Semiconductor quantum rods as single molecule fluorescent biological labels. *Nano Lett.* **7**, 179–182.

49. Yong, K.-T., Roy, I., Pudavar, H. E., Bergey, E. J., Tramposch, K. M., Swihart, M. T., Prasad, P. N. (2008) Multiplex imaging of pancreatic cancer cells by using functionalized quantum rods. *Adv. Mater.* **20**, 1412–1417.

50. Deka, S., Quarta, A., Lupo, M. G., Falqui, A., Boninelli, S., Giannini, C., Morello, G., De Giorgi, M., Lanzani, G., Spinella, C., Cingolani, R., Pellegrino, T., Manna, L. (2009) CdSe/CdS/ZnS double shell nanorods with high photoluminescence efficiency and their exploitation as biolabeling probes. *J. Am. Chem. Soc.* **131**, 2948–2958.

51. Quarta, A., Ragusa, A., Deka, S., Tortiglione, C., Tino, A., Cingolani, R., Pellegrino, T. (2009) Bioconjugation or rod-shaped fluorescent nanocrystals for efficient targeted cell labeling. *Langmuir.* **25**, 12614–12622.

52. Tomczak, N., Jańczewski, D., Han, M.-Y., Vancso, G. J. (2009) Designer

polymer-quantum dot architectures. *Progr. Polym. Sci.* **34**, 393–430.

53. Hines, M. A., Guyot-Sionnest, P. (1996) Synthesis and characterization of strongly luminescing ZnS-capped CdSe nanocrystals. *J. Phys. Chem.* **100**, 468–471.
54. Peng, X., Schlamp, M. C., Kadavanich, A. V., Alivisatos, A. P. (1997) Epitaxial growth of highly luminescent CdSe/CdS core/shell nanocrystals with photostability and electronic accessibility. *J. Am. Chem. Soc.* **119**, 7019–7029.
55. Mulvaney, P., Liz-Marzán, L. M., Giersig, M., Ung, T. (2000) Silica encapsulation of quantum dots and metal clusters. *J. Mater. Chem.* **10**, 1259–1270.
56. Gerion, D., Pinaud, F., Williams, S. C., Parak, W. J., Zanchet, D., Weiss, S., Alivisatos, A. P. (2001) Synthesis and properties of biocompatible water soluble silica-coated CdSe/ZnS semiconductor quantum dots. *J. Phys. Chem. B* **105**, 8861–8871.
57. Darbandi, M., Thomann, R., Nann, T. (2005) Single quantum dots in silica spheres by microemulsion synthesis. *Chem. Mater.* **17**, 5720–5725.
58. Zhu, M.-Q., Chang, E., Sun, J., Drezek, R. A. (2007) Surface modification and functionalization of semiconductor quantum dots through reactive coating of silanes in toluene. *J. Mater. Chem.* **17**, 800–805.
59. Parak, W. J., Gerion, D., Zanchet, D., Woerz, A. S., Pellegrino, T., Micheel, C., Williams, S. C., Seitz, M., Bruehl, R. E., Bryant, Z., Bustamante, C., Bertozzi, C. R., Alivisatos, A. P. (2002) Conjugation of DNA to silanized colloidal semiconductor nanocrystalline quantum dots. *Chem. Mater.* **14**, 2113–2119.
60. Schroedter, A., Weller, H., Eritja, R., Ford, W. E., Wessels, J. M. (2002) Biofunctionalization of silica-coated CdTe and gold nanocrystals. *Nano Lett.* **2**, 1363–1367.
61. Selvan, S. T., Tan, T. T., Ying, J. Y. (2005) Robust, non-cytotoxic, silica-coated CdSe quantum dots with efficient photoluminescence. *Adv. Mater.* **17**, 1620–1625.
62. Chan, W. C. W., Nie, S. (1998) Quantum dot bioconjugates for ultrasensitive nonisotopic detection. *Science* **281**, 2016–2018.
63. Tomczak, N., Jańczewski, D., Tagit, O., Han, M. Y., Vancso, G. J. Surface engineering of quantum dots with designer ligands. In *Surface Design: Applications in Bioscience and Nanotechnology*. Förch, R., Schönherr, H., Jenkins, A. T. A. Eds. Wiley, 2009.
64. Susumo, K., Mei, B. C., Mattoussi, H. (2009) Multifunctional ligands based on dihydrolipoic acid and polyethylene glycol to promote biocompatibility of quantum dots. *Nature Protocols* **4**, 424–436.
65. Mei, B. C., Susumu, K., Medintz, I. L., Mattoussi, H. (2009) Polyethylene glycol-based bidentate ligands to enhance quantum dot and gold nanoparticle stability in biological media. *Nature Protocols.* **4**, 412–423.
66. Carion, O., Mahler, B., Pons, T., Dubertret, B. (2007) Synthesis, encapsulation, purification and coupling of single quantum dots in phospholipid micelles for their use in cellular and in vivo imaging. *Nature Protocols.* **2**, 2383– 2390.
67. Lin, C.-A. J., Sperling, R. A., Li, J. K., Yang, T.-Y., Li, P.-Y., Zanella, M., Chang, W. H., Parak, W. J. (2008) Design of an amphiphilic polymer for nanoparticle coating and functionalization. *Small* **4**, 334–341.
68. Pellegrino, T., Manna, L., Kudera, S., Liedl, T., Koktysh, D., Rogach, A. L., Keller, S., Radler, J., Natile, G., Parak, W. J. (2004) Hydrophobic nanocrystals coated with an amphiphilic polymer shell: A general route to water soluble nanocrystals. *Nano Lett.* **4**, 703–707.
69. Gao, X., Cui, Y., Levenson, R. M., Chung, L. W. K., Nie, S. (2004) In vivo cancer targeting and imaging with semiconductor quantum dots. *Nature Biotechnol.* **22**, 969–976.
70. Jańczewski, D., Tomczak, N., Khin, Y. W., Han, M. Y. Vancso, G. J. (2009) Designer multi-functional comb-polymers for surface engineering of quantum dots on the nanoscale. *Eur. Polym. J.* **45** (1), 3–9.
71. Yang, L., Mao, H., Wang, Y. A., Cao, Z., Peng, X., Wang, X., Duan, H., Ni, C., Yuan, Q., Adams, G., Smith, M. Q., Wood, W. C., Gao, X., Nie, S. (2009) Single chain epidermal growth factor receptor antibody conjugated nanoparticles for in vivo tumor targeting and imaging. *Small.* **5**, 235–243.
72. Dubertret, B., Skourides, P., Norris, D. J., Noireaux, V., Brivanlou, A. H., Libchaber, A. (2002) In vivo imaging of quantum dots encapsulated in phospholipid micelles. *Science.* **298**, 1759–1762.
73. Jańczewski, D., Tomczak, N., Liu S. H., Han, M. Y., Vancso G. J. (2010) Covalent assembly of functional inorganic nanoparticles by "click" chemistry in water. *Chem. Commun.* 46 (19), 3253–3255.
74. Mulder, W. J. M., Koole, R., Brandwijk, R. J., Storm, G., Chin, P. T. K., Strijkers, G. J., de Mello Donega, C., Nicolay, K., Griffioen, A. W. (2006) Quantum dots with a paramagnetic coating as a bimodal molecular imaging probe. *Nano Lett.* **6**, 1–6.

75. Bakalova, R., Zhelev, Z., Aoki, I., Kanno, I. (2007) Designing quantum-dot probes. *Nature Photonics* **1**, 487–489.

76. Wang, S., Jarrett, B. R., Kauzlarich, S. M., Louie, A. Y. (2007) Core/shell quantum dots with high relaxivity and photoluminescence for multimodality imaging. *J. Am. Chem. Soc.* **129**, 3848–3856.

77. Cormode, D. P., Skajaa, T., van Schooneveld, M. M., Koole, R., Jarzyna, P., Lobatto, M. E., Calcagno, C., Barazza, A., Gordon, R. E., Zanzonico, P., Fisher, E. A., Fayad, Z. A., Mulder, W. J. M. (2008) Nanocrystal core high-density lipoproteins: A multimodality contrast agent platform. *Nano Lett.* **8**, 3715–3723.

78. Bruns, O. T., Ittrich, H., Peldschus, K., Kaul, M. G., Tromsdorf, U. I., Lauterwasser, J., Nikolic, M. S., Mollwitz, B., Merkel, M., Bigall, N. C., Sapra, S., Reimer, R., Hohenberg, H., Weller, H., Eychmuller, A., Adam, G., Beisiegel, U., Heeren, J. (2009) Real-time magnetic resonance imaging and quantification of lipoprotein metabolism in vivo using nanocrystals. *Nature Nanotechnol.* **4**, 193–201.

79. Duconge F., Pons, T., Pestourie, C., Herin, L., Theze, B., Gombert, K., Mahler, B., Hinnen, F., Kuhnast, B., Dolle, F., Dubertret, B., Tavitian, B. (2008) Fluorine-18-labeled phospholipid quantum dot micelles for in vivo multimodal imaging from whole body to cellular scales. *Bioconj. Chem.* **19**, 1921–1926.

80. Schipper, M. L., Cheng, Z., Lee, S. W., Bentolila, L. A., Iyer, G., Rao, J. H., Chen, X. Y., Wul, A. M., Weiss, S., Gambhir, S. S. (2007) MicroPET-based biodistribution of quantum dots in living mice. *J. Nucl. Med.* **48**, 1511–1518.

81. Nel, A., Xia, T., Li, N. (2006) Toxic potential of materials at the nanolevel. *Science* **311**, 622–627.

82. Nel, A. E., Madler, L., Velegol, D., Xia, T., Hoek, E. M. V., Somasundaran, P., Klaessig, F., Castranova, V., Thompson, M. (2009) Understanding biophysicochemical interactions at the nano-bio interface. *Nature Mater.* **8**, 543–557.

83. Hussain, S. M., Braydich-Stolle, L. K., Schrand, A. M., Murdock, R. C., Yu, K. O., Mattie, D. M., Schlager, J. J., Terrones, M. (2009) Toxicity evaluation for safe use of nanomaterials: Recent achievements and technical challenges. *Adv. Mater.* **21**, 1549–1559.

84. Derfus, A. M., Chan, W. C. W., Bhatia, S. N. (2004) Probing the cytotoxicity of semiconductor quantum dots. *Nano Lett.* **4**, 11–18.

85. Kirchner, C., Liedl, T., Kudera, S., Pellegrino T., Munoz Javier, A., Gaub, H. E., Stolzle, S., Fertig, N., Parak, W. J. (2005) Cytotoxicity of colloidal CdSe and CdSe/ZnS nanoparticles. *Nano Lett.* **5**, 331–338.

86. Lovric, J., Bazzi, H. S., Cuie, Y., Fortin, G. R. A., Winnik, F. M., Maysinger, D. (2005) Differences in subcellular distribution and toxicity of green and red emitting CdTe quantum dots. *J. Mol. Med.* **83**, 377–385.

87. Alsharif, N. H., Berger, C. E. M., Varanasi, S. S., Chao, Y., Horrocks, B. R., Datta, H. K. (2009) Alkyl-capped silicon nanocrystals lack cytotoxicity and have enhanced intracellular accumulation in malignant cells via cholesterol-dependent endocytosis. *Small* **5**, 221–228.

88. Bharali, D. J., Lucey, D. W., Jayakumar, H., Pudavar, H. E., Prasad, P. N. (2005) Folate-receptor-mediated delivery of InP quantum dots for bioimaging using confocal and two-photon microscopy. *J. Am. Chem. Soc.* **127**, 11364–11371.

89. Hussain, S., Won, N., Nam, J., Bang, J., Chung, H., Kim, S. (2009) One-pot fabrication of high-quality InP/ZnS (core/shell) quantum dots and their application to cellular imaging. *ChemPhysChem.* **10**, 1466–1470.

90. Hoshino, A., Fujioka, K., Oku, T., Suga, M., Sasaki, Y. F., Ohta, T., Yasuhara, M., Suzuki, K., Yamamoto, K. (2004) Physicochemical properties and cellular toxicity of nanocrystal quantum dots depend on their surface modification. *Nano Lett.* **4**, 2163–2169.

91. Jiang, W., Kim, B. Y. S., Rutka, J. T., Chan, W. C. W. (2008) Nanoparticle-mediated cellular response is size-dependent. *Nature Nanotechnol.* **3**, 145–150.

92. Liu, W., Choi, H. S., Zimmer, J. P., Tanaka, E., Frangioni, J. V., Bawendi, M. (2007) Compact cysteine-coated CdSe(ZnCdS) quantum dots for in vivo applications. *J. Am. Chem. Soc.* **129**, 14530–14531.

93. Liu, W., Howarth, M., Greytak, A. B., Zheng, Y., Nocera, D. G., Ting, A. Y., Bawendi, M. G. (2008) Compact biocompatible quantum dots functionalized for cellular imaging. *J. Am. Chem. Soc.* **130**, 1274–1284.

94. Howarth, M., Liu, W., Puthenveetil, S., Zheng, Y., Marshall, L. F., Schmidt, M. M., Wittrup, K. D., Bawendi, M. G., Ting, A. Y. (2008) Monovalent, reduced-size quantum dots for imaging receptors on living cells. *Nature Meth.* **5**, 397–399.

95. Zaman, M. B., Baral, T. N., Zhang, J., Whitfield, D., Yu, K. (2009) Single-domain antibody functionalized CdSe/ZnS quantum dots for cellular imaging of cancer cells. *J. Phys. Chem. C* **113**, 496–499.

96. Chang, E., Thekkek, N., Yu, W. W., Colvin, V. L., Drezek, R. (2006) Evaluation of quantum dot cytotoxicity based on intracellular uptake. *Small.* **2**, 1412–1417.
97. Tarantola, M., Schneider, D., Sunnick, E., Adam, H., Pierrat, S., Rosman, C., Breus, V., Sonnichsen, C., Basche, T., Wegener, J., Janshoff, A. (2009) Cytotoxicity of metal and semiconductor nanoparticles indicated by cellular micromotility. *ACS Nano.* **3**, 213–222.
98. Ipe, B. I., Lehnig, M., Niemeyer, C. M. (2005) On the generation of free radical species from quantum dots. *Small.* **1**, 706–709.
99. Samia, A. C. S., Chen, X., Burda, C. (2003) Semiconductor quantum dots for photodynamic therapy. *J. Am. Chem. Soc.* **125**, 15736–15737.
100. Bakalova, R., Ohba, H., Zhelev, Z., Nagase, T., Jose, R., Ishikawa, M., Baba, Y. (2004) Quantum dot anti-CD conjugates: Are they potential photosensitizers or potentiators of classical photosensitizing agents in photodynamic therapy of cancer? *Nano Lett.* **4**, 1567–1573.
101. Tsay, J. M., Trzoss, M., Shi, L., Kong, X., Selke, M., Jung, M. E., Weiss, S. (2007) Singlet oxygen production by peptide-coated quantum dot-photosensitizer conjugates. *J. Am. Chem. Soc.* **129**, 6865–6871.
102. Mattoussi, H., Mauro, J. M., Goldman, E. R., Anderson, G. P., Sundar, V. C., Mikulec, F. V., Bawendi, M. G. (2000) Self-assembly of CdSe-ZnS quantum dot bioconjugates using an engineered recombinant protein. *J. Am. Chem. Soc.* **122**, 12142–12150.
103. Goldman, E. R., Balighian, E. D., Mattoussi, H., Kuno, M. K., Mauro, J. M., Tran, P. T., Anderson, G. P. (2002) Avidin: A natural bridge for quantum dot-antibody conjugates. *J. Am. Chem. Soc.* **124**, 6378–6382.
104. Jaiswal, J. K., Goldman, E. R., Mattoussi, H., Simon, S. M. (2004) Use of quantum dots for live cell imaging. *Nature Methods.* **1**, 73–78.
105. Bhang, S. K., Won, N., Lee, T.-J., Jin, H., Nam, J., Park, J., Chung, H., Park, H.-S., Sung, Y.-E., Hahn, S. K., Kim, B.-S., Kim., S. (2009) Hyaluronic acid-quantum dot conjugates for in vivo lymphatic vessel imaging. *ACS Nano.* **3**, 1389–1398.
106. Akerman, M. E., Chan, W. C. W., Laakkonen, P., Bhatia, S. N., Ruoslahti, E. (2002) Nanocrystals targeting in vivo. *Proc. Natl. Acad. Sci. USA.* **99**, 12617–12621.
107. Pinaud, F., King, D., Moore, H.-P., Weiss, S. (2004) Bioactivation and cell targeting of semiconductor CdSe/ZnS nanocrystals with phytochelatin-related peptides. *J. Am. Chem. Soc.* **126**, 6115–6123.
108. Delehanty, J. B., Medintz, I. L., Pons, T., Brunel, F. M., Dawson, P. E., Mattoussi, H. (2006) Self-assembled quantum dot-peptide bioconjugates for selective intracellular delivery. *Bioconj. Chem.* **17**, 920–927.
109. Winter, J. O., Liu, T. Y., Korgel, B. A., Schmidt, C. E. (2001) Recognition molecule directed interfacing between semiconductor quantum dots and nerve cells. *Adv. Mater.* **13**, 1673–1677.
110. Xing, Y., Chaudry, Q., Shen, C., Kong, K. Y., Zhau, H. E., Chung, L. W., Petros, J. A., O'Regan, R. M., Yezhelyev, M. V., Simons, J. W., Wang, M. D. (2007) Bioconjugated quantum dots for multiplexed and quantitative immunohistochemistry. *Nature Protocols.* **2**, 1152–1165.
111. Rosenthal, S. J., Tomlinson, I., Adkins, E. M., Schroeter, S., Adams, S., Swafford, L., McBride, J., Wang, Y., DeFelice, L. J., Blakely, R. D. (2002) Targeting cell surface receptors with ligand-conjugated nanocrystals. *J. Am. Chem. Soc.* **124**, 4586–4594.
112. Chen, X.-C., Deng, Y.-L., Lin, Y., Pang, D.-W., Qing, H., Qu, F., Xie, H.-Y. (2008) Quantum dot-labeled aptamer nanoprobes specifically targeting glioma cells. *Nanotechnology.* **19**, 235105.
113. Khorev, O., Stokmaier, D., Schwardt, O., Cutting, B., Ernst, B. (2008) Trivalent, Gal/GalNAc-containing ligands designed for the asialoglycoprotein receptor. *Bioorg. Med. Chem.* **16**, 5216–5231.
114. Kikkeri, R., Lepenies, B., Adibekian, A., Laurino, P., Seeberger, P .H. (2009) In vitro imaging and in vivo liver targeting with carbohydrate capped quantum dots. *J. Am. Chem. Soc.* **131**, 2110–2112.
115. Babu, P., Sinha, S., Surolia, A. (2007) Sugar-quantum dot conjugates for a selective and sensitive detection of lectins. *Bioconj. Chem.* **18**, 146–151.
116. Howarth, M., Takao, K., Hayashi, Y., Ting, A. Y. (2005) Targeting quantum dots to surface proteins in living cells with biotin ligase. *Proc. Natl. Acad. Soc.* USA. **102**, 7583–7588.
117. Gussin, H. A., Tomlinson, I. D., Little, D. M., Warnement, M. R., Qian, H., Rosenthal, S. J., Pepperberg, D. R. (2006) Binding of muscimol-conjugated quantum dots to GABAc receptors. *J. Am. Chem. Soc.* **128**, 15701–15713.
118. Rajan, S. S., Vu, T. Q. (2006) Quantum dots monitor TrkA receptor dynamics in the interior of neural PC12 cells. *Nano Lett.* **6**, 2049.
119. Luccardini, C., Yakovlev, A., Gaillard, S., Vant Hoff, M., Alberola, A. P., Mallet, J.-M., Parak, W. J.,

Feltz, A., Oheim, M. (2007) Getting across the plasma membrane and beyond: intracellular uses of colloidal semiconductor nanocrystals. *J. Biomed. Biotechnol.* **2007**, 68963.

120. Chen, F., Gerion, D. (2004) Fluorescent CdSe/ZnS nanocrystal-peptide conjugates for long-term, nontoxic imaging and nuclear targeting in living cells. *Nano Lett.* **4**, 1827–1832.
121. Ruan, G., Agrawal, A., Marcus, A. I., Nie, S. (2007) Imaging and tracking of Tat peptide-conjugated quantum dots in living cells: New insights into nanoparticle uptake, intracellular transport, and vesicle shedding. *J. Am. Chem. Soc.* **129**, 14759–14766.
122. Medintz, I. L., Pons, T., Delehanty, J. B., Susumu, K., Brunel, F. M., Dawson, P. E., Mattoussi, H. (2008) Intracellular delivery of quantum dot-protein cargos mediated by cell penetrating peptides. *Bioconj. Chem.* **19**, 1785–1795.
123. Anas, A., Okuda, T., Kawashima, N., Nakayama, K., Itoh, T., Ishikawa, M., Biju, V. (2009) Clathrin-mediated endocytosis of quantum dot-peptide conjugates in living cells. *ACS Nano.* **3**, 2419–2429.
124. Delehanty, J. B., Mattoussi, H., Medintz, I. L. (2009) Delivering quantum dots into cells: strategies, progress and remaining issues. *Anal. Bioanal. Chem.* **393**, 1091–1105.
125. Zhang, Y., So, M. K., Rao, J. (2006) Protease-modulated cellular uptake of quantum dots. *Nano Lett.* **6**, 1988–1992.
126. Mok, H., Bae, K. H., Ahn, C.-H., Park, T. G. (2008) PEGylated and MMP-2 specifically dePEGylated quantum dots: Comparative evaluation of cellular uptake. *Langmuir.* **25**, 1645–1650.
127. Barua, S., Rege, K. (2009) Cancer-cell-phenotype-dependent differential intracellular trafficking of unconjugated quantum dots. *Small.* **5**, 370–376.
128. Duan, H., Nie, S. (2007) Cell-penetrating quantum dots based on multivalent and endosome-disrupting surface coatings. *J. Am. Chem. Soc.* **129**, 3333–3338.
129. Xue, F. L., Chen, J. Y., Guo, J., Wang, C. C., Yang, W. L., Wang, P. N., Lu, D. R. (2007) Enhancement of intracellular delivery of CdTe quantum dots (QDs) to living cells by Tat conjugation. *J. Fluoresc.* **17**, 149–154.
130. Jablonski, A. E., Humphries IV, W. H., Payne, C. K. (2009) Pyrenebutyrate-mediated delivery of quantum dots across the plasma membrane of living cells. J. *Phys. Chem. B* **113**, 405–408.
131. Biju, V., Muraleedharan, D., Nakayama, K., Shinohara, Y., Itoh, T., Baba, Y., Ishikawa, M. (2007) Quantum dot-insect neuropeptide conjugates for fluorescence imaging, transfection, and nucleus targeting of living cells. *Langmuir.* **23**, 10254–10261.
132. Ballou, B., Ernst, L. A., Andreko, S., Harper, S., Fitzpatrick, J. A. J., Waggoner, A. S., Bruchez, M. P. (2007) Sentinel lymph node imaging using quantum dots in mouse tumor models. *Bioconj. Chem.* **18**, 389–396.
133. Larson, D. R., Zipfel, W. R., Williams, R. M., Clark, S. W., Bruchez, M. P., Wise, F. W., Webb, W. W. (2003) Water-soluble quantum dots for multiphoton fluorescence imaging in vivo. *Science.* **300**, 1434–1436.
134. Dong, W., Guo, L., Xu, S. (2009) CdTe QDs-based prostate-specific antigen probe for human prostate cancer cell imaging. *J. Lumin.* **129**, 926–930.
135. Yong, K.-T., Ding, H., Roy, I., Law, W.-C., Bergey, E. J., Maitra, A., Prasad, P. N. (2009) Imaging pancreatic cancer using bioconjugated InP quantum dots. *ACS Nano.* **3**, 502–510.
136. Yong, K.-T., Roy, I., Ding, H., Bergey, E. J., Prasad, P. N. (2009) Biocompatible near-infrared quantum dots as ultrasensitive probes for long-term in vivo imaging applications. *Small.* **5**, 1997–2004.
137. So, M.-K., Xu, C., Loening, A. M., Gambhir, S. S., Rao, J. (2006) Self-illuminating quantum dot conjugates for in-vivo imaging. *Nature Biotechnol.* **24**, 339–343.
138. Gomez, N., Winter, J. O., Shieh, F., Saunders, A. E., Korgel, B. A., Schmidt, C. E. (2005) Challenges in quantum dot-neuron active interfacing. *Talanta.* **67**, 462–471.
139. Vu, T. Q., Maddipati, R., Blute, T. A., Nehilla, B. J., Nusblat, L., Desai, T. A. (2005) Peptide-conjugated quantum dots activate neuronal receptors and initiate downstream signaling of neurite growth. *Nano Lett.* **5**, 603–607.
140. Hoshino, A., Manabe, N., Fujioka, K., Hanada, S., Yasuhara, M., Kondo, A., Yamamoto, K. (2008) GFP expression by intracellular gene delivery of GFP-coding fragments using nanocrystal quantum dots. *Nanotechnology.* **19**, 495102.

Chapter 17

The Role of Nanophotonics in Regenerative Medicine

Maria F. Garcia-Parajo

Abstract

Cells respond to biochemical and mechanical stimuli through a series of steps that begin at the molecular, nanometre level, and translate finally in global cell response. Defects in biochemical- and/or mechanical-sensing, transduction or cellular response are the cause of multiple diseases, including cancer and immune disorders among others. Within the booming field of regenerative medicine, there is an increasing need for developing and applying nanotechnology tools to bring understanding on the cellular machinery and molecular interactions at the nanoscale. Nanotechnology, nanophotonics and in particular, high-resolution-based fluorescence approaches are already delivering crucial information on the way that cells respond to their environment and how they organize their receptors to perform specialized functions. This chapter focuses on emerging super-resolution optical techniques, summarizing their principles, technical implementation, and reviewing some of the achievements reached so far.

Key words: Super-resolution optical microscopy, Nanophotonics, Near-field optical microscopy, Cell membrane organization

1. Introduction

Many innovations in biomedicine and in our understanding of how cells work have been made possible through the last decades by a series of new technologies, from biochemistry starting in the 1970s, to biotechnology, and finally through the deciphering of the human genome at the end of the 1990s. However, despite our in-depth molecular knowledge, how the cell achieves in organizing individual cellular functions at the cell surface and subsequent downstream signalling is largely unknown. Similarly, in disease, we still do not understand how tissues become abnormal, such as in cancer or cardiac hypertrophy. In general, cells respond to biochemical and mechanical cues through a series of steps that begin at the molecular (nanometre) level. Defects in biochemical- and/ or mechanical-sensing, transduction, or cellular response at this

Melba Navarro and Josep A. Planell (eds.), *Nanotechnology in Regenerative Medicine: Methods and Protocols*, Methods in Molecular Biology, vol. 811, DOI 10.1007/978-1-61779-388-2_17,

level underlie many diseases, such as cancer, immune disorders, genetic deformation, etc. In this context, new advances in nanotechnology and nanophotonics are emerging nowadays and capable of analyzing cellular processes at the molecular level with unprecedented detail. At the core of these novel technologies, single-molecule fluorescence techniques and nanophotonics hold great potential since they allow to visualize and to identify molecular interactions between different multimolecular cellular components providing deep understanding in cell biology processes that can help to discriminate health from disease.

Regenerative medicine is a field devoted in great part to treatments in which stem cells are induced to differentiate into the specific cell type required to repair damaged or destroyed cell populations or tissues. Unfortunately, stem cells have also been found to be the source of some and possibly most cancers, even if they normally make up only a very small percentage of the total tumour mass. Like most cells, stem cells display a characteristic set of protein receptors on their cell surface and stem cell homing is triggered by interactions between cell surface adhesion molecules (such as selectins, integrins, and ICAMs) and the surrounding environment of the cell. However, as already mentioned, despite great advances identifying molecular components, very little is known on how these receptors organize on the cell surface and how biochemical and mechanical inputs translate into cell responses. This chapter focuses on advanced fluorescence microscopy techniques that are able to probe these interactions at the molecular scale.

2. Fluorescence Microscopy for Live Cell Investigation

To enhance our understanding of biological processes as they occur in living organisms, imaging strategies have been developed and refined that reveal cellular and molecular events of biology in real time. In particular, fluorescence microscopy has become one of the most prominent and versatile research tools used in modern cell biology and in principle ideal to investigate cell membrane organization in living cells (1). The reasons for it are essentially twofold. First, light-based microscopy allows the study of living specimens in their native environment in a non-invasive manner. Additionally, fluorescence microscopy offers chemical specificity by exploiting polarization, lifetime, and spectral contrast (2). Furthermore, progress in detector technology has recently pushed fluorescence microscopy to its ultimate level of sensitivity: the detection of individual molecules (3, 4). Second, enormous progress on the development of specific and highly efficient fluorescent probes for exogenous labelling has been achieved. In parallel to external antibody labelling, the advent of green fluorescent protein (GFP) technology has revolutionized live cell imaging because autofluorescent proteins can be genetically encoded as a fusion with the c-DNA of interest (5). Indeed, the spectral variants of GFP and the unrelated red fluorescent protein (DsRed) make it possible to perform nowadays multicolour imaging in living cells (5, 6).

In the last decade, a number of fluorescent-based techniques have been applied to study molecular processes in living cells. In particular, confocal, wide field, and total internal reflection microscopy can resolve structures on the cell membrane and track proteins and other biomolecules in living cells (Fig. 1). However, a major drawback of standard light microscopy is the fundamental limit of the attainable spatial resolution, which is dictated by the laws of diffraction. This diffraction limit originates from the fact that it is impossible to focus light to a spot smaller than half its wavelength. In practice, this means that the maximal resolution in optical microscopy is ~250–300 nm. Since a large body of evidence indicates that dynamic cell-signalling events start by oligomerization and interaction of individual proteins (i.e., on the molecular scale), the need for imaging techniques that have a higher resolution is growing.

Traditionally, high-resolution cell biology has been the arena of transmission electron microscopy (TEM) (Fig. 1), which offers superb resolution but lacks the above-mentioned advantages of fluorescence microscopy. Unfortunately, since TEM requires extensive sample preparation it cannot be extended to live cell

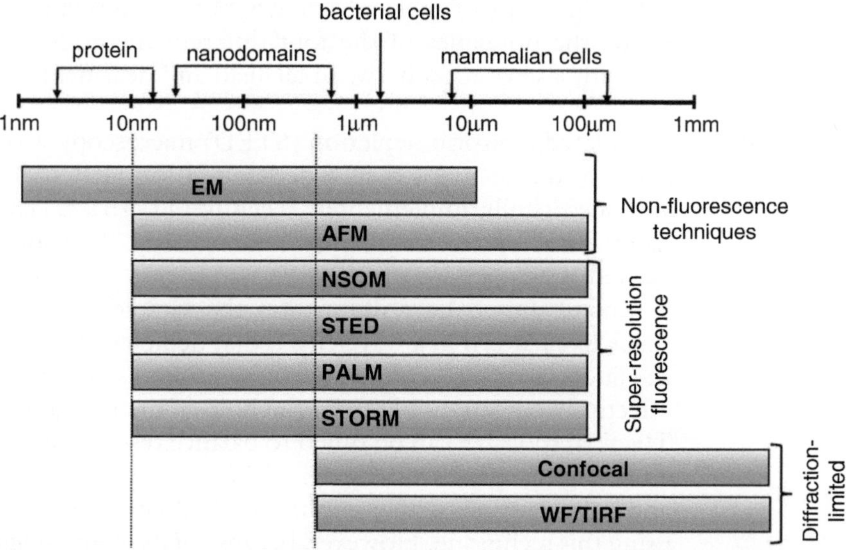

Fig. 1. Comparison of spatial resolution techniques for biological imaging. *WF* wide field microscopy, *TIRF* total internal reflection fluorescence microscopy, *STED* stimulated emission depletion, *PALM* photoactivated localization microscopy, *STORM* stochastic optical reconstruction microscopy, *EM* electron microscopy, *AFM* atomic force microscopy. STED, PALM, and STORM belong to far-field super-resolution techniques while NSOM is a near-field super-resolution technique.

imaging. The advent of scanning probe microscopy (Fig. 1), and especially atomic force microscopy (AFM), in which an atomically sharp probe attached to a cantilever is scanned over the surface of interest, has made nanometre resolution also attainable on living cells (7, 8). However, although AFM produces a high-resolution topographical image of the sample, it lacks biochemical specificity. Hence, although individual molecules can be seen, their identities cannot be defined. This seriously limits the usefulness of AFM for high-resolution imaging on cells. A promising way around the problem relies on specific labelling of the AFM probe with biomolecules (e.g., with antibodies or ligands). This introduces a contrast mechanism based on specific interactions between the probe and a certain type of molecules in the specimen (9). More recently, molecular recognition imaging using AFM and biofunctionalized probes has been successfully implemented by the Hinterdorfer's group (10). Although extremely sensitive, the experimental approach is, so far, restricted to a single type of interaction being probed.

High-resolution fluorescence microscopy is compatible with live cell imaging, provides excellent spectral contrast and in combination with sensitive detectors allows the detection of individual molecules. Until only a few years ago, near-field scanning optical microscopy (NSOM) has been the only optical technique able to provide resolution beyond the diffraction limit of light (Fig. 1). However, recently developed far-field methods have also demonstrated optical resolution in the nanometre range, not only laterally but also in 3D (Fig. 1). Each of the methods is briefly discussed below in terms of their advantages but also limitations. Figure 2 shows the principles of the four different methods developed so far, with a separation between far-field and near-field approaches.

2.1. Far-Field Optical Nanoscopy

Stimulated emission depletion (STED) microscopy was conceptually introduced more than a decade ago by Hell and colleagues, and successfully implemented recently (11–15). STED creates a nanometric optical region by first exciting fluorophores to an excited state over a diffraction-limited region using a pulsed laser. A second pulsed laser illuminates the sample with a doughnut-shape like pattern in a wavelength that depletes the excited state of the fluorescent molecules back to the ground state. Fluorescence is effectively detected only from the hole of the doughnut (Fig. 2a). The final spot size can be tuned to balance resolution against signal and imaging speed by controlling the power of the depleting laser, and indeed images with a resolution of ~30 nm have been reported using this technique. However, because of its mere principle, STED requires precise control of the position, phase and amplitude of two laser beams (for single colour fluorescence), and its best resolution is restricted to certain dyes able to withstand repeated cycles of excitation and depletion at extremely high intensities. So far, the

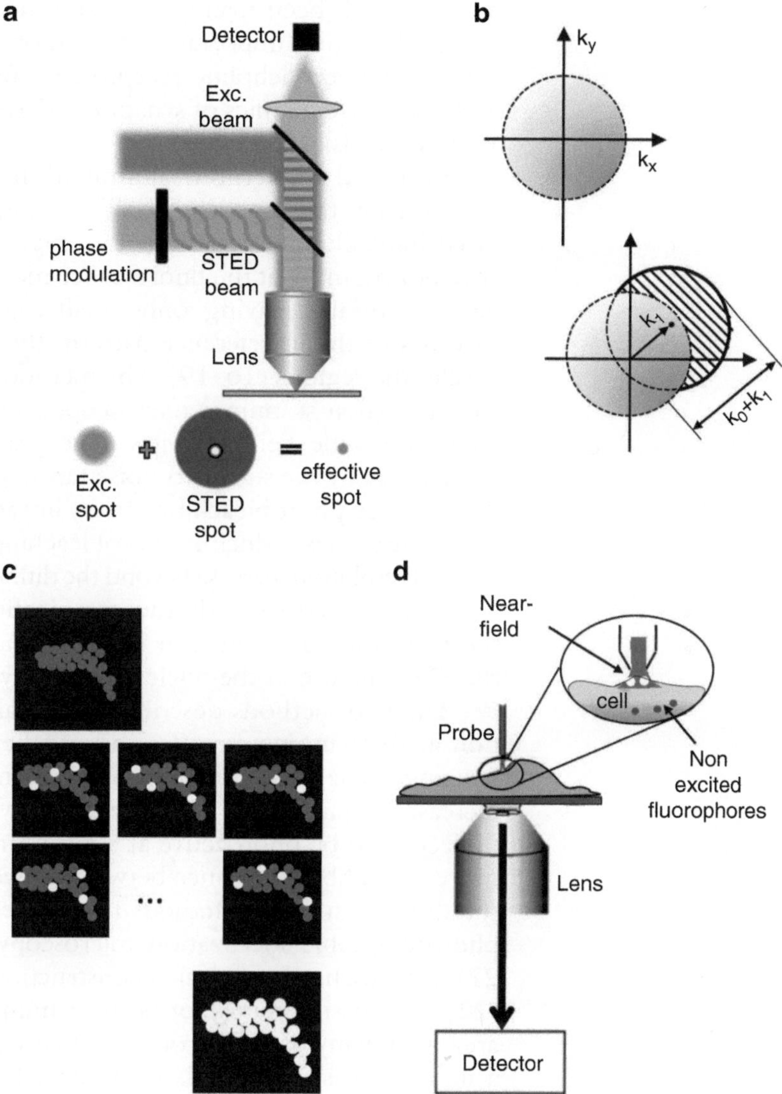

Fig. 2. Different schemes for super-resolution imaging microscopy. (**a**) Stimulated emission depletion (STED) microscopy, adapted from *ref.* 11. (**b**). Structured illumination concept. *Top*: circular observable region of radius k_0 in frequency space observed by a conventional microscope. *Bottom*: New set of information available in the form of moiré fringes (*hatched circle*) provided that the excitation light contains a spatial frequency k_1. The new region has the same shape as the normal observable region but it is centred at k_1. The maximum spatial frequency that can be detected in this direction is $k_0 + k_1$, adapted from *ref.* 16. (**c**) Principle of PALM/FPALM and STORM. The techniques use the stochastic photoactivation of single molecules (set to the dark state at the beginning of the experiment as shown in the *top panel*) and their subsequent nanometric localization over thousands of widefield image frames (series of *small panels*) to construct a super-resolution image (*bottom panel*). (**d**) NSOM uses a sub-wavelength aperture (~50–100 nm) probe to locally excite the sample surface and to generate point-by-point a super-resolution image related to the size of the probe. Only fluorophores at the cell surface are effectively excited (*bright dots* close to the near-field region) reducing the contribution of background fluorescence from other regions of the cell (*dark dots* in the interior of the cell).

technique has been mainly applied in neurobiology by the Hell's group, delivering important information on the study of syntaxin clusters and acetylcholine receptors on fixed cultured neurons, as well as to the dynamics of synaptic vesicles over small fields of view at high speeds (12–15).

Saturated structured illumination microscopy (SSIM) is conceptually the opposite of STED (Fig. 2b). By using a structured light illumination from two high-intensity power interference beams, most of the fluorescence molecules in the illuminating beams saturate, leaving only small regions unsaturated at the shadows of the interference pattern: the higher the intensity, the smaller the regions (16–19). The method can be implemented in a widefield (non-scanning) microscope and is capable of high frame rates over wide fields of view. The practical resolving power is determined by the signal-to-noise ratio, which is in turn limited by fluorescence photobleaching. In its linear form, SIM can work at lower intensities reducing photobleaching but can provide only a twofold resolution increase beyond the diffraction limit. The technique has been applied to study the organization of specific proteins at the neuromuscular junction in Drosophila (19) and to elucidate the 3D structure of the nuclear periphery (17).

The two methods described above allow truly optical resolution at the nanometre scale, and can be readily extended to 3D imaging. The resolution in fluorescence microscopy can be increased even further by allowing only a subset of fluorescent molecules to be photoactive at a given time and ensuring that the nearest-neighbour distance between active molecules is larger than the diffraction limit. Methods that make use of this principle are photoactivatable localization microscopy (PALM/FPALM) (20, 21) and stochastic optical reconstruction microscopy (STORM) (22). The basic premise of both techniques is to fill the imaging area with many dark fluorophores that can be photoactivated into a fluorescing state by a flash of light. Because photoactivation is stochastic, only a few well-separated molecules "turn on." Then, Gaussians are fit to their point spread functions (PSF) to high precision. After the few bright dots photobleach, another flash of the photoactivating light activates random fluorophores again and the PSFs are fit of these different well-spaced objects. This process is repeated many times, building up an image molecule-by-molecule; and because the molecules were localized at different times, the "resolution" of the final image can be much higher than that limited by diffraction (Fig. 2c). The main difference between PALM and STORM resides on the type of fluorophores used for photoactivation: PALM relies on autofluorescent proteins while STORM uses organic switchable dyes (from the cyanine family). The ascertainable localization accuracy depends strongly on the total number of photons being detected. Especially PALM/FPALM can quantitatively map relative molecular densities with very high localization accuracy over wide fields and in living cells. As already mentioned, these forms of nanoscale image reconstruction methodologies rely on photoswitchable fluorophores, and therefore imaging conditions should be consistent with single-molecule detection and require so far long acquisition times. PALM/FPALM has been used to reconstruct images of various proteins in thin cellular sections and near the surfaces of whole, fixed cells (20), to study the organization of different proteins within adhesion complexes (23) and to track large populations of single proteins molecules in the plasma membrane of living cells (24–26).

2.2. Super-resolution NSOM

In far-field optical microscopy, the diffraction limit implies that the minimum distance Δx required to resolve independently two distinct objects is dependent on the wavelength λ of the light used to observe the specimen, and by the condenser and objective lens system, through their refractive indices n and angle of acceptance α, such that $\Delta x = \lambda / 2n \sin\alpha$. This implies that Δx exceeds 300 nm in the case of visible light. When an object, such as microscopic specimen, is illuminated with a monochromatic plane wave, the transmitted or reflected light is collected by a lens and projected onto a detector to form the image. Usually, for convenience and practicality, the detector is placed in the far-field so that the far-field component of the light, which propagates in an unconfined way, is the only component used to generate the image. On the other hand, the interaction between the imaging light and the specimen also generates a near-field component, which consists of a non-propagating (evanescent) field existing only near the object at distances less than the wavelength of the light. Because the near-field decays exponentially within a distance less than the wavelength, usually it cannot be collected by the lens, thus it is not detected. This effect leads to the well-known Abbé's diffraction limit. By detecting the near-field component before it undergoes diffraction, NSOM allows non-diffraction limited high-resolution optical imaging (27–29). This is achieved in NSOM by placing a probe tip in close proximity to the sample in order to illuminate and/or detect in the near-field. Thus, in NSOM microscopes, the resolution Δx no longer depends on λ but instead on the aperture diameter of the probe (typically between 50 and 100 nm). In contrast to the previously described super-resolution techniques that are restricted to fluorescence and rely on the photophysical properties of the probes used, NSOM can exploit many other optical contrast mechanisms (i.e., absorption, polarization, and spectroscopy) in addition to fluorescence.

In its most commonly implemented mode, a sub-wavelength aperture probe is scanned in close proximity (<10 nm) to the specimen under study (Fig. 2d) to generate an image. Using the probe as a near-field excitation source, the interaction with the sample surface induces changes in the far-field radiation, which is collected in the far-field by conventional optics and directed to highly sensitive detectors to provide an optical image (27–30). An independent mechanism is used to control the distance separation between the tip and the sample and to simultaneously generate a topographic image (31, 32). In this way, a singular feature pertaining to NSOM is produced: correlative optical and topographical imaging with a spatial resolution determined by the probe configuration. Another unique characteristic of near-field excitation is given by the finite size of the probe itself: decreasing the area of illumination obviously reduces the interaction volume and background scatter, which is of major importance in enhancing the sensitivity for spectroscopic applications (fluorescence, Raman, etc.).

Instead of using the probe to illuminate the sample, one can employ far-field optics to illuminate the sample and use the probe to collect the evanescent field in close proximity to the sample surface. Although perfectly suitable for some photonic applications, its use in fluorescence imaging is less appropriate since far-field illumination translates in unnecessary sample photobleaching. A different experimental strategy to NSOM is based on the use of metallic tips, known in the literature as apertureless NSOM when the tip is used as passive scatterer (33), or tip-enhanced NSOM when the metallic tip is excited to enhance the electromagnetic field at the end of the tip apex (34). In both cases, the sample is illuminated in the far-field and a metal probe is placed in the tight focus of the illumination beam. The local interaction with the sample surface is subsequently detected as a modulation in the scattered far-field. Extreme sensitivity is required to observe the weakly scattered light from the nanometre-sized tip in the presence of the light scattered by the sample. When combined with fluorescence, and the tip is properly excited with radial fields along the tip axis, optical resolutions in the order to 30 nm can be achieved (35–37). This method is however accompanied by a large fluorescence background generated from far-field illumination of the sample, requiring therefore modulation techniques to recover the high-resolution signal (38). On the positive side of the balance, this method is free from the associated practical difficulties of fabricating circular apertures.

2.3. Implementation of NSOM for Quantitative Bioimaging

For biological applications, the most widely used configuration is an aperture-type NSOM, incorporated into an inverted optical microscope, with near-field excitation and far-field detection (see Fig. 3a). This scheme preserves most of the conventional imaging modes (confocal microscopy for instance), which remain available in combination with the near-field approach. NSOM aperture probes are commonly fabricated by the heating-and-pulling method or chemical etching of glass fibres, followed by aluminium evaporation. A representative NSOM probe with an aperture of 70 nm in diameter is shown in Fig. 3b. Light that is emitted by the

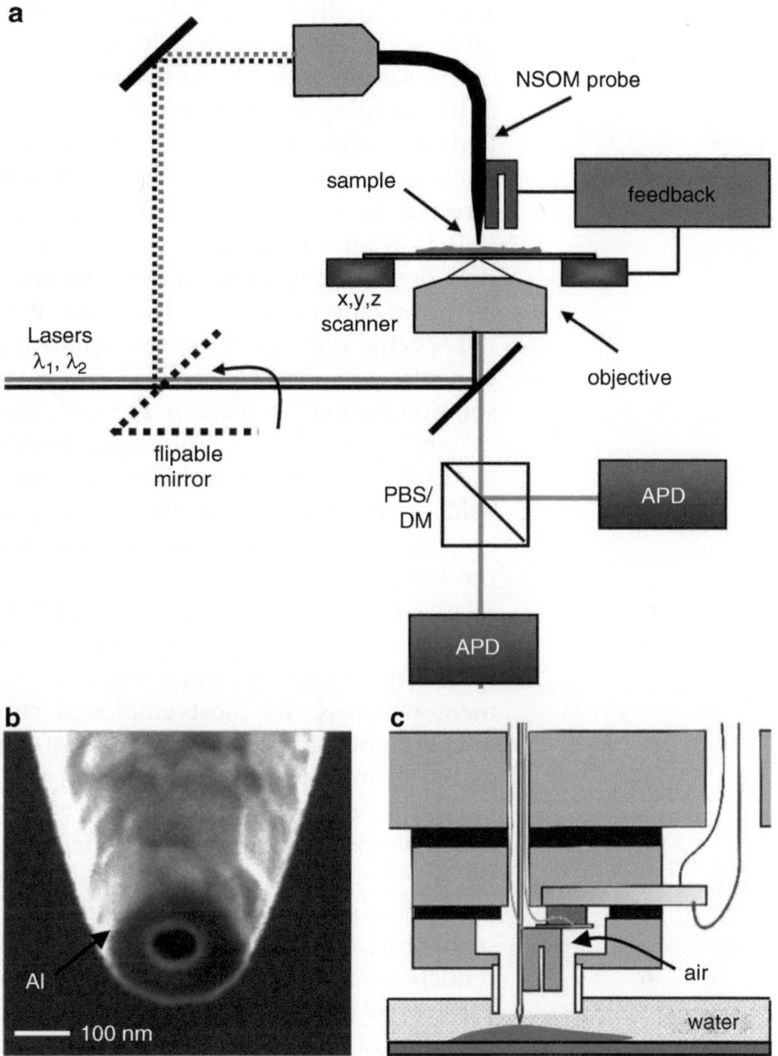

Fig. 3. (**a**) Schematics layout of the combined confocal/NSOM set-up. Two laser lines can be simultaneously coupled into the microscope using the confocal or NSOM excitation configuration modes. Fluorescence light is collected using a high NA objective and selected using appropriate filters. The fluorescence signal is then separated according its polarization (using a polarizing beam splitter, PBS) or spectral (using a dichroic mirror, DM) properties and sent to two APDs. (**b**). Focused ion beam image of a 70 nm diameter NSOM probe. The aperture functions as a local light source, and its diameter primarily determines the optical resolution of the microscope. Because of the evanescent character of the light exiting the probe, the optical near-field excitation has significant intensity only in a layer of <100 nm away from the aperture. This essentially means that lower lying fluorophores are not excited resulting in an effective suppression of background fluorescence. (**c**). Schematic drawing of the diving bell concept implemented for NSOM operation in liquid conditions. Only the tip end (~200 μm) is immersed in liquid while the tuning fork piezo-electric element is vibrating in air.

aperture locally interacts with the sample. It may be absorbed, phase shifted, or used to locally excite fluorescent markers, depending on the sample and the contrast mechanisms employed. In any case, light emerging from the imaging zone must be collected with the highest possible efficiency. For this purpose, high NA (oil immersion) microscope objectives are usually employed. The collected light is directed to sensitive detectors, such as avalanche photo-diodes (APD) or photo-multiplier tubes (PMT), via suitable dichroic mirrors for spectral splitting or through a polarizing beam splitter cube for polarization detection. Filters are also commonly used to select the spectral regions of interest removing unwanted spectral components, and inverted optical microscopes are an advantageous solution for light collection, redistribution, and filtering.

In Fig. 3a, the excitation light from one or more laser sources is coupled into the optical fibre. The tip is maintained in the near-field of the specimen by the feedback system operating in close loop that precisely controls the separation between the probe and the sample. In addition, a 3D scanner is employed to control the relative positioning of sample and probe. Depending upon design and applications, in principle the scanner may either move the specimen or the probe. In the case of the scanner locked to the specimen, which is the most employed configuration for biological imaging, the sample is moved in a raster pattern. The image is generated from the signal arising from the tip–specimen interaction under the probe, which is fixed and aligned confocally to the objective. The size of the area imaged depends uniquely on the range of the scanner. During raster scanning, data obtained both from the feedback system and from the optical detectors are simultaneously stored by a computer point by point. Finally, the PC compiles and renders the acquired data and furnishes simultaneously topography and optical image of the specimen.

One of the major obstacles that have restricted the use of NSOM in cell biology has been related to its difficulty to operate in liquid conditions, a crucial step towards live cell imaging. Successful control of the tip-sample distance has been routinely achieved in air by using tuning forks as sensing elements and driven at resonance (31, 32). However, this approach systematically failed once the tuning fork was immersed in a liquid. Our group has demonstrated that, in aqueous environments, sensitivity of the surface topography can be regained by keeping the tuning fork dry in a "diving bell" enclosure just above the probe (Fig. 3c) (39, 40). Alternatively, Höppener and colleagues used the tuning fork with the tip placed perpendicular to the prongs of the fork and protruding about ~2 mm below the fork. The configuration works thus as "tapping-mode" with the tip immerse in solution and the tuning fork kept dry above the liquid (41). An alternative method for position control is based on ion conductance. The method relies on the use of sharp micropipettes. As the probe approaches the sample, ion conduction is partially blocked and the change in conductivity is used as a measure of the tip-sample distance (42). This mechanism has been coupled to NSOM to obtain images in living cells (42).

Some examples of high-resolution NSOM imaging of different receptors expressed on cell membranes are shown in Fig. 4. The technique has been successfully applied for cell membrane quantitative imaging, and in particular to investigate the degree of clustering of different receptors on the cell membrane. For instance, NSOM has been used to elucidate clustering of the pathogen recognition receptor DC-SIGN on dendritic cells and relate clustering to the capability of DC-SIGN to bind with very high efficiency to a multitude of different pathogens (40, 43). Recently, Chen et al. used NSOM in combination with quantum dots to label the T cell receptor (TCR) of T cells in live animals before and after cell stimulation (44). In these experiments, it was shown that TCR organization plays a significant role in antigen recognition and cytokine production (44). Dual colour NSOM has also been used to measure the association of host and parasite proteins in malaria-infected erythrocytes (45), association between different members of the interleukin receptor family in T cells (46), colocalization of β-adrenergic receptors, and caveolae on the cell surface of cardiac myocytes (47) and more recently, the association of integrin receptors and lipid rafts on monocytes (48). It is well worthy of mentioning that dual colour NSOM has the inherent advantage of focussing different wavelength excitations through the same sub-wavelength aperture and as such, it does not suffer from chromatic aberrations, such as commonly occurring in other fluorescence-based techniques.

Although NSOM provides nanometric optical resolution together with simultaneous topographic information using a multitude of different optical contrast mechanisms, one should also be aware of the current limitations of the technique. For instance, NSOM is prone to artefacts generated from the topographic signal used to control the separation between the tip and sample. Therefore, relatively flat samples (with height differences below 1 μm) are preferred for imaging. As scanning technique, NSOM is inherently slow, and thus not suitable for recording fast dynamic processes in living cells. On the other hand, recent developments on NSOM combined with FCS might provide truly dynamic information at the nanometre scale (49). Finally, aperture probes have low throughput efficiencies (typically 10^{-4} to 10^{-6}), limiting the number of photons that can be forced out the tip. Current developments using optical nanoantennas to concentrate and enhance the electric field at the antenna end hold great promise for the use of these engineered probes in bioimaging (50) and are further described below.

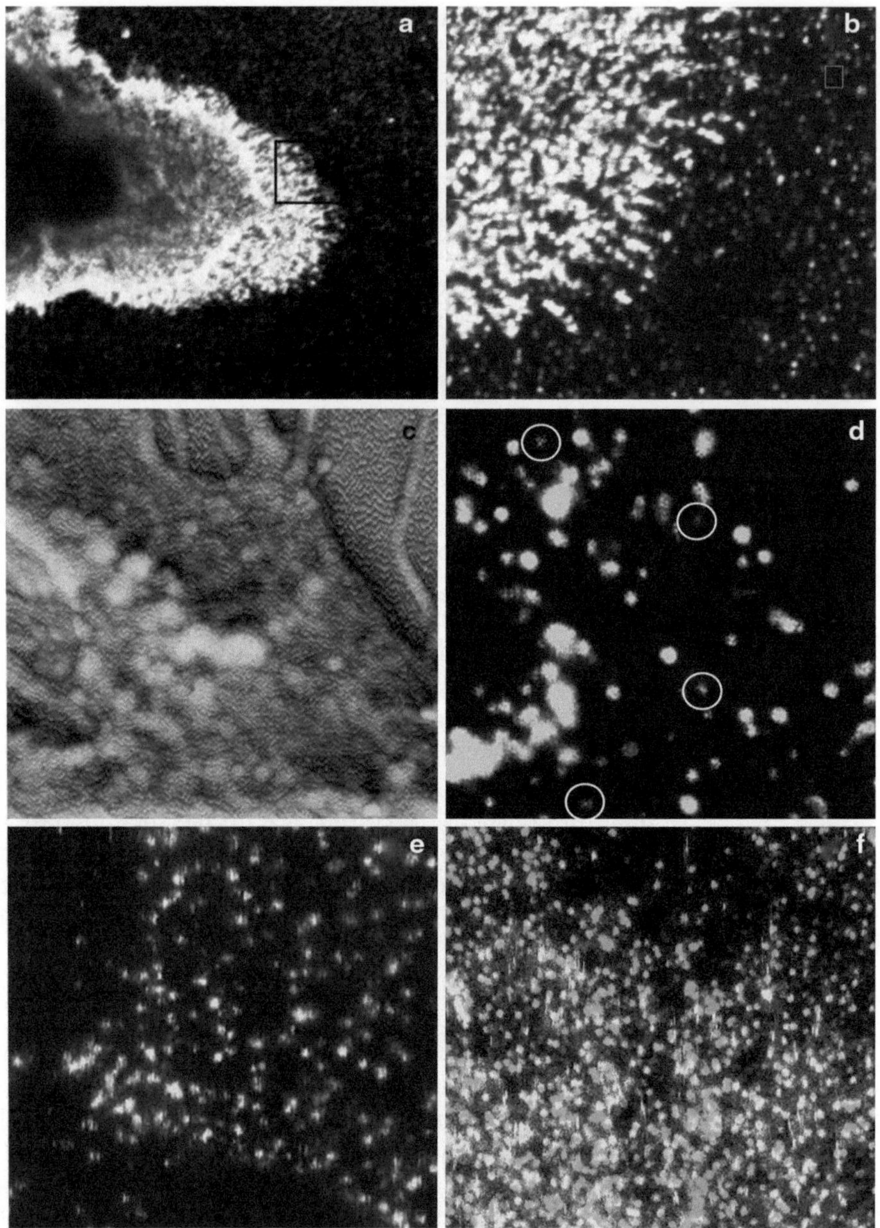

Fig. 4. Gallery of NSOM images on cell membranes in aqueous conditions. (**a**) Confocal image of ganglioside GM1 labelled with CTxB on monocytes. Image is 40 × 40 μm². (**b**) NSOM fluorescence image of the part highlighted in A, showing the higher spatial resolution obtained by NSOM. The less intense region (*right side*) corresponds to some individual CTxB on the glass surface, while the more intense region (*left side*) corresponds to GM1 nanodomains (<100 nm diameter) expressed on monocytes. Image is 7 × 7 μm². (**c** and **d**). Simultaneously obtained topography (**c**) and fluorescence image (**d**) as obtained by NSOM on immature dendritic cells (DC) expressing the pathogen recognition receptor DC-SIGN. The difference in spot intensity and size implies that DC-SIGN organizes in clusters on DC with a remarkably heterogeneous manner. The circles on the image correspond to individual DC-SIGN molecules. Image is 6 × 6 μm². (**e**) Single-molecule NSOM image of DC-SIGN in a less clustered sample, with the left side of the image corresponds to glass. Image is 7 × 7 μm². (**f**) Single-molecule image of the transferring receptor CD71 expressed on monocytes. This receptor organizes largely as monomers on monocytes, as derived from single spot intensity analysis. However, the large number of spots covering the cell membrane corresponds to the high expression level of the receptor. Image is 7 × 7 μm².

2.4. Optical Nanoantennas for Super-resolution NSOM Imaging

The main idea of optical antennas is to localize and enhance the optical radiation to a nanometric region, similar to electromagnetic antennas, which convert propagating radiation into a confined zone. In the biological context, gold nanoparticles attached to glass tips have been exploited as nanoantennas (36) and used to image single Ca^{2+} channels on erythrocyte plasma membranes with 50 nm optical resolution (37). Unfortunately, the method relies on far-field illumination to excite the antenna adding therefore a significant background contribution to the antenna response and requiring modulation techniques to reduce the background (38). A different excitation scheme that suppresses background illumination was first proposed by Frey et al. (51) and more recently, refined by Taminiau et al. (52). In these tip-on-aperture antennas, the local illumination properties of aperture-type NSOM are used to drive the antenna to resonance. Using this configuration, single-molecule detection with 30 nm resolution and virtually no background has been recently demonstrated (51, 52). Although nanoscale imaging of biological samples should be one of the most promising applications of this approach (50) its use in intact cell membranes in physiological conditions has not been explored until very recently.

Our group has recently demonstrated the potential of optical antennas for nanobioimaging of individual receptors and nanodomains on intact cells of the immune system (53). The probe-based monopole optical antennas were fabricated by carving of the antenna on the tip apex of conventional NSOM probes at the glass-metal interface using (Ga+)-FIB milling (53). The geometry, i.e., length, width, and radius of curvature of the antennas can be carefully controlled during FIB to maximize their response in liquid conditions. In our case, the dimensions of the fabricated antennas varied from 50 to 60 nm in width, ~20 nm of radius of curvature and lengths between 90 and 135 nm. These probes were then used under appropriate excitation antenna conditions to image individual antibodies in liquid conditions with an unprecedented resolution of 26 ± 4 nm and virtually no surrounding background. On intact cell membranes in physiological conditions, the obtained resolution is currently 30 ± 6 nm. Importantly, the method allowed us to distinguish individual proteins from nanodomains and to quantify the degree of clustering by directly measuring physical size and intensity of individual fluorescent spots (53). Improved antenna geometries by carefully reducing the width and adjusting the length to optimum resonance in liquid conditions should lead to true live cell imaging below 10-nm resolution with position accuracy in the sub-nanometric range.

3. High-Resolution Imaging in Tissue

While extremely useful, the techniques described above are mostly restricted so far to the study of single cells in in vitro and/or ex vivo conditions. During the past two decades, new optical microscopy techniques that use non-linear light–matter interactions to generate signal contrast have been developed. Non-linear optical microscopy techniques have special features that make them less sensitive to scattering and thus well suited for high resolution in tissues. In particular, two-photon-excited fluorescence laser-scanning microscopy (2PLSM) (54), combined with in vivo fluorescence-labelling techniques, has opened a rapidly expanding field of imaging studies in intact tissues and living animals. Specimens, such as lymphatic organs (55, 56), brain (57, 58), and deep tissue (59), have been examined in detail at depths of up to 1 mm, without damaging of the specimen. Moreover, the technique has also been used to study the development, progression, and potential treatment of pathological disorders, such as tumours (60) and Alzheimer disease (61–63).

Two-photon excitation occurs when two-low energy photons (normally in the infrared region) together excite a fluorophore and generate fluorescence. There are two main advantages of two-photon excitation as compared to conventional fluorescence (one-photon) excitation. First, the light scattering cross-section of living tissue is less in the infrared so that examination of fluorophores deep in living samples becomes possible. Second, two-photon absorption requires a high local concentration of photons at the focal point of excitation. This essentially means that two-photon excitation has build-in optical sectioning and does not require out-of-focus rejection like the one needed in confocal microscopy. Because of this, sample excitation occurs selectively only at the sample plane, reducing problems of sample photodamage and/or fluorophore photobleaching that would occur from comparable excitation with confocal microscopy. From the above, it can be inferred that 2PLSM is ideally suited for in vivo imaging at depth having no real advantage for transparent or thin imaging applications, for which confocal or more traditional fluorescence microscopes are superior.

The tight link existing between stem cells fate and their surrounding physiological environment requires imaging cells in living organisms or in fixed thick tissue slices while maintaining the native histological architecture. As a consequence, two-photon microscopy is now massively entering the field of stem cells and regenerative medicine. Besides increased penetration depth, the use of infrared-pulsed sources for imaging provides superb performances in functional microscopy assays. Two-photon absorption also opens the possibility to excite fluorescence in the UV range, even at relevant depths, making visible and quantifiable the spatial distribution of a wide variety of endogeneously expressed metabolic products as NADH, a relevant readout of the glycolysis activity of the cell. The ability to browse through the spectrum of emitted light by spectral analysis allows to simultaneously quantifying fluorescence from different molecular sources excited in the same wavelength range and where different components are then detected based on their photophysical properties.

4. Outlook

Cell-based therapies, in particular those based on stem cells, have generated much excitement in the scientific community, and are one of the most promising and active areas of research in regenerative medicine. While cancer-related research is a large part of the nanomedicine effort, there is great potential for applying nanotechnology in cell-based therapies for regenerative medicine. For instance, with the enormous self-repair potential of stem cells, it is important to be able to locate, recruit, and signal these cells to begin the regeneration processes. Improving non-invasive monitoring methods is particularly desirable since current approaches of evaluating cell treatments typically involve destructive or invasive techniques, such as tissues biopsies. Traditionally, non-invasive methods, such as magnetic resonance imaging (MRI) and positron emission tomography (PET), which rely heavily on contrast agents, lack the specificity or resident time to be a viable option for cell tracking. However, in vitro and in vivo visualization of nanoscale systems using techniques, such as the ones described in this chapter, will advance the field of regenerative nanomedicine. Combined with the use of suitable nanoparticles as fluorescence markers and in conjunction with nanofabrication techniques, nanotechnology, and nanophotonics are already providing powerful new tools for non-invasive tracking of cells in in vitro and in vivo conditions. The field is just booming.

Acknowledgments

The author would like to thank B. I. de Bakker, M. Koopman and T.S. van Zanten for NSOM images and fruitful discussions. Financial support has been provided by the EC-RTN-IMMUNANOMAP and EC-NEST-BIO-LIGHT-TOUCH, Spanish Ministry of Science and Technology MAT2007-66629-C02-01 and Generalitat de Catalunya 2009 SGR 597.

References

1. Stephens, D. J., and Allan, V. J. (2003) Light microscopy techniques for live cell imaging. *Science* **300**, 82–86.
2. Michalet, X., Kapanidis, A. N., Laurence, T., Pinaud, F., Doose, S., Pfughoefft, M., and Weiss, S. (2003) The power and prospects of fluorescence microscopies and spectroscopies. *Annu. Rev. Biophys. Biomol. Struct.* **32**, 161–182.
3. Betzig, E., and Chichester, R. J. (1993) Single molecules observed by near-field scanning optical microscopy. *Science* **262**, 1422–1425.
4. Weiss, S. (1999) Fluorescence spectroscopy of single biomolecules. *Science* **283**, 1676–1683.
5. Tsien, R. Y. (1998) The green fluorescent protein *Annu. Rev. Biochem.* **67**, 509–544.
6. Lippincott-Schwartz, J., and Patterson, G. H. (2003) Development and use of fluorescence protein markers in living cells. *Science* **300**, 87–91.
7. Hansma, P. K., Cleveland, J. P., Radmacher, M., Walters, D. A., Hillner, P. E., Bezanilla, M., Fritz, M., Vie, D., Hansma, H. G., Prater, C. B. (1994) Tapping mode atomic force microscopy in liquids *Appl. Phys. Lett.* **64**, 1738–1740.
8. Putman, C. A., van der Werf, K. O., de Groot, B. G., van Hulst, N. F., and Greve, J. (1994) Tapping mode atomic force microscopy in liquid *Appl. Phys. Lett.* **64**, 2454–2456.
9. Willemsen, O. H., Snel, M. M., Cambi, A., Gerve, J., de Groot, B. G., and Figdor, C. G. (2000) Biomolecular interactions measured by atomic force microscopy *Biophys. J.* **79**, 3267–3281.
10. Stroh, C., Wang, H., Bash, R., Ashcroft, B., Nelson, J., Gruber, H., Lohr, D., Lindsay, S. M., and Hinterdorfer, P (2004) Single-molecule recognition imaging microscopy *Proc. Natl. Acad. Sci. USA* **101**, 12503–12507.
11. Donnert, G., Keller, J., Medda, R., Andrei, M. A., Rizzoli, S. O., Luhrmann, R., Jahn, R., Eggeling, C., and Hell, S. W. (2006) Macromolecular-scale resolution in biological fluorescence microscopy *Proc. Natl. Acad. Sci. USA* **103**, 11440–11445.
12. Willig, K. I., Rizzoli, S. O., Wesphal, V., Jahn, R., and Hell, S. W. (2006) STED microscopy reveals hat synaptotagmin remains clustered after synpatic vesicle exocytosis *Nature* **440**, 935–939.
13. Sieber, J. J., Willig, K. I., Kutzner, C., Gerding-Reimers, C., Harke, B., Donnert, G., Rammner, B., Eggeling, C., Hell, S. W., Grubmuller, H., and Lang, T. (2007) Anatomy and dynamics of a supramolecular membrane protein cluster *Science* **317**, 1072–1076.
14. Kellner, R. R., Baier, C. J., Willig, K. I., Hell, S. W., and Barrantes, F. J. (2007) Nanoscale organization of nicotinic acetylcholine receptors revealed by stimulated emission depletion microscopy *Neuroscience* **144**, 135–143.
15. Wesphal, V., Rizzoli, S. O., Lauterbach, M. A., Kamin, D., Jahn, R., and Hell, S. W. (2008) Video-rate far-field optical nanoscopy dissects synaptic vesicle movement *Science* **320**, 246–249.
16. Gustafsson, M. G. L. (2005) Non-linear structured-illumination microscopy: wide-field fluorescence imaging with theoretically unlimited resolution *Proc. Natl. Acad. Sci. USA* **102**, 13081–13086.
17. Schermelleh, L., Carlton, P. M., Haase, S., Shao, L., Winoto, L., Kner, P., Burke, B., Cardoso, M. C., Agard, D. A., Gustafsson, M. G. L., Leonhardt, H., and Sedat, J. W. (2008) Subdiffraction multicolor imaging of the nuclear periphery with 3D structured illumination microscopy *Science* **320**, 1332–1336.
18. Gustaffson, M. G. L., Shao, L., Carlton, P. M., Wang, C. J. R., Golubovskaya, I. N., Cande, W. Z., Agard, D. A., and Sedat, J. W. (2008) Three-dimensional resolution doubling in wide-field fluorescence microscopy by structured illumination *Biophys. J.* **94**, 4957–4970.
19. Pielage, J., Cheng, L., Fetter, R. D., Carlton, P. M., Sedat, J. W., and Davis, G. W. (2008) A presynaptic giant ankrin stabilizes the NMJ through regulation of presynaptics microtubules and transsynaptic cell adhesion *Neuron* **58**, 195–209.
20. Betzig, E., Patterson, G. H., Sougrat, R., Lindwasser, O. W. S., Olenuch, Bonifacino, J. S., Davidson, M. W., Lippincott-Schwartz, J., and Hess, H. F. (2006) Imaging intracellular fluorescent proteins at nanometer resolution *Science* **313**, 1642–1645.
21. Hess, S. T., Girirajan, T. P. K., and Mason, M. D. (2006) Ultra-high resolution imaging by fluorescence photoactivation localization microscopy *Biophys. J.* **91**, 4258–4272.
22. Rust, M. J., Bates, M., and Zhuang, X. (2006) Subdiffraction limit imaging by stochastic optical reconstruction microscopy (STORM) *Nat. Methods* **3**, 793–796.
23. Schroff, H., Galbraith, C. G., Galbraith, J. A., White, H., Gillete, J., Olenych, S., Davidson, M. W., and Betzig, E. (2007) Dual-color superresolution imaging of genetically expressed probes within adhesion complexes *Proc. Natl. Acad. Sci. USA* **104**, 20308–20313.
24. Manley, S., Gillete, J. M., Patterson, G. H., Schroff, H., Hess, H. F., Betzig, E., and

Lippincott-Schwartz, J. (2008) High-density mapping of single molecule trajectories with photoactivated localization microscopy *Nat. Methods* **5**, 155–157.

25. Hess, S. T., Gould, T. J., Gudheti, M. V., Maas, S. A., Mills, K. D., and Zimmerberg, J. (2007) Dynamic clustered distribution of hemagglutinin resolved at 40nm in living cell membranes discriminates between raft theories *Proc. Natl. Acad. Sci. USA* **104**, 17370–17375.
26. Schroff, H., Galbraith, C. G., Galbraith, J. A., and Betzig, E. (2008) Live cell photoactivated localization microscopy of nanoscale adhesion dynamics *Nat. Methods* **5**, 417–423.
27. Pohl, D. W., Denk, W., and Lanz, M. (1984) Optical stethoscopy: image recording with resolution l/20 *Appl. Phys. Lett.* **44**, 651–653.
28. Betzig, E., Trautman, J. K., Harris, T. D., Weiner, J. S., Kostelak, R. L. (1991) Breaking the diffraction barrier: optical microscopy on a nanometric scale *Science* **251**, 1468–1470.
29. van Hulst, N. F., Garcia-Parajo, M. F., Moers, M. H. P., Veerman, J. A., and Ruiter, A. G. T. (1997) Near-field fluorescence imaging of genetic material: towards the molecular limit *J. Struct. Bio.* **119**, 222–231.
30. Hecht, B., Sick, B., Wild, U. P., Deckert, V., Zenobi, R., Martin O. J. F. and Dieter, D. W. (2000) Scanning near-field optical microscopy with aperture probes: fundamentals and applications *J. Chem. Phys.* **112**, 7761–7774.
31. Karrai, K., and Grober, R. D. (1995) Piezoelectric tip-sample distance control for near field optical microscopes *Appl. Phys. Lett.* **66**, 1842–1844.
32. Ruiter, A. G., Veerman, J. A., van der Werf, K. O., van Hulst, N. F. (1997) Dynamic behavior of tuning fork shear-force feedback *Appl. Phys. Lett.* **71**, 28–30.
33. Zenhausern,F.,Martin,Y.,andWickramasinghe, H. K. (1995) Scanning interferometric apertureless microscopy: Optical imaging at 10 Angstrom resolution *Science* **269**, 1083–1085.
34. Novotny, L., and Stranick, S. J. (2006) Near-field optical microscopy and spectroscopy with pointed probes *Annu. Rev. Phys. Chem.* **57**, 303–331.
35. Sanchez, E. J., Novotny, L., Sunney-Xie, X. (1999) Near-field fluorescence microscopy based on two-photon excitation with metal tips *Phys. Rev. Lett.* **82**, 4014–4017.
36. Anger, P., Bharadwaj, P., and Novotny, L. (2006) Enhancement and quenching of single-molecule fluorescence *Phys. Rev. Lett.* **96**, 1130021–1130024.
37. Höppener, C., and Novotny, L. (2008) Antenna-based optical imaging of single Ca^{2+} transmembrane proteins in liquids *Nano Lett.* **8**, 642–646.
38. Höppener, C., Beams, R., Novotny, L. (2009) Background suppression in near-field optical imaging *Nano Lett.* **9**, 903–908.
39. Koopman, M., de Bakker B. I., Garcia-Parajo, M. F., and van Hulst, N. F. (2003) Shear force imaging of soft samples in liquid using a diving bell concept *Appl. Phys. Lett.* **83**, 5083–5085.
40. Koopman, M., Cambi, A., de Bakker, B. I., Joosten, B., Figdor, C. G., van Hulst, N. F., and Garcia-Parajo, M.F. (2004) Near-field scanning optical microscopy in liquid for high resolution single molecule detection on dendritic cells *FEBS Lett.* **573**, 6–10.
41. Höppener, C., Siebrasse, J. P., Peters, R., Kubitscheck, U., and Naber, A. (2005) High-resolution near-field optical imaging of single nuclear pore complexes under physiological conditions *Biophys. J.* **88**, 3681–3688.
42. Korchev, Y. E., Raval, M., Lab, M. J., Gorelik, J., Edwards, C. R. W., Rayment, T., and Klenerman, D. (2000) Hybrid scanning ion conductance and scanning near-field optical microscopy for the study of living cells *Biophys. J.* 78, 2675–2679.
43. de Bakker, B. I., de Lange, F., Cambi, A., Korterik, J. P., van Dijk, E. M. H. P., van Hulst, N. F., Figdor, C. G., and Garcia-Parajo, M. F. (2007) Nanoscale organization of the pathogen receptor DC-SIGN mapped by single-molecule high-resolution fluorescence microscopy *Chemphyschem* **8**, 1473–1480.
44. Chen, Y., Shao, L., Ali, Z., Cai, J., and Chen, Z. W. (2008) NSOM/QD-based nanoscale immunofluorescence imaging of antigen-specific T-cell receptor responses during an in vivo clonal Vγ2Vδ2 T-cell expansion **Blood** *111*, 4220–4232.
45. Enderle, T., Ha, T., Ogletree, D. F., Chemla, D. S., Magowan, C., and Weiss, S. (1997) Membrane specific mapping and colocalization of malarial and host skeletal proteins in the Plasmodium falciparum infected erythrocyte by dual-color near-field scanning optical microscopy *Proc. Natl. Acad. Sci. USA* **94**, 520–525.
46. de Bakker, B. I., Bodnar, A., van Dijk, E. P., Vámosi, G., Damjanovich, S., Waldmann, T. A., van Hulst, N. F., Jenei A., and Garcia-Parajo, M. F. (2008) Nanometer-scale organization of the alpha subunits of the receptors for IL2 and IL15 in human T lymphoma cells *J. Cell Sci.* **121**, 627–633.
47. Ianoul, A., Grant, D. D., Rouleau, Y., Bani-Yaghoub, M., Johnston, L. J., and Pezacki, J. P. (2005) Imaging nanometer domains of

β-adrenergic receptor complexes on the surface of cardiac myocytes *Nat. Chem. Biol.* **1**, 196–202.

48. van Zanten, T. S., Cambi, A., Koopman, M., Joosten, B., Figdor, C. G., and Garcia-Parajo, M. F. (2009) Hotspots of GPI-anchored proteins and integrin nanoclusters function as nucleation sites for cell adhesion *Proc. Natl. Acad. Sci. USA* **106**, 18557–18562.
49. Vobornik, D., Banks, D. S., Lu, Z., Fradin, C., Taylor, R., and Johnston, L. J. (2008) Fluorescence correlation spectroscopy with sub-diffraction-limited resolution using near-field optical probes *Appl. Phys. Lett.* **93**, 163904–163906.
50. Garcia-Parajo, M. F. (2008) Optical antennas focus in on biology *Nat. Photonics* **2**, 201–203.
51. Frey, H. G., Witt, S., Felderer, K., Guckenberger, R. (2004) High-resolution imaging of single fluorescent molecules with the optical near-field of a metal tip *Phys. Rev. Lett.* **93**, 200801–200804.
52. Taminiau, T. H., Moerland, R. J., Segerink, F. B., Kuipers, L. van Hulst, N. F. (2007) λ/4 Resonance of an optical monopole antenna probed by single molecule fluorescence *Nano Lett.* 7, 28–33.
53. van Zanten, T. S., Lopez, M. J., Garcia-Parajo, M. F. (2009) Imaging individual proteins and nanodomains on intact cell membranes with a probe-based optical antenna *Small* **6**, 270–275.
54. Denk, W., Strickler, J. H., and Webb, W. W. (1990) Two-photon laser scanning fluorescente microscopy *Science* **248**, 73–76.
55. Miller, M. J., Wei, S. H., Cahalan, M. D., and Parker, I. (2003) Autonomous T cell trafficking examined *in vivo* with intravital two-photon microscopy *Proc. Nat. Acad. Sci. USA* **100**, 2604–2609.
56. Bousso, P., and Robey, E. A. (2004) Dynamic behavior of T cells and thymocytes in lymphoid organs as revealed by two-photon microscopy *Immunity* **21**, 349–355.
57. Nimmerjahn, A., Kirchhoff, F., and Helmchen, F. (2005) Resting microglial cells are highly dynamic surveillants of brain parenchyma *Science* **308**, 1314–1318.
58. Davalos, D. et al. (2005) ATP mediates rapid microglial response to local brain injury *in vivo* *Nat. Neurosci* **8**, 752–758.
59. Helmchen, F., and Denk, W. (2005) Deep tissue two-photon microscopy *Nat. Methods* **2**, 932–940.
60. Jain, R. K., Munn, L. L., and Fukumura, D. (2002) Dissecting tumor pathophysiology using infravital microscopy *Nat. Rev. Cancer* **2**, 266–276.
61. Skoch, J., Hickey, G. A., Kajadasz, S. T., Hyman, B. T., and Bacskai, B. J. (2005) In-vivo imaging of amyloid-beta deposits in mouse brain with multiphoton microscopy *Methods Mol. Bio.* **299**, 349–363.
62. Tsai, J., Grutzendler, J., Duff, K., and Gan, W. (2004) Fibrillar amyloid deposition leads to local synaptic abnormalities and breakage of neuronal branches *Nat. Neurosci.* 7, 1181–1183.
63. Bacskai, B. J. et al. (2003) Four-dimensional multiphoton imaging of brain entry, amyloid binding, and clearance of an amyloid-beta ligand in transgenic mice *Proc. Natl. Acad. Sci. USA* **100**, 12462–12467.

Chapter 18

Molecular Dynamics Methods for Modeling Complex Interactions in Biomaterials

Antonio Tilocca

Abstract

The molecular dynamics method is a powerful computer simulation technique which provides access to the detailed time evolution (trajectory) of a system in specified conditions, such as a particular temperature or pressure. The full trajectory of the system can be analyzed using statistical mechanics tools to obtain thermodynamical quantities and dynamical properties; the mechanism of chemical reactions and other time-dependent processes, such as diffusion, can also be revealed in high detail. When applied to model extended and complex system such as biomaterials, MD simulations represent an invaluable tool to discover structure–activity relationships and rationalize biomedical applications.

Key words: Computer simulations, Molecular dynamics, Force fields, Ab-initio, Car–Parrinello, Biomaterials

1. Introduction

Understanding the properties of biomaterials at the microscopic level is an extremely challenging task, due to the complexity of the diverse physicochemical interactions that control their activity. For instance, bioactive implants used for bone repair or replacement are typically inorganic materials (glasses, ceramics, or a suitable combination) which are able to interact and integrate with a biological environment (1–3). A number of different interactions between the synthetic material and the living body take place simultaneously, dynamically modifying the material–tissue interface until the implant is either firmly anchored to the tissue by strong chemical bonds, or has completely dissolved after having accomplished its task (for instance, after releasing a drug or an antibacterial agent in-situ) (2). Interactions between ionic inorganic species dominate the initial stages of the bioactive fixation mechanism

Melba Navarro and Josep A. Planell (eds.), *Nanotechnology in Regenerative Medicine: Methods and Protocols*, Methods in Molecular Biology, vol. 811, DOI 10.1007/978-1-61779-388-2_18,

of bioglasses; these interactions are eventually supplemented and gradually replaced by further interactions between the inorganic surface and biomolecules such as collagen or tissue growth factors, leading to the formation of a bond between the artificial implanted material and the natural tissues (3). Surface-analytical experimental techniques are able to reveal macroscopic effects and processes occurring at the interface between an implant and its biological environment, sometimes with very good time and space resolution, providing an explicit measurement of the different response of specific biomaterials in realistic conditions (4). However, to obtain a deeper insight into the way in which biomaterials work, one needs to look at processes which occur on an "atomistic" scale (that is, interatomic separations as small as 0.1 nm and a time resolution of $\sim 10^{-6}$ μs or less): unfortunately, even the most powerful experimental methods cannot reach this resolution. Computer simulations provide a direct route to explore structural and dynamical properties of assemblies of atoms, and therefore represent an invaluable tool to probe the properties of biomaterials at the atomistic level. In particular, with the advent of powerful supercomputers and the availability of parallel simulation codes optimized to make the most of the unprecedented computational power on hand, today we can carry out accurate and reliable computer experiments of systems and processes which were absolutely impractical only few years ago (5, 6). The molecular dynamics (MD) method is currently the most common approach to model relatively large and complex systems, including periodic solids, liquids, and biomolecules in solution, with atomistic resolution (7, 8). An MD simulation provides a series of "snapshots" of the system at different times: this *trajectory* represents the natural time evolution of the system under specified operating conditions, for instance room temperature and atmospheric pressure. As we illustrate in the following sections, an MD trajectory provides access to the thermodynamic properties of the system, but also, unlike other computational methods such as Monte Carlo, allows us to directly look at *dynamical* processes, driven by the finite temperature of the simulation.

On the one hand, classical MD using predefined force fields (6) can be used to investigate biological system and materials containing up to millions of atoms (6, 9), for time lengths up to the microsecond scale, with an accuracy which is completely and exclusively determined by the force field adopted. On the other hand, ab-initio (AI) MD (10) can be used to tackle systems that are intrinsically difficult or impossible to treat with classical MD and predefined potentials, such as those involving chemical reactions, breaking and formation of chemical bonds, and in general electronic effects which can hardly be incorporated in a classical interatomic potential.

Whereas the size and time scales (for instance, millimeters and seconds) characterizing real, macroscopic samples of biomaterials are still far beyond those accessible by atomistic MD, the information obtained by performing MD simulation on smaller samples is generally relevant to describe these systems, except in selected cases where there is a serious mismatch between the intrinsic scale of the phenomenon under study and the MD scale. This is the case of activated processes, such as chemical reactions involving high energy barriers, or phase transitions: a very active area of research involves the development of computational methods designed to overcome these complications (11), thus further extending the applicability of MD to realistic problems in material science.

The following sections illustrate the molecular dynamics method in detail, focusing on the basic approximations and algorithms needed to set up and run an MD simulation. The features and limitations of common interatomic potentials used in classical MD are then discussed, followed by a description of the Car–Parrinello method of ab-initio MD. A brief outline of some recent applications to the field of bioactive glasses conclude the chapter, showing how the interplay between classical and ab-initio MD is often beneficial for improving our knowledge of these systems.

2. The Molecular Dynamics Method

2.1. The Basics

Given a system composed of N atoms or interaction sites, all molecular dynamics methods involve the numerical integration of *classical* equations of motion:

$$\mathbf{F}_i = m_i \mathbf{a}_i ; \mathbf{F}_i = -\partial V / \partial \mathbf{r}_i \quad i = 1, \ldots, N, \tag{1}$$

where the index i runs from 1 to N; $\mathbf{r}_i$ and $\mathbf{a}_i$ are the position and acceleration vectors of atom i, m_i is the atomic mass, $\mathbf{F}_i$ is the total force acting on i, and $V = V(\mathbf{r}_1, \ldots, \mathbf{r}_N)$ is the *potential* function. The latter is the key ingredient of an MD simulation: once V has been defined, eq. 1 allows us to propagate an initial configuration $\Gamma(0) = \{\mathbf{R}(t=0); \mathbf{P}(t=0)\}$, where $\mathbf{R} = \{\mathbf{r}_1, \ldots, \mathbf{r}_N\}$ and $\mathbf{P} = \{\mathbf{p}_1, \ldots, \mathbf{p}_N\}$ are the set of individual positions $\mathbf{r}_i$ and momenta $\mathbf{p}_i$, respectively, along discrete subsequent time points Δt, separated by the *time step* Δt. The appropriate Δt for an MD run depends on the fastest motions characterizing the system, such as high-frequency vibrations: Δt should be short enough to limit errors in the (finite-difference) integration of these fast motions, avoiding an unnecessarily short Δt, which will only waste computer time. Depending on the mass of the lightest atoms in the system, typical values for classical MD simulations range between 1 and 20 fs; shorter values (0.1–0.2 fs) can be needed in AIMD (see below).

2.2. Statistical Mechanics: Microstates and Ensembles

A single configuration (or microstate) Γ of our N-particle system is represented by the $6N$ components of the $\{\mathbf{R}\}$ and $\{\mathbf{P}\}$ position and momentum vectors; the set of all the possible $\Gamma = \{\mathbf{R};\mathbf{P}\}$ microstates is named *phase space*. A standard MD simulation of N_{run} time steps will produce a *trajectory* $\{\Gamma(0), \Gamma(\Delta t), \Gamma(2\Delta t), \ldots, \Gamma(N_{\text{run}}\Delta t)\}$ in phase space, which is the sequence of N_{run} successive instantaneous configurations visited by the system during its dynamical evolution. More specifically, given that the total energy E is a conserved quantity in standard MD, an MD run will describe a trajectory in a relatively small portion of the phase space, where all microstates share the same total energy E. In other words, whereas the individual configurations of an MD trajectory are all unique and different from each other, they belong to a common constant-energy portion of the phase space. The collection of *all* microstates sharing the same total energy (or some other macroscopic quantity) is the statistical *ensemble*: the NVE ensemble, where all microstates have the same number of particles, volume and total energy is the "natural" ensemble of MD simulations, but extensions to the method (mimicking an external thermostat or barostat) can be made to probe other ensembles, such as the canonical (NVT) or the isothermal–isobaric NPT ensemble. The choice of the appropriate statistical ensemble for the simulation depends of the phenomena that one wants to model: for instance, the NPT ensemble may be more appropriate to simulate processes in the laboratory, which normally occur at constant pressure and temperature rather than at constant volume, or to study any process which involve dynamical changes in the size and shape of the simulation box.

2.3. Periodic Boundary Conditions

No matter how large the number of atoms N in the simulated system is, it will always be much smaller than the number of atoms (of the order of the Avogadro number, $\sim 6 \times 10^{23}$) in a *real*, macroscopic sample. The most immediate consequence is that the ratio of the number of atoms found at the boundary and in the bulk of the system will be much higher in the simulated than in the real system: *everything would be surface*. The common solution to remove this artifact in MD simulations is to apply periodic boundary conditions (PBC) to a central box containing the system: the box is replicated infinitely along each direction, such that a particle leaving the central box will be mirrored by the image of the same particle in the adjacent box, which will enter the box from the opposite side. In this way, boundaries are completely removed and every atom in the central box is embedded in a bulk-like environment. This approach is natural for crystalline solids, where the central box has the symmetry of the periodic unit cell, but it is also effective for liquids and disordered solids, where the artificial periodicity introduced by the PBC is only a small drawback compared to the advantages of removing fictitious surface effects. The absence of boundaries entails that to model processes that depend on the presence of surfaces, such as evaporation of a liquid sample, or gas adsorption and reactivity at a solid surface, one will have to reintroduce the exposed surface by using a slab geometry, where PBC are removed or altered along one direction and only act in two directions (12).

2.4. Calculating Thermodynamic Quantities

Provided that a specific property A can be defined as a function of the microscopic configuration Γ, i.e., $A = A(\Gamma)$, then the macroscopic observable $<A>$ is the average of the values $A(\Gamma)$ corresponding to *all* the M microstates belonging to the statistical ensemble, weighted with the ensemble probability function $\rho(\Gamma)$.

$$\langle A \rangle = \sum_{k=1}^{M} \rho(\Gamma_k) A(\Gamma_k). \tag{2}$$

For complex systems, performing the summation in eq. 2 is not practical, because there are a very large number of microstates, most of which with a very low associated probability: how can we filter out the low-probability states, and include only the representative, high-probability states in the sum? Molecular dynamics provides a way to perform this task: in the ergodic hypothesis that, during an MD run, the system will visit all the representative configurations of the ensemble (visiting several times and more frequently microstates with a high associated probability), one can then calculate the property A by simply averaging over the instantaneous values assumed by A at each timestep of an MD trajectory (13):

$$\langle A \rangle = \frac{1}{N_{\text{run}}} \sum_{j=1}^{N_{\text{run}}} A\left[\Gamma(j\Delta t)\right] = \frac{1}{N_{\text{run}}} \sum_{j=1}^{N_{\text{run}}} A(j) \tag{3}$$

In other words, provided that an "instantaneous" definition $A(j)$ of the macroscopic variable A can be formulated, the ensemble average of statistical mechanics is replaced by a *time average* in molecular dynamics. Because the ergodic hypothesis is in principle satisfied only in the limit of an infinitely long trajectory, whereas the MD average is necessarily taken over a comparatively small number of microstates, care must be taken to check the convergence of the averaged quantity over the finite length of the MD run: a drift in the running average typically indicates that the run was too short, making the sampling of the property of interest ineffective.

A typical macroscopic property which is straightforwardly calculated along an MD run is the temperature (from the average kinetic energy K of the system):

$$T = \frac{2\langle K \rangle}{3Nk_{\text{B}}} = \frac{2}{3Nk_{\text{B}}}\left[\frac{1}{N_{\text{run}}} \sum_{j=1}^{N_{\text{run}}} K(j)\right] \tag{4}$$

where k_B is the Boltzmann constant, N is the number of atoms, and $K(j)$ is the instantaneous value of the total kinetic energy, defined as:

$$K(j) = \sum_{i=1}^{N} \frac{\mathbf{p}_i^2(j)}{2m_i} \quad (5)$$

The calculation of the system pressure is only slightly more involved, and it makes use of the virial equation (6):

$$P = \frac{1}{V}\left[Nk_B T + \frac{1}{3}\frac{1}{N_{\text{run}}}\sum_{j=1}^{N_{\text{run}}} W(j) \right] \quad (6)$$

where V is the volume of the system and $W(j)$ is the instantaneous virial function, defined as:

$$W(j) = \sum_{i=1}^{N} \mathbf{r}_i(j) \cdot \mathbf{F}_i(j) \quad (7)$$

where $\mathbf{F}_i$ is the internal force acting on atom i.

These examples show how, based on the ergodic hypothesis, some macroscopic properties of the system can be extracted from direct time averages of appropriate instantaneous quantities, such as the kinetic energy and the virial function, over a sufficiently long MD trajectory. Not only their mean value, but also the *fluctuations* (i.e., the variance) of the instantaneous quantities can be linked to thermodynamic observables: for instance, the heat capacity can be calculated in the NVE ensemble from the kinetic energy fluctuations of the system (7).

2.5. Calculating Structural and Dynamical Properties

The previous section showed how the extremely detailed description of an MD trajectory can be processed to obtain measurable thermodynamic properties of a system. However, in most cases, it *is* this highly detailed description, and its correspondingly high space and time resolution, which make MD simulations an invaluable tool, providing an unique route to investigate structural and dynamical properties which are not easily *measurable* with standard experimental techniques.

For instance, radial distribution functions (RDF) are a powerful tool to analyze the atomistic structure of liquids and solids: these functions show maxima at the typical interatomic distances which characterize the system (7). The individual contributions of specific atomic pairs to the total radial distribution function of disordered solids and liquids are hard to identify in the experimental RDF obtained from diffraction experiments, and advanced (high-energy and isotopic) techniques are often needed to separate the individual patterns (14). On the contrary, the straightforward postprocessing analysis of an MD trajectory can easily provide each individual contribution to the total RDF, thus enormously helping the assignment of the peaks in the experimental curve, and the interpretation of the underlying atomic structure (15). In fact, it is quite straightforward to calculate *partial* RDFs from an MD trajectory, simply by focusing on the distance r_{AB} separating all atoms of species A and B during the trajectory: the partial A–B RDF(r) is the probability of finding an A–B pair separated by r, relative to a homogeneous (random) distribution of A and B:

$$\mathrm{RDF}_{AB}(r) = \frac{\left\langle \frac{1}{N_A} \sum_{i=1}^{NA} \sum_{j=1}^{NB} \delta(r - r_{ij}) \right\rangle}{\rho_{AB}^{\mathrm{hom}}(r)}, \tag{8}$$

where the <...> brackets denote a time average, N_A and N_B are the total number of atoms of species A and B, respectively, and the ideal density of a corresponding homogeneous system is at the denominator, whereas the numerator is essentially the cumulative (running) coordination number of species B found in a sphere of radius r centered on atom A.

The dynamical behavior of a system can be captured using the time-correlation function formalism (6, 11): the time autocorrelation function $C(t)$ of a dynamical variable $x(t)$ measures the agreement between two values of x taken at two different times t_1 and t_2, separated by an interval t:

$$C(t) = \langle x(0) \cdot x(t) \rangle \tag{9}$$

where the < ... > brackets denote in this case an average done over all pairs of configurations separated by the time interval t present in the MD trajectory.

Using time-correlation functions, MD simulations provide a direct route to dynamical quantities such as the diffusion coefficient of each species i, which can be extracted from the slope of its time-dependent mean square displacement $\mathrm{MSD}_i(t)$, easily calculated as:

$$\mathrm{MSD}_i(t) = \langle | \mathbf{r}_i(t) - \mathbf{r}_i(0) |^2 \rangle, \tag{10}$$

where the <...> brackets here denote an average over all atoms of type i, and over all pairs of configurations separated by the time interval t present in the MD trajectory.

Further useful time-correlation functions are the velocity autocorrelation function, which can be Fourier-transformed to yield the vibrational (power) spectrum (16), and orientational correlation functions, which reflect the rotational dynamics of molecules (17). Once again, it is straightforward to extract the individual contributions of groups of atoms (for instance, belonging to a specific functional group or to a specific region of the system) from the MD trajectory, and use this information to perform difficult tasks, such as assigning the peaks corresponding to individual functional groups in the vibrational spectrum (18).

Space *and* time correlation functions, such as the self- and distinct part of the van Hove functions (19), are extremely powerful tools to analyze complex dynamical features difficult to investigate, such as the nature of correlated hops characterizing ion migration in ionic solids and liquids (20).

It should be remarked that, provided that the characteristic time of a dynamical event is compatible (i.e., significantly shorter) with the length of the MD trajectory, such that several instances of the event will occur *spontaneously* in the course of the simulation, perhaps the most important feature of MD simulations is the possibility to reveal the mechanism of dynamical processes which can only be guessed using alternative (indirect) probes. For instance, the mechanism of spontaneous (barrierless) chemical reactions in solution or at the surface of a solid catalyst or biomaterial, of ion migration in amorphous materials, and of proton transfer in biomolecules, have all been highlighted by standard (either classical or ab-initio) MD (11, 21–24). In cases where a substantial energy barrier prevents the process to occur with a significant probability during the scale of molecular dynamics, special techniques have been developed to enhance the sampling of the low-probability region nearby the transition state, and therefore permit its study through MD (10, 25).

3. Classical MD with Empirical Potentials

3.1. The Potential Function

Methods to advance atomic positions through the numerical integration of the equations of motion eq. 1, such as the Verlet algorithm (6), require the force $\mathbf{f}_i$ acting on each atom i, which must be recalculated at each timestep, that is, for each new atomic configuration $\mathbf{R} = \{\mathbf{r}_1, \ldots, \mathbf{r}_N\}$ visited along the trajectory. In classical MD, $\mathbf{f}_i$ is obtained as the analytical gradient (partial derivative with respect to $\mathbf{r}_i$) of an $\mathbf{R}$-dependent potential function V: $\mathbf{F}_\mathrm{i} = -\partial V(\mathbf{R})/\partial \mathbf{r}_i$.

What does V represent exactly? In the Born–Oppenheimer approximation, the electrons adapt instantaneously to the motion of the heavier and slower nuclei, and are always found at the minimum (ground state) for each nuclear configuration visited during the dynamics. The total, ground-state electronic energy E can then be considered dependent on the instantaneous nuclear positions, that is $E = E(\mathbf{R})$: if we were able to calculate and map E for *every* possible value of $\mathbf{R}$, we would have built the potential energy function governing the motion of the nuclei. The potential function $V(\mathbf{R})$ is thus the ground-state electronic energy obtained by solving the quantum-mechanical Schrödinger equation for a fixed nuclear configuration $\mathbf{R}$, in the Born–Oppenheimer approximation.

In MD simulations, the nuclei move in the potential energy surface $V(\mathbf{R})$. For most systems, it is not possible or practical to obtain $V(\mathbf{R})$ through ab-initio energy minimizations: in fact, the *relevant* fraction of the configurations space that the system will cross during the MD run is not known a priori, and therefore, one will have to obtain and interpolate $V(\mathbf{R})$ for a very large number of configurations $\mathbf{R}$, spanning all the (potentially) relevant microstates, and possibly much more. Ab-initio MD, with *on-the-fly* quantum-mechanical calculation of forces (discussed below), represents a more efficient way to evolve the system on an ab-initio potential energy surface, because the ab-initio forces in AIMD are calculated *only* for the relevant fraction of microstates effectively visited by the system during its "natural" evolution. However, notwithstanding this improvement, the computational cost inherent in the explicit full quantum-mechanical calculation of forces is still very high and limits the system size and the trajectory length which can be simulated by AIMD.

A computationally much cheaper approach, at the cost of a lower accuracy, is classical MD using predefined, empirical potentials. In classical MD, one selects an analytical potential function suitable for the system under study *before* starting the simulation, and then calculates the forces as analytical gradients of this potential function. No matter how sophisticated the classical potential can be made, the main drawback is that, by preselecting the interatomic potential, one is somewhat biasing the dynamical evolution of the system: by constraining the forces acting between atomic species, one is also applying constraints to the range of physical phenomena that can be observed in the simulation. Nonetheless, provided that an adequate interatomic potential is selected using chemical and physical insight on the system under study, classical MD represents an extremely powerful tool to explore systems and phenomena whose investigation requires size and time scales out of the reach of ab-initio MD: systems containing millions of atoms can *only* be targeted using classical MD (5, 6).

The interatomic potential function for classical MD is typically written as a sum of different terms, each one representing a different kind of interaction. For instance, short-range, non-bonded interactions can be described through a term of the form:

$$V^{NB}(\mathbf{R}) = \sum_{i}\sum_{j>i} v(r_{ij}), \tag{11}$$

where the double sum runs over all different pairs of atoms and $v(r_{ij})$ is the ***pair potential***, representing the contribution of the individual interaction of atom i with atom j. A typical choice for the pair potential is the Buckingham form:

$$v(r) = Ae^{-r/\rho} - C / r^6, \tag{12}$$

where the first term represents the effect of the short-range repulsion due to the overlap of electronic clouds at short interatomic distances, and the second term represents the attraction between i and j due to London dispersion forces arising from fluctuating induced dipoles. It can be immediately noticed that, through an appropriate choice of the potential parameters A, ρ, and C, one can approximately incorporate average electronic effects (such as Pauli repulsion and dispersion) in the potential, without explicitly including the instantaneous electronic degrees of freedom in the model. The Buckingham pair potential (complemented by long-range, electrostatic interactions between the charges of each ion) can be successfully used to model ionic systems, and also partially ionic systems containing strong covalent bonds, such as silicates (26). In this case the total potential function will be:

$$V(\mathbf{R}) = \sum_i \sum_{j>i} A_{ij} e^{-r_{ij}/\rho_{ij}} - C_{ij} / r_{ij}^6 + \frac{q_i q_j}{4\pi\varepsilon_0 r_{ij}} \tag{13}$$

where q_i is the charge of ion i and the last term is the Coulomb electrostatic interaction.

For systems dominated by covalent interactions, such as organic species and biomolecules, many force fields complement the non-bonded interaction terms with additional terms, more suited to represent the individual bonds and functional groups. In this case, the potential function can be made dependent not only of the bonding distance between pairs of atoms, but also on the bending angle between groups of three atoms and on the dihedral (torsion) angle involving chains of four atoms:

$$V^{\text{bonded}}(\mathbf{R}) = \sum_{\substack{\text{bonds} \\ i-j}} k_{ij}(r_{ij} - r_{ij}^0)^2 + \sum_{\substack{\text{angles} \\ i-j-k}} k_{ijk}(\theta_{ijk} - \theta_{ijk}^0)^2 + \sum_{\substack{\text{dihedrals} \\ i-j-k-l}} k_{ijkl}(\phi_{ijkl} - \varphi_{ijkl}^0)^2 \tag{14}$$

where the superscript "0" labels the reference values of distances and angles, which do not necessarily correspond to typical values found in reference structures: in fact, it is the combination of the effects of all the different terms of the potential, rather than each individual term, which ensures that reliable models are obtained. Note also that many other mathematical forms can be used for each term in eq. 14, where the common harmonic form has been employed for ease of notation.

3.2. Potential Parameters and Transferability

Having selected a suitable analytical form, the selection of appropriate parameters is the most critical task in the building a model potential for MD simulations. Because each individual interaction term in eqs. 13 and 14 requires specific parameters, which are different for bonds, angles, dihedral, van der Waals, and electrostatic interactions involving different pairs or groups of atoms, it is easy to see that the number of potential parameters needed to start the simulation can be quite substantial, for multicomponent systems. Different strategies can be followed to obtain the set of parameters: one can variationally identify the best combination of parameters which, when used in the potential, yields a model which reproduces experimental or ab-initio properties of systems related to the one under study (27). For instance, cell parameters, elastic constants and phonon frequencies of crystalline structures, obtained from experimental data or from ab-initio calculations, can all be used to parameterize a potential aimed at modeling solid, amorphous and liquid phases. Alternatively, as mentioned before, one can fit the force field to the potential energy surface calculated for a relatively large set of significant configurations: modern approaches involve the use of artificial neural networks to perform the highly complex task of identify the potential parameters linking every input configuration to the corresponding ab-initio energy (28). Moreover, the recent *force-matching* methods aim at producing a potential which directly reproduces ab-initio *forces*, rather than energies, calculated for several reference configurations (29).

A critical issue is the *transferability* of the potential: how well a forcefield developed to reproduce the structure of crystalline phases of silica, for example, will perform in the simulation of the corresponding liquid? Or, likewise, can the same potential be used to model liquid water and ice, and can we model their surface as well as the bulk with the same potential? Analogously, there is no guarantee that, if a potential can be used to successfully predict the structural properties of a system, it will also provide a reliable prediction of vibrational and other dynamical properties. These intrinsic limitations all arise from the predefined nature of an empirical forcefield and the necessarily limited range of systems and properties which have been used for its parameterization. While hard to predict, the transferability of a potential depends to a great extent on the degree of similarity of the (short-range) atomic environments in the fitting datasets and in the target system: for instance, it has been recently shown that amorphous silica (a-SiO_2) can be modeled using a potential originally fitted to crystalline quartz, whose basic building block (the SiO_4 tetrahedron) is the same as a-SiO_2 (23).

3.3. Atomic Charges and Polarizability

Amongst all the potential parameters, a particularly critical choice in calculations employing empirical potentials involves the atomic charges q_i, which affect the calculation of long-range electrostatic forces: whereas in some cases it is possible to adopt formal ionic charges, many standard forcefields employ partial (fractional) charges, for instance obtained from quantum mechanical calculations, which may provide a more accurate representation of the actual charge *on average* surrounding an atom in a partially covalent structure. However, because – as mentioned above in relation to the reference values of distances and angles – the performance of a potential ultimately only depends on the balance between the different (short-range, electrostatic, bonded, etc.) terms, rather than on the physical significance of each individual term or parameter, accurate potentials using full atomic charges have been developed for ionic and partially covalent systems (23).

A more subtle issue is the use of *fixed* charges in the potential: this is equivalent to assuming that the charge q_i of atom i can be approximated by the average of the *local* charge in all the different coordination and bonding environments experienced by i in the fitting dataset, as this averaging will be the effective result of the fitting procedure. This same mean charge will then be used in the MD simulation, regardless of the specific local structure around atom i. While this approximation may work well for systems and atomic species whose charge distribution is relatively rigid, many systems contain highly polarizable species, whose charge distribution will depend strongly on the *local* electric field, which can change often in the course of the dynamic run. These systems are better described by complementing the potential with additional terms that take into account the local electric field and the effect on the charge distribution of polarizable species, such as oxide ions. The discussion of how to incorporate polarizability in an empirical potential is beyond the general scope of this chapter, and the reader is referred to one of the many excellent reviews available on the subject (30). We only mention that the inclusion of polarization effects in the potential has been shown to be essential to model amorphous materials, ionic solutions and biomolecules, and it is recommended that polarization terms are incorporated in the force field whenever possible (31). It should also be remarked that electronic polarization is naturally accounted for in AIMD simulations.

4. Ab-Initio MD: The Car–Parrinello Method

All issues related to the transferability of the potential are removed if the interatomic forces are calculated “on-the-fly” from first-principles, that is through a rigorous quantum-mechanical approach for each new configuration visited along the MD trajectory. This is the spirit of ab-initio MD (AIMD): whereas the nuclei still move following the classical equations of motion 18.1, their motion is now controlled by highly accurate ab-initio forces. Because the calculation of ab-initio forces involves an explicit optimization of the electronic structure at each time step, this enormously increases the computational requirements of AIMD. In 1985, Car and Parrinello devised a method to advance the nuclear positions using ab-initio forces, but without the need of repeating an explicit, self-consistent electronic minimization at each time step (32). This remarkable goal was achieved by introducing the electrons as additional (fictitious) *dynamical* degrees of freedom, which led to a set of coupled differential equations for the (real) motion of nuclei and that (fictitious) of the electrons. The electronic degrees of freedom need to be explicitly minimized to the ground state only for the starting configuration; thereafter, the coupled ion-electron dynamics ensures that electrons remain close to the ground state for each new nuclear configuration; this is achieved by assigning a small fictitious mass μ to the electrons, such that their high-frequency oscillatory motion allows them to respond almost instantaneously to the motion of the slower nuclei (33). Their small kinetic energy is such that electrons are always slightly off the ground state during the dynamics, and therefore the *instantaneous* forces on the ions $f_i(t)$ can show small and oscillatory deviations from the true forces, but the fast oscillation around *exact* average forces ensures that the CP dynamics is accurate (30). Obviously, if the electrons drifted away from the ground state to a significant extent, the average ionic forces and the nuclear dynamics would no longer be accurate. The success of the CP method relies on preserving an adiabatic separation between the "hot" nuclei, which move at a certain temperature, and the "cold" electrons, which move with a very low kinetic energy. The equipartition principle would in principle tend to clear this metastable state, through a continuous flow of energy from the hot to the cold subsystem, ultimately matching the temperatures of the two subsystems and thus shifting the electrons away from the ground state. However, in CPMD simulations this undesired energy transfer is inefficient, due to the mechanical separation between the two subsystems with very different masses: by using a small fictitious mass for the electronic degrees of freedom, the lowest frequency in the electronic spectrum, ω_{min}, is made much larger than the highest frequency in the ionic (phonon) spectrum, and no energy transfer is normally observed during the relatively short timescale of a CPMD simulation (10).

In order to integrate the high-frequency electronic motion, one has to use a time step which is about an order of magnitude smaller than typical time steps used in classical MD, for instant 0.15 fs in CPMD vs. 1–10 fs in classical MD. As the fictitious electronic mass μ is an arbitrary parameter, one can increase it as much as possible to reduce ω_{min} and use the largest possible time step, but care must be taken not to shifit ω_{min} within the range of the ionic frequencies. Typical values of μ in CPMD are between 300 and 1,000 atomic units, depending on the system under study.

Car–Parrinello MD was originally developed in the framework of solid-state physics, and is therefore naturally suited to model periodic systems: however, the ubiquitous use of PBC in molecular dynamics is such that the CPMD framework can be applied successfully to study many different systems, periodic and aperiodic, ranging from active sites of enzymes, to biomaterials, to liquids and solutions, to amorphous materials. The parameter-free nature of the method poses virtually no limits to the range of systems that can be simulated, as long as nuclear quantum effects are negligible. In fact, the use of *classical* equations of motion to advance the nuclei using quantum-mechanical forces is not fully justified for systems and conditions where quantum effects are important, such as low temperatures and light atoms (note that the same consideration applies to classical MD simulations). Another difficult case for CPMD is represented by metallic systems: a small or zero energy gap between occupied and empty electronic states always leads to an irreversible energy transfer into the electronic degrees of freedom, which are shifted away from the ground state causing the method to fail, unless the energy transfer can be counterbalanced, for instance using a thermostat which keeps the electrons "cold" (9).

Possibly the most remarkable feature of CPMD simulations is that, besides providing a very accurate and unbiased view of complex systems, the system is actually free to evolve following spontaneous transformations: in other words, whereas in classical MD the accessible bonding patterns and structural environments are essentially preset from the beginning, in CPMD spontaneous events such as chemical reactions and processes involving formation and breaking of chemical bonds, changing the interatomic connectivity, can occur, exactly as they would do in a "real" system. The only limitation for this to happen is the short timescale of the CPMD simulations: only processes involving a small or no energy barrier can be observed spontaneously during a standard CPMD run. However, the extensions to the MD method developed to enhance the sampling of activated processes, mentioned before, can also be easily implemented in CP simulations.

5. Some Applications to Inorganic Biomaterials

In order to illustrate the versatility of MD method, and the interplay between classical and ab-initio simulations, we discuss here some recent applications in the field of biomaterials. The bioactivity of Hench-type bioglasses can be rationalized, at least to some extent, through the connectivity of the glass network, which can be calculated from the distribution of Q^n species, where Q^n is a SiO_4 or PO_4 tetrahedron linked to n other tetrahedra through n shared oxygen atoms (1). Whereas the Q^n distribution of glasses can be estimated only *indirectly* from NMR and vibrational data, it can be directly accessed by postprocessing an MD trajectory.

In this case, an adequate statistics on medium-range intertetrahedral properties such as the Q distribution can be attained using relatively large ($N > 1{,}000$ atoms) system sizes, only affordable with classical MD. The Q^n distribution and other medium-range structural features of several Hench bioglasses has been recently calculated using classical MD simulations (14, 34): the results of the calculations helped to interpret the composition–bioactivity relationships of these materials by providing direct access to atomistic structural data not easily available otherwise. Whereas CPMD simulations cannot access the large system sizes needed to explore medium-range features, they represent an excellent tool to obtain *local* properties of glasses and liquids, using supercells containing around 100 atoms (15, 35): the time-correlation analysis of a CPMD trajectory has recently allowed the assignment of bands in the vibrational spectrum of bulk 45S5 Bioglass® (17). Because the mechanism by which biomaterial implants integrate with living tissues involves the partial or complete dissolution of the material in a biological, aqueous medium, the focus of MD simulations of bioactive glasses has recently shifted toward the interface of the biomaterial with water. As discussed before, CPMD is the natural choice when dealing with chemisorption, hydrolysis and in general chemistry at surfaces, where the frequent and substantial electronic rearrangements occurring at room temperature cannot be modeled with a classical interatomic potential. CPMD simulations of surface models of 45S5 Bioglass® interacting with water have recently permitted the characterization of surface sites in terms of their relative strength, and highlighted the physicochemical processes in the first stages of the glass dissolution (11, 36). Using the accurate and highly specific information gathered in these CPMD simulations, classical MD runs of large slab models were then carried out to compare the surface structure of different compositions, whose different bioactivity was rationalized on the basis of the availability of the reactive surface sites characterized in the CPMD runs (37).

These recent examples highlight the power and efficacy of molecular dynamics to explore complex systems such as biomaterials; in particular, classical and ab-initio MD should not be considered alternative, but *complementary* computational approaches, whose combination is often necessary to reveal new features of the material and its activity in realistic conditions.

Acknowledgments

Financial support from the UK's Royal Society (University Research Fellowship) is gratefully acknowledged.

References

1. Tilocca, A. (2009) Review: Structural Models of Bioactive Glasses from Molecular Dynamics simulations *Proc. Royal Soc. A* **465**, 1003–27.
2. Hench, L. L. (1998) Bioceramics *J. Am. Ceram. Soc.* **81**, 1705–28.
3. Hertz, A., and Bruce, I. J. (2007) Inorganic materials for bone repair or replacement applications *Nanomedicine* **2**, 899–918.
4. Cerruti, M., Greenspan, D., and Powers, K. (2005) Effect of pH and ionic strength on the reactivity of Bioglass 45S5. *Biomaterials* **26**, 1665–74.
5. Sanbonmatsu, K. Y.; Tung C.-S. (2007) High performance computing in biology: Multimillion atom simulations of nanoscale systems *J. Struc. Biol.* **157**, 470–80.
6. Schulz, R., Lindner, B., Petridis, L., and Smith, J. C. (2009) Scaling of Multimillion-Atom Biological Molecular Dynamics Simulation on a Petascale Supercomputer *J. Chem. Theory Comput.* **5**, 2798–2808.
7. Allen, M. P., and Tildesley, D. J. (1989) *Computer Simulations of Liquids* Oxford University Press.
8. Frenkel, D., and Smit, B. J. (2001) *Understanding Molecular Simulation: From Algorithms to Applications* Academic Press, San Diego, California.
9. Freddolino, P. L., Arkhipov, A. S., Larson, S. B., McPherson, A., and Schulten, K. (2006) Molecular dynamics simulations of the complete satellite tobacco mosaic virus. *Structure* **14**, 437–449.
10. Marx, D., and Hütter, J. (2000) Ab initio molecular dynamics: theory and implementation. In *Modern methods and algorithms of quantum chemistry, vol. 1* (ed. J. Grotendorst). NIC Series, pp. 301–449. Julich, Germany: John von Neumann Institute for Computing.
11. Kanai, Y., Tilocca, A., Selloni, A., and Car, R. (2004) First-principles string molecular dynamics: An efficient approach for finding chemical reaction pathways *J. Chem. Phys.* **121**, 3359–67.
12. Tilocca, A., and Cormack, A. N., Exploring the Surface of Bioactive Glasses: Water Adsorption and Reactivity *J. Phys. Chem. C* **112**, 11936–45.
13. Chandler, D. (1987) *Introduction to Modern Statistical Mechanics.* Oxford University Press.
14. Moss, R. M., Pickup, D. M., Ahmed, I., Knowles, J. C., Smith, M. E., Newport, R. J. (2008) Structural Characteristics of Antibacterial Bioresorbable Phosphate Glass Adv. Fun. Mater. **18**, 634–39.
15. Tilocca, A., Cormack, A. N., de Leeuw, N. H. (2007) The structure of bioactive silicate glasses: new insight from molecular dynamics simulations *Chem. Mater.* **19**, 95–103.
16. Tilocca, A. (2007) Structure and dynamics of bioactive phosphosilicate glasses and melts from *ab initio* molecular dynamics simulations *Phys. Rev. B* **76**, 224202.
17. Fois, E., Gamba, A., and Tilocca, A. (2002) Structure and dynamics of the flexible triple helix of water inside VPI-5 molecular sieves *J. Phys. Chem. B* **106**, 4806–4812.
18. Tilocca, A., and de Leeuw, N. H, (2006) Ab-initio molecular dynamics study of 45S5 bioactive silicate glass *J. Phys. Chem. B* **110**, 25810–16.
19. Rahman, A. (1964) Correlations in the Motion of Atoms in Liquid Argon *Phys. Rev.* **136**, 405–11.
20. Smith, W., Forester, T. R., Greaves, G. N., Hayter, S., and Gillan, M. J. (1997) Molecular dynamics simulations of alkali-metal diffusion in alkali-metal disilicate glasses *J. Mater. Chem.* 7, 331–36.
21. Ivanov, I, and Klein, M. L. (2005) Dynamical Flexibility and Proton Transfer in the Arginase Active Site Probed by ab Initio Molecular Dynamics *J. Am. Chem. Soc.* **127**, 4010–20.
22. Molteni, C., and Parrinello, M. (1998) Glucose in Aqueous Solution by First Principles Molecular Dynamics *J. Am. Chem. Soc.* **120**, 2168–71.
23. Minary, P., and Tuckerman, M. E. (2004) Reaction Pathway of the (4 + 2) Diels–Alder Adduct Formation on Si(100)-2×1 *J. Am. Chem. Soc.* **126**, 13920–21.
24. Meyer, A., Horbach, J., Kob, W., Kargl, F., and Schober, H. (2004) Channel Formation and Intermediate Range Order in Sodium Silicate Melts and Glasses *Phys. Rev. Lett.* **93**, 027801.
25. Laio, A., and Parrinello, M. (2002) Escaping free energy minima *Proc Natl. Acad. Sci. USA*, **99**, 12562–66.
26. Tilocca, A., de Leeuw, N. H., and Cormack, A. N. (2006) Shell-model molecular dynamics calculations of modified silicate glasses *Phys. Rev. B* **73**, 104209.
27. Gale, J. D., and Rohl, A. L. The General Utility Lattice Program (2003) *Mol. Simul.*, **29**, 291.
28. Pukrittayakamee, A., Malshe, M., Hagan,M., Raff, L. M., Narulkar,R., Bukkapatnum, S., and Komanduri, R. (2009) Simultaneous fitting of a potential-energy surface and its corresponding force fields using feedforward neural networks *J. Chem. Phys.* **130**, 134101.
29. Izvekov S., Parrinello M., Burnham C. J., and Voth G. A. (2004) Effective force fields for condensed phase systems from ab initio molecular dynamics simulation: A new method for force-matching *J. Chem. Phys.* **120** 10896–10913.

30. Lopes, P. E., Roux, B., and MacKerell Jr., A. D. (2009) Molecular modeling and dynamics studies with explicit inclusion of electronic polarizability: theory and applications *Theor. Chem. Acc.* **124**, 11–28.
31. Tilocca, A. (2008) Short- and medium-range structure of multicomponent bioactive glasses and melts: An assessment of the performances of shell-model and rigid-ion potentials *J. Chem. Phys.* **129**, 084504.
32. Car, R. and Parrinello, M. (1985) Unified Approach For Molecular Dynamics and Density Functional Theory *Phys. Rev. Lett.* **55**, 2471.
33. Pastore, G., Smargiassi, E., and Buda, F. (1991) Theory of ab-initio molecular dynamics calculations *Phys. Rev. A* **44**, 6334–47.
34. Tilocca, A., and Cormack, A. N. (2007) Structural effects of phosphorus inclusion in bioactive silicate glasses *J. Phys. Chem. B* **111**, 14256–64.
35. Donadio, M., Bernasconi, M., and Tassone, F. (2004) Photoelasticity of sodium silicate glass from first principles *Phys. Rev. B* **70**, 214205.
36. Tilocca, A., and Cormack, A. N. (2009) Modeling the water-bioglass interface by ab-initio Molecular Dynamics simulations *ACS Applied Materials & Interfaces* **1**, 1324–33.
37. Tilocca, A., and Cormack, A. N. (2010) Surface signatures of bioactivity: MD simulations of 45S and 65S silicate glasses *Langmuir* **26**, 545–51.

Chapter 19

Regenerative Nanomedicine: Ethical, Legal, and Social Issues

Linda MacDonald Glenn and Jeanann S. Boyce

Abstract

Advances in regenerative nanomedicine raise a host of ethical, legal, and social questions that healthcare providers and scientists will need to consider. These questions and concerns include definitions, appropriate applications, dual use, potential risks, regulations, and access. In this chapter, we provide an overview of the questions and concerns and recommend proactive consideration and solutions.

Key words: Nanomedicine, Regenerative medicine, Regenerative nanomedicine, Ethical, legal, and social issues, ELSI, Therapy, Enhancement, Alterations, Health care disparities

If there were no regeneration there could be no life. If everything regenerated there would be no death. All organisms exist between these two extremes. – Richard J. Goss, 1992.

1. Introduction

Regenerative medicine is at the forefront of the evolution of medical therapies and treatments; the promise of enabling scientists to regenerate damaged tissues and organs in vitro or in vivo has the potential of developing therapies for previously untreatable diseases and conditions, combating rising health care costs, solving the shortage of transplantable organs, and offering improved quality of life for an aging population (1). Advances in material design at the nanoscale, merged with new insights in the fields of molecular and cell biology, have allowed for the burgeoning of a new cross-disciplinary field: regenerative nanomedicine. It has the capacity to reconstitute significant aspects of medicine, medical practice, and health care systems. In so doing, regenerative medicine is giving rise to a host of ethical issues associated with adaptation

Melba Navarro and Josep A. Planell (eds.), *Nanotechnology in Regenerative Medicine: Methods and Protocols*, Methods in Molecular Biology, vol. 811, DOI 10.1007/978-1-61779-388-2_19, © Springer Science+Business Media, LLC 2012

and transition (2). As this field continues to develop, this chapter will examine the potential ethical, legal, and social implications as this realm evolves.

2. Definitions

2.1. Regenerative Medicine

A variety of definitions of the term "regenerative medicine" has been proffered by different institutions: In the USA, the National Institute of Health (NIH) uses the terms regenerative medicine and tissue engineering interchangeably: "Regenerative medicine/ tissue engineering is a rapidly growing multidisciplinary field involving the life, physical and engineering sciences that seeks to develop functional cell, tissue, and organ substitutes to repair, replace or enhance biological function that has been lost due to congenital abnormalities, injury, disease, or aging" (3).

However, other researchers have proposed a more comprehensive definition:

Regenerative medicine is an interdisciplinary field of research and clinical applications focused on the repair, replacement or regeneration of cells, tissues or organs to restore impaired function resulting from any cause, including congenital defects, disease, trauma, and aging. It uses a combination of several converging technological approaches, both existing and newly emerging, that moves it beyond traditional transplantation and replacement therapies. The approaches often simulate and support the body's own self-healing capacity. These approaches may include, but are not limited to, the use of soluble molecules, gene therapy, stem and progenitor cell therapy, tissue engineering and the reprogramming of cell and tissue types (4).

The NIH definition is succinct, while the Daar and Greenwood definition is more detailed; however, both definitions are not incompatible with one another. The difficulty arises when we attempt to define regenerative nanomedicine.

2.2. Nanotechnology in Regenerative Medicine/ Nanomedicine

The term *nanotechnology* has been hard to define, partly because the field is multidisciplinary and different disciplines use the term in myriad ways. Some researchers claim that use of natural occurring nanoparticles should be treated different than their synthetically prepared analogs, but exposure to naturally occurring particles can create significant health problems, problems areas which are often regulated by laws and government agencies (e.g., coal dust, talc, second-hand smoke). As of this writing, the ISO (International Organization for Standards) has no ratified definition of nanotechnology.

In the USA, the National Nanotechnology Initiative (NNI) declared that "Nanotechnology is the understanding and control of matter at dimensions of roughly 1–100 nm, where unique phenomena enable novel applications"(5).

How small is a nanometer? To give a meaningful, comparative illustration of a nanometer (one billionth of a meter), a human hair measures between 50,000 and 100,000 nm wide. A red blood cell measures between 5,000 and 8,000 nm in diameter; a DNA molecule is about 2.5-nm wide. Ten atoms of hydrogen, side-by-side, equal 1 nm (6).

Despite the lack of consensus of the definition of nanotechnology, nanomedicine has been broadly defined as the approach of science and engineering at the nanometer scale toward biomedical applications (7).

2.3. Regenerative Nanomedicine

By forging the definitions of the aforementioned fields, it would seem that an appropriate and useful definition of regenerative nanomedicine would be: "the interdisciplinary field of research and clinical applications focused on the repair, replacement, or regeneration of cells, tissues or organs to restore, replace, or enhance biological function that has been lost due to congenital abnormalities, injury, disease, or aging, utilizing nanoscale technology or materials." This definition would be broad enough so it could be a subspecialty of regenerative medicine, yet avoid (for the time being) the difficulties associated with defining and regulating nanotechnologies in general. With the USA, medical drugs and devices are regulated by the Food and Drug Agency (FDA) and any research or clinical application of regenerative nanomedicine would be subject to the FDA's jurisdiction. While the FDA has not established its own formal definition of "nanotechnology," to facilitate the regulation of nanotechnology products, the Agency has formed a NanoTechnology Interest Group (NTIG), which relies upon the NNI's definition, as set forth above. Some of the legal and ethical difficulties that this lack of definition poses are explored in Subheading 4, below.

3. Applications

While the malleability of nanomaterials allows to them to be a perfect platform for the convergence of other technologies, such as interacting with biological systems at the molecular level, they also have novel electronic, optical, magnetic, and structural properties that cannot be obtained from either individual molecules or bulk materials (5). These novel properties represent a double edged sword; the emergent properties of the nanoparticles can behave in an unexpected way; although nanoparticles generally act in accordance with the laws of classical physics they can, at times, exhibit principles of quantum physics. Quantum physics, unlike classical physics, is undeterministic; that is, outcomes are measured probabilistically, rather than with certainty. As nanoscale engineered devices become smaller, this may affect the predictability of outcomes in their use, as a well as a consideration in risk assessment and toxicity screens.

Notwithstanding potential quantum effects, several conditions must be meticulously met and considered in designing nanomaterials or nanoparticles into medical devices. They should (1) be designed to interact with proteins and cells without perturbing their biological activities, (2) maintain their physical properties after the surface conjugation chemistry, and (3) be biocompatible and nontoxic (8, 9). While these are the standards for incorporation into medical uses or devices, these standards do not yet take into account the further downstream effects of bodily disposal or elimination as medical waste and environmental impact, which are discussed as a potential ethical issue later in this chapter.

Three strategies are being pursued in the development of a regenerative medicine:

1. The replacement of damaged tissue by transplanting cell suspensions or aggregates.
2. The implantation of bioartificial tissues or organs constructed in the laboratory to replace the natural ones ("tissue engineering").
3. The induction of regeneration in situ (10).

Although these strategies have yet to achieve broad success, here are a few examples of some bench to bedside projects in the works utilizing nanomaterials:

- In replacing damaged heart tissue, stem cells have been injected to regenerate damaged myocardium (11). However, some of the difficulties of this approach have been differentiation of the mesenchymal stem cells (MSC) into cardio myocytes, as well as, targeting, tracking, and retaining of cells to the damaged area to viability and function. Recent research has shown that stem cell nanotechnology is developing *toward imaging, active tracing, and controllable regulation of proliferation and differentiation of stem cells* (12). Note: Keep in mind that the biocompatibility of carbon nanotubes (CNTs) in cardiac muscle has already been established (13).
- Nanoscale magnet particles have been used to bring magnetically tagged endothelial progenitor cells – a type of stem cell shown to be important in vascular healing processes – to sites of cardiac damage and vascular injury, stimulating repair and regenerative growth (14).

- Recent and ongoing efforts to engineer whole organs involve natural extracellular matrix molecules, commonly utilized in vitro, being used as effective, natural, acellular allografts, and being integrated into nanoscale scaffolds and matrices with programmable responsiveness (15).
- Coined as "nanosuturing," researchers in Boston have been able to close wounds utilizing a light activated dye (Rose Bengal – a stain commonly used to highlight corneal lesions) plus a laser to "knit" layers of collagen together. The wound is painted with the dye, and then the laser illumination serves to activate to the dye, helping to transfer electrons between the dye molecules and the collagen, causing the molecular chains of the collagen to chemically bond or link together nearly instantaneously. This technique could be used for skin closures, eye surgeries, and tendon, blood vessel and nerve repairs (16, 17).
- To repair spinal cord injuries, injured optical pathways, and other central nervous system injuries, researchers have created "nanoneural knitting," where transplanted neural progenitor cells have been cultured on self-assembled nanofiber scaffold (SAPNS), creating a three-dimension environmental "bridge" that allows for living cells to migrate (18).
- In yet another tool in the arsenal for erectile dysfunction, researchers have announced a nanoparticle cream that will allow delivery of nitric oxide locally, thereby avoiding systemic side effects (19). Scientists hope this delivery system will be of use in treatment of female sexual arousal disorder as well, since there are currently very few, if any, effective treatments.

4. Ethical, Legal, and Social Concerns

4.1. Health Risks and Regulatory Issues

In general, because engineered nanoparticles are still such a relatively new technology, there are many questions about occupational, consumer, environmental safety and health. While there have been some attempts to assess the risks of engineered nanoparticles (20), part of the difficulty, which was mentioned in Subheading 2, lay in the lack of an agreed upon definition; without a definition, nanotechnology cannot be regulated. Although the National Nanotechnology Initiative (NNI) set forth a general definition, it has spread the responsibility of regulating nanotechnology between 25 different federal agencies. One might think that because medical devices and drugs go through clinical trials to test effectiveness and toxicity and must be approved by the FDA, that the products/inventions of regenerative nanomedicine may escape the concerns of environmental risks. However, as example of potential difficulty, the FDA must determine the medical product falls within the category of "drug" or "device" or "cosmetic"; depending on the classification, treatment by the FDA will be very different. Depending on how medical nanodevices are classified, this distinction could have significant consequences.

Under FDA regulations, a "drug" is defined as an article intended for use in diagnoses, cure, mitigation, treatment, or prevention of disease or an article, other than food, intended to affect the structure or function of the body (21). Applicants for new "drugs" must demonstrate the product's safety and efficacy; the FDA's multi-step approval process requires drug manufacturers to submit studies to the agency to evaluate the safety and efficacy of the product and can take several years (22). These studies include preclinical research, synthesis and purification, as well as animal testing. Additionally, clinical testing on human subjects in randomized clinical trials (RCTs) is part of the process, which requires review and approval of Institutional Review Boards (IRBs) (22).

A "medical device" is defined, in part, as any health care product that does not achieve its primary intended purposes by chemical action or by being metabolized (21 CFR Part 812, 814). Unlike drugs, the approval process for premarket approval of a medical "device" is less rigorous; the FDA requires that the manufacturer provide "reasonable assurance that the device is safe and effective" for its intended use. By contrast, the time-consuming three-phased system of RCTs is required for drug approval the FDA will accept any "valid scientific evidence" regarding the safety of the device, including partially or noncontrolled studies or any other reports of its safety. Unlike the standards for drug approval, there is no requirement that it be shown to be more effective to currently available alternatives (22).

A "cosmetic" is defined by its intended use, as in an "article[s] intended to be rubbed, poured, sprinkled, or sprayed on, introduced into, or otherwise applied to the human body … for cleansing, beautifying, promoting attractiveness, or altering the appearance" (23). Among some of the products included in this definition are skin moisturizers, perfumes, lipsticks, fingernail polishes, eye and facial makeup preparations, shampoos, toothpastes, and deodorants, as well as any material intended for use as a component of a cosmetic product. Currently, the FDA does not have a premarket approval system for cosmetic products or ingredients.

The difficulty is that, as mentioned previously, nanoscale materials can exhibit chemical, mechanical, electrical, optical, and magnetic properties that are not present at macro levels; so it becomes impossible to distinguish "chemical" from "electrical" or "mechanical." For example, picture a mouthwash solution with nanobots that can scrub away plaque, prevent gingivitis, and whiten teeth. This product may appear to be a "device" for the purposes of FDA regulation because of the scrubbing action of the nanobots; however, if touted as primarily a cosmetic product because of its whitening properties, then no premarket approval is required.

In short, the FDA standards and approval process for the approval of drugs is "significantly more burdensome, time-consuming and scrupulous than that for devices" (22). Products designated as "drugs" undergo far more careful preapproval review, delaying public access to the remedies. By contrast, the designation of "device" means that the product will reach the bedside for clinical use in a much shorter period with less scrupulous review. And if the product is touted as "cosmetic," no premarket approval is required. To add another layer of complexity, there is no "bright line" between the presence of nanoparticles in products and an understanding and appreciation of possible uses of nanotechnology in medicine (24).

In general, the current formal FDA approval process provides some assurance that nanomedical products are and will be evaluated for safety and efficacy before being marketed and distributed for human use. Researchers and manufacturers would be wise to emphasize informed consent and avoid a situation that could set gene-therapy back substantially, as in the Jesse Gelsinger case (25). At least one bioethicist has recommended post-approval registries (22) for cohort studies and long term observational studies are needed to ensure long term safety.

4.2. Environmental and Epidemiological Concerns

Because nanomaterials will be used for many things other than nanomedicine, there has been much concern expressed about effects on the environment and nature (26). As mentioned previously, the disagreement of a precise definition, and the unique, sometimes unexpected properties of nanomaterials has led to difficulties in risk assessment. While there have studies initiated to study the impact of nanomaterials in the environment nature, only time will tell as to the long-term impact. Similarly, epidemiological studies have been proposed only recently (27). Once a classification schemata has been agreed upon, the next logical step would be to have researchers, industry, physicians, lawyers and ethicists meet to establish principles, guidelines, and recommendations similar to those set up at the Asilomar Conference on Recombinant DNA (28–30).

In terms of nanomedicine, manufacturers, clinical investigators and health care providers will likely face a scenario similar to a current one, where pharmaceutical residues and hormones are ending up in drinking water supplies and causing harm (31–33). Several possible mitigation program and remedies have been proposed in the last year, including a hospital and pharmacy monitoring system, a bioremediation tank for medical waste water, and incineration, but such practices are far from being widespread. Although epidemiological studies will likely result in recommendation and adoption of federal regulations of medical pharmaceutical waste in the next decade, manufacturers, clinical investigators and health care providers would be well advised to consider incorporation of mitigational and remedial programs into their standard best practices.

4.3. Sources of Stem Cells

Some of the most promising research in regenerative is based upon the use of stem cells in conjunction with nanotechnology; the ability to engineer whole organs as replacements for allografts and xenografts is the Holy Grail of regenerative medicine (15).

The primary ethical conundrum in using stem cells is the source from which they are obtained. Adult stem cells, which have been identified in many organs and tissues, including brain, bone marrow, peripheral blood, blood vessels, skeletal muscle, skin, teeth, heart, gut, liver, ovarian epithelium, and testis, to do not pose an ethical challenge; however, even though they are pluripotent, they are somewhat limited in their capacity to differentiate (34). Embryonic stem cells, by contrast, are totipotent, able to differentiate into any cell in the body. Embryonic stem cells are controversial because of differing beliefs in the moral status of embryos. Those opposed to human Embryonic Stem Cell (hESC) research believe that the fertilized egg has the moral rights of personhood, and that the destruction of blastocysts devalues human life and dignity (35, 36). Of course, moral and philosophical arguments have been made in favor of embryonic stem cell research, too; the argument being that embryos are not persons, and that the moral rights of sentient persons suffering from diseases that could be helped by hESC research trumps any moral right the blastocyst may have (37, 38). In the center of the spectrum of beliefs are people who do not object to the use of existing (or future) frozen blastocysts that are destined to be destroyed, rather used for reproductive purposes, to derive new hESCs for research, but they do not want new blastocysts created specifically for experimentation. Currently, in an attempt to follow this middle path, the policy in both the USA and Canada is that the federal government will only pay for research conducted with stem cells harvested from leftover embryos developed in vitro fertilization procedures, and only with the donors' consent (34).

To avoid this ethical controversy, researchers have found ways of producing what are being called "induced pluripotent cells" (iPCs). These are stem cells that they hope will demonstrate the same key properties of regeneration and unrestricted differentiation that human embryonic stem cells (hESCs) possess, but which are derived from skin cells not from embryos. In other words, these scientists have succeeded in reprogramming skin cells to behave like hESCs (39). iPC research is attractive for several reasons: first, it does not require the gathering and use of human eggs; secondly, it is genuinely gender-neutral because it does not impose a disproportionate burden on one gender over the other, in risks and burdens of egg collection/donation (39). Additional research is needed before iPCs are to be in widespread use, partly because viruses are currently used to introduce reprogramming factors into adult cells, and this process must be carefully controlled and tested before the technique can lead to useful treatments for humans.

Additionally, the possibility of cross-species grafting raises ethical concerns; human embryonic cells can be grafted to the embryos of other species. Picture, for example, "an experiment in which the whole brain of a spontaneously aborted human embryo is grafted in place of the brain of an embryonic nonhuman primate" (10). Arguably, such as creature, if allowed to develop and brought to term, could be recognized as human and thus raise questions about their moral and legal status (40). Most countries in Europe and Canada have banned the production of such chimeras and there have been federal laws proposed in the USA banning the creation of chimeras.

4.4. Therapy Versus Enhancement Versus Alteration

Regenerative medicine, by the NIH definition, includes both therapy and enhancement of biological function "that has been lost due to congenital abnormalities, injury, disease, or aging" (3). Among the most interesting and controversial research studies within the field of regenerative nanomedicine involve nanoneural interfaces: the potential not only for brain repair or axon regeneration, but the merger of artificial intelligence with human capabilities (41, 42). Futurist and scientist Ray Kurzweil predicts within the next 25 years, "the union of human and machine, in which the knowledge and skills embedded in our brains will be combined with the vastly greater capacity, speed, and knowledge-sharing ability of our own creations;" and that the merger between human and machine will be so complete that distinctions will be impossible or rendered meaningless (43). The ethical, legal, and societal concern is that by merging with technology, we change what it means to be "human" or a "person" and change basic human nature (44, 45). Brain implants that create interfaces between human neural systems and computers will allow for (1) the improvement and augmentation of human capabilities, (2) the advent of "human" immortality through cloning and implantation of bioelectronic chips with the uploaded emotions, memories and knowledge of the source human, and (3) the possibility that humanity may be replaced by the next stage in guided evolution (46).

Similar to the embryonic stem cell debate, there is a spectrum of beliefs. At one end of the spectrum, there are people who argue that technology should be embraced as freeing and transcendent (43). At the other end of the spectrum are those who argue that we have already done harm to ourselves and to the earth and should put the brakes on further technological development (47). In the center of the spectrum are those who recognize that technology is changing/challenging our relationships and that notions of personhood are evolving, and the laws must adapt to recognize the changes (48).

Most ethicists recognize that new technologies present the challenges of weighing and balancing potential benefits vs. potential harms. At least one ethicist argues that distinctions should be made between therapy, enhancement and alteration:

- *Level 1 Therapy.* restoration (partial or complete) of neurological functions through nanodevices. Therapeutic applications could enhance certain traits or abilities ... [A]n example at this level would be neural prosthetics.
- *Level 2 Enhancement.* enhancing mental capacities such as the ability to process more data and faster...[Neural] enhancement...Level 2 is less concerned with patients than with individuals wanting to improve their mental capacities for nonhealth-related purposes.
- *Level 3 Alteration.* altering neurobiological functions. This level is the most controversial since the goal is to transcend biological boundaries through technological means to alter human capacities (species atypical). Here the concern is to perform neural alteration of the human brain using, for instance, nanorobots replacing neurons. This means adding new features to brain functions (brain to brain interface, web access, etc.) through technological means unknown so far (49).

However, making such distinctions is not an easy task; many therapeutic interventions can and are used later on for nontherapeutic purposes; for example, sildenafil citrate (popularly known as Viagra) may have started off as a therapy, but has quickly found a place in society as a drug to enhance romantic encounters. Or Minoxidil, originally used to treat high blood pressure, now advertised as the drug of choice for thinning hair or hair loss in men and women (45). Such would be the envisioned mouthwash mentioned in Section 4.1 that brightens and whitens teeth.

Even though enhancements and alterations are not likely to be banned, most ethicists in the field advocate a thoughtful, cautious approach, with discourse recognizing public input as well as political considerations. Even with extensive analysis, there are unintended consequences that may occur. Rather than a reactive approach to potential problems, a new creative proactive approach – "a thorough reflection on the implications of the brain computer interface" is needed to address ethical and moral conundrums (49).

4.5. Access to Health Care and Health Care Disparities

With a worldwide aging population, regenerative medicine will be in peak demand in the decades to come. Within the USA, health disparities are well documented in minority populations such as African Americans, Native Americans, Asian Americans, and Latinos (50). When compared to whites, these minority groups have higher incidence of chronic diseases, higher mortality, and poorer health outcomes (50). Although there are many reasons for disparities in health care, such lack of financial resources, health literacy, linguistic barriers and mistrust due to past tragedies (such as the Tuskegee Syphilis experiments or the United Radiation experiments), the primary reason is lack of insurance coverage. Without health insurance, patients are more likely to postpone medical care, more likely to go without needed medical care and more likely to go without prescription medicines; and minority groups in the USA lack insurance coverage at higher rates than whites.

Further access questions arise: "For example, is it appropriate to commit research funding, expertise, personnel, and infrastructure resources on developing and disseminating technologies with enhancement potentials when more than 50 million people are without adequate health care in the USA and hundreds of millions are without even basic health care globally? Are such technologies likely to exacerbate or reduce unjust inequalities? What rights (or obligations) do parents have with respect to adopting such technologies for their children?" (2).

As of the writing of this, the drama of pending health care reform and accessibility is playing out in front of the US Congress. How the issue of health care access and reform plays out will determine who will, or who will, not be able to avail themselves to the modern miracles of regenerative medicine.

5. Recommendations

Progress in regenerative medicine is being made around the world and breakthroughs in regenerative nanomedicine are expected to follow suit. To address the potential ethical issues threatening to slow down medical advances, several steps would be useful and appropriate:

1. A proactive approach including continuing dialog and collaboration among all those involved in the cross disciplinary field of regenerative medicine, scientists, ethicists, lawmakers, economists, futurists, as well members of the public.
2. Establish a comprehensive definition or classification schemata of nanomaterials in a way that is suitable for scientific, regulatory, and policy purposes.
3. Have stakeholders (e.g., researchers, industry, physicians) and lawyers, lawmakers, economists and ethicists meet to establish principles, guidelines, and recommendations similar to those set up at the Asilomar Conference on Recombinant DNA.

4. Come to terms that our creations can have unintended or unforeseen consequences and consider who will decide issues of regulation and liability. Among considerations, should there be international oversight or federal government oversight or will individual jurisdictions be called upon to decide enact statutes or decide on a case-by-case basis?
5. Continue explorations and discussions of the Property – Personhood Continuum, issues of personal identity, and consider whether current law is sufficient or will new laws be needed?

As Abraham Lincoln once said, "The best way to predict your future is to create it." As new developments in regenerative nanomedicine arise, and when both humans and nonhumans receive therapeutic benefits, enhancements, or alterations, it will be up to the stakeholders, as well as lawmakers, policymakers, and courts, to configure guidelines that will serve as an ethical framework for future advances in regenerative nanomedicine.

References

1. Goss, R.J. (1992) The natural history (and mystery) of regeneration. In: Dinsmore CE (ed.). A History of Regeneration Research. Cambridge: Cambridge University Press, 7–23.
2. Sandler, R. (2009) Nanomedicine and nanomedical ethics, *American Journal of Bioethics* **9**(10), 16–17.
3. 2020: A new vision: A future for regenerative medicine, (2009) Department of Health and Human Services. Available at http://www.hhs.gov/reference/newfuture.shtml, last accessed October 23, 2009.
4. Daar, A.S., Greenwood, H.L. (2007) A proposed definition of regenerative medicine. *Journal of Tissue Engineering and Regenerative Medicine*, **1**, 179–84.
5. National Nanotechnology Initiative (NNI). Available at http://www.nano.gov/html/about/home_about.html, http://www.nano.gov/html/facts/whatIsNano.html, http://www.nano.gov/Understanding_Risk_Assessment.pdf, last accessed November 9, 2009.
6. Alpert, S. (2008) Neuroethics and nanoethics: do we risk ethical myopia? *Neuroethics* **1**(1), 55–68. Available at http://www.springerlink.com/content/v6352402k048qq53, last accessed November 6, 2009.
7. Lee, K.B., Solanki, A., Kim, J.D., and Jung, J. (2009) *Nanomedicine: Dynamic Integration of Nanotechnology with Biomedical Science*. In Zhang, M and Xi, N: Nanomedicine, a systems approach, Pan Stanford Publishing Pte Ltd.
8. Freitas, R.A. (1999) Nanomedicine, Vol 1: Basic Capabilities, Landes Bioscience.
9. Mooney, E., Dockery, P., Greiser, U., Murphy, M., Barron V. (2008) Carbon Nanotubes and Mesenchymal Stem Cells: Biocompatibility, Proliferation and Differentiation *Nano Letters* 8(8), 2137–2143.
10. Stocum, D.L. (2009) *Regenerative Biology and Medicine*, Elsevier Press, San Diego, 28
11. *Stem Cell Information*. Bethesda, M.D: National Institutes of Health, U.S. Department f Health and Human Services, November 06, 2009. Available at: http://stemcells.nih.gov/info/basics/basics10, last accessed November 9, 2009.
12. Wang, Z., Ruan, J., Cui, D. (2009) Advances and prospect of nanotechnology in stem cells. *Nanoscale Res Letters*, **4**(7), 593–605.
13. Garibaldi, S., Brunelli, C., Bavastrello, V., Ghigliotti, G., Nicolini, C. (2006) Carbon nanotube biocomptibility with cardiac muscle cells, *Nanotech* **17**, 391–397.
14. Panagiotis, G.K., Lehtolainen, P., Junemann-Ramirez, M., Garcia-Prieto, A., Price, A., Martin, J.F., Gadian, D.G., Pankhurst, A.Q., Lythgoe, M.F. (2009) Magnetic tagging increases delivery of circulating progenitors in vascular injury. *JACC Cardiovascular Interventions*, **2**(8), 794.

15. Traphagen, S., Yelick, P.C. (2009) Reclaiming a natural beauty: whole-organ engineering with natural extracellular materials, *Regenerative Medicine*, **4**(5), 747–758
16. Chang,W.C., Kliot, M., Sretavan, D.W. (2008) Microtechnology and nanotechnology in nerve repair, *Neurol Res.* **30**, 10, 1053–62.
17. Gravitz, L. (2009) *Laser* Show in the Surgical Suite, *Technology Review*, March/April 2009, MIT Press, available at http://www.technologyreview.com/biomedicine/22088/, last accessed October 25, 2009.
18. Guo, J., Su, H., Zeng, Y., Liang, Y.X., Wong, W.M., Ellis-Behnke, R.G., So, K.F., Wu, W. (2007) Reknitting the injured spinal cord by self-assembling peptide nanofiber scaffold, *Nanomedicine*, **3**(4), 311–21.
19. *Science Daily*, Nanoparticle-based Erectile Dysfunction Therapy Shows Promise, Sept. 20, 2009. Available at http://www.sciencedaily.com/releases/2009/09/090918181456.htm, last accessed October 30, 2009.
20. Woodrow Wilson Report, Project *on Emerging Nanotechnologies*. Available at http://www.nanotech-project.us/process/assets/files/2741/18_nanotechnologyhumanhealthimpactframeworkstrategicresearch.pdf, last accessed November 9, 2009.
21. *United States Code of Federal Regulations*, Part 312 and 314.
22. Jotterand, F. (Ed) (2009) *Emerging Conceptual, Ethical, and Policy Issues in Bionanotechnology*, 2009, Springer, 183–192.
23. U.S. Federal *Food, Drug and Cosmetic Act* of 1962 *Section 201 (f)*.
24. White, G.B. (2009) Missing the Boat on Nanoethics. *The American Journal of Bioethics*, **9**(10), 18–19. Available at http://www.informaworld.com/10.1080/15265160903162642, last accessed November 08, 2009
25. Note 1: Jesse Gelsinger was an 18 year old patient who participated in a gene therapy trial for Ornithine transcarboxylase deficiency (OTCD); he died from multiple organ failure 4 days after starting treatment. His death was believed to have been triggered by a severe immune response to the adenovirus carrier and was a major setback in gene therapy. Available at http://www.ornl.gov/sci/techresources/Human_Genome/medicine/genetherapy.shtml, last accessed November 08, 2009
26. Allhoff, F., Lin, P. (2008) Nanotechnology *& Society: Current and Emerging Ethical Issues.* (Dordrecht: Springer).
27. Schmidt, C.W. (2009) Nanotechnology-related environment, health, and safety research: examining the national strategy. *Environ Health Perspect*, **117**(4), A158–61. Available at http://www.ehponline.org/members/2009/117-4/EHP117pa158PDF.PDF, last accessed November 4, 2009.
28. Note 2: The Asilomar Conference on Recombinant DNA was an influential conference organized to discuss the potential hazards and regulation of biotechnology, held in February 1975 at a conference center Asilomar State Beach. A group of around 140 professionals participated in the conference to draw up voluntary guidelines to ensure the safety of recombinant DNA technology. Available at http://www.wikidoc.org/index.php/Asilomar_conference_on_recombinant_DNA last accessed August 25, 2010.
29. Fredrickson, D.S. (2001) *The recombinant DNA controversy: A memoir: science, politics, and the public interest 1974–1981.* Washington, D.C.: ASM Press.
30. Paradise, J. (2008) Developing oversight frameworks for nanobiotechnology. *Minn. J.L. Sci. & Tech.* 9(1), 399–416.
31. Kristof, N. (2009) Chemicals and our health, *New York Times*, July 15, 2009, available at http://www.nytimes.com/2009/07/16/opinion/16kristof.html?_r=2&ref=opinion, last accessed November 5, 2009
32. Diamanti-Kandarakis, E., Bourguignon, J.P., Giudice, L.C., Hauser, R., Prins, G.S., Soto, A.M., Zoeller, R.T., Gore, A.C. (2009) Endocrine-disrupting chemicals: an Endocrine Society scientific statement. *Endocr Rev.* **30**(4), 293–342.
33. Jones, O.A.H., Voulvoulis, N., Lester, J.N. (2001) Human pharmaceuticals in the aquatic environment: a review. *Environ Technol* **22**, 1383–1395.
34. *Stem Cell Information.* Bethesda, MD: National Institutes of Health, U.S. Department of Health and Human Services, 2009, Available at http://stemcells.nih.gov/info/scireport/2006chapter6, last accessed October 23, 2009
35. Munn, D. (2001) Moral Issues of Human Embryo Research. *Science* **293,** 211.
36. Pellegrino, E. (2000) Testimony, in National Bioethics Advisory Commission, *Ethical Issues in Human Stem Cell Research*, vol. III (Washington: NBAC, 2000), F-1-F-5.
37. Guenin, L. (2001) Morals and Primordials. *Science* **292**, 1659–1660. Available at http://www.sciencemag.org/cgi/content/full/292/5522/1659, last accessed November 6, 2009.
38. Brannigan, M.C. (2001) What if we assign moral status to human embryos? *Genetic Eng News* **12,** 6.

39. Chan, S., Harris, J. (2008) Adam's fibroblast? The (pluri)potential of iPCs. *J MedEthics.* **34** (2), 64–6.

40. Glenn, L.M. (2003) Biotechnology at the margins of personhood: An evolving legal paradigm, *Journal of Evolution and Technology*, **13**, 110.

41. Ellis-Behnke, R.G., Liang, Y.X., You, S.W., Tay, D.K., Zhang, S., So, K.F., Schneider, G.E. (2006) Nano neuro knitting: peptide nanofiber scaffold for brain repair and axon regeneration with functional return of vision. *Proc Natl Acad Sci* USA, **103**(13), 5054–9.

42. Sclabassi, R.J. (2009) *Brain Implantable Computer Platforms*, Presentation at Unither Technology Conference, Magog, Quebec. Available at http://www.unithertechnology-conference.com/downloads09/SessionsDayOne/SCLABASSI_web.pdf, last accessed November 6, 2009

43. Kurzweil, R. (2005) *The Singularity is Near: When Humans Transcend Biology*, Viking Press.

44. Glenn, L.M. (2003) Biotechnology at the margins of personhood: An evolving legal paradigm, *Journal of Evolution and Technology*, **13**, 110.

45. Glenn, L.M., Boyce, J.S. (2008) Nanotechnology: Considering the Complex Ethical, Legal, and Societal Issues with Parameters of Human Performance, *Nanoethics*, **2**, 265–275.

46. McGee, E. (2009) Nanomedicine: Ethical concerns beyond diagnostics, drugs, and techniques. *American Journal of Bioethics*, **9**(10), 14–15.

47. McKibben, B. (2003) Enough: Staying human in an engineered age, Times Books, New York.

48. McGee, E., (2007) Should there be a law: Brain chips: Ethical and policy issues. *Thomas M. Cooley Law Review*, **24**(1), 81–97.

49. Jotterand, F. (2008) Beyond Therapy and Enhancement: The Alteration of Human Nature, *NanoEthics*, **2**, 15–23

50. Goldberg, J., Hayes, W., Huntley, J. (2004) *Understanding Health Disparities*, publication of the Health Policy Institute of Ohio. Ohio.

Index

Melba Navarro and Josep A. Planell (eds.), *Nanotechnology in Regenerative Medicine: Methods and Protocols*, Methods in Molecular Biology, vol. 811, DOI 10.1007/978-1-61779-388-2, © Springer Science+Business Media, LLC 2012

MIX
Papier aus verantwortungsvollen Quellen
Paper from responsible sources
FSC® C105338

If you have any concerns about our products,
you can contact us on
ProductSafety@springernature.com

In case Publisher is established outside the EU,
the EU authorized representative is:
Springer Nature Customer Service Center GmbH
Europaplatz 3, 69115 Heidelberg, Germany

Printed by Libri Plureos GmbH
in Hamburg, Germany